S0-CAH-178

Reactivity of Solids:
Past, Present and Future

INTERNATIONAL UNION OF PURE AND APPLIED CHEMISTRY

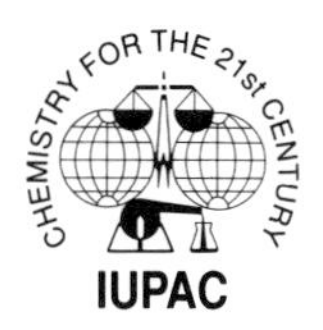

Reactivity of Solids: Past, Present and Future

A 'Chemistry for the 21st Century' monograph

EDITED BY
V.V. BOLDYREV

Institute of Solid State Chemistry
Siberian Branch of the Russian Academy of Sciences
Novosibirsk, Russia

Blackwell
Science

Editorial Offices:
Osney Mead, Oxford OX2 0EL
25 John Street, London WC1N 2BL
23 Ainslie Place, Edinburgh EH3 6AJ
238 Main Street, Cambridge
Massachusetts 02142, USA
54 University Street, Carlton
Victoria 3053, Australia

Other Editorial Offices:
Arnette Blackwell SA
224, Boulevard Saint Germain
75007 Paris, France

Blackwell Wissenschafts-Verlag GmbH
Kurfürstendamm 57
10707 Berlin, Germany

Zehetnergasse 6
A-1140 Wien, Austria

First published 1996

Set by Semantic Graphics, Singapore
Printed and bound in Great Britain at the University Press, Cambridge

DISTRIBUTORS

Marston Book Services Ltd
PO Box 269
Abingdon
Oxon OX14 4YN
(*Orders*: Tel: 01235 465500
Fax: 01235 465555)

USA
Blackwell Science, Inc.
238 Main Street
Cambridge, MA 02142
(*Orders*: Tel: 800 215-1000
617 876-7000
Fax: 617 492-5263)

Canada
Copp Clark, Ltd
2775 Matheson Blvd East
Mississauga, Ontario
Canada, L4W 4P7
(*Orders*: Tel: 800 263-4374
905 238-6074)

Australia
Blackwell Science Pty Ltd
54 University Street
Carlton, Victoria 3053
(*Orders*: Tel: 03 9347 0300
Fax: 03 9349 3016)

A catalogue record for this title is available from the British Library

ISBN 0-86542-687-2

Library of Congress
Cataloging-in-Publication Data

Reactivity of solids : past, present, and future /
edited by V.V. Boldyrev.
p. cm. — (A 'chemistry for the 21st century' monograph)
At head of title: International Union of Pure and Applied Chemistry.
Includes bibliographical references and index.
ISBN 0-86542-687-2
1. Solid state chemistry.
2. Reactivity (Chemistry)
I. Boldyrev, V. V. (Vladimir Viacheslavovich)
II. International Union of Pure and Applied Chemistry.
III. Series.
QD478.R43 1996
541′.0421—dc20 96-13655
CIP

Contents

Contributors

M.A. BLESA *Departamento Química de Reactores, Comisión Nacional de Energía Atómica, Avenida del Libertador 8250, (1429) Buenos Aires, Argentina* [43]

V.V. BOLDYREV *Institute of Solid State Chemistry, Siberian Branch of the Russian Academy of Sciences, Kutateladze, 18 Novosibirsk-128, 630128, Russia* [267]

E.V. BOLDYREVA *Institute of Solid State Chemistry, Siberian Branch of the Russian Academy of Sciences, Kutateladze, 18 Novosibirsk-128, 630128, Russia* [141]

G.R. DESIRAJU *School of Chemistry, University of Hyderabad, Hyderabad 500 134, India* [223]

L.C. DUFOUR *Laboratoire de Recherches sur la Réactivité des Solides, URA 23 CNRS, Université de Bourgogne, BP 138, 21004 Dijon cedex, France* [69]

A.K. GALWEY *School of Chemistry, The Queen's University of Belfast, Belfast BT9 5AC, Northern Ireland (UK)* [15]

B.S. GOUD *School of Chemistry, University of Hyderabad, Hyderabad 500 134, India* [223]

V. GUTMANN *Institute of Inorganic Chemistry, Technical University of Vienna, Getreidemarkt 9, A-1060 Vienna, Austria* [1]

N. LYAKHOV *Institute of Solid State Chemistry, Siberian Branch of the Russian Academy of Sciences, Kutateladze, 18 Novosibirsk-128, 630128, Russia* [121]

M. MARTIN *Institut für Physikalishe Chemie, Elektrochemie, Technishe Hochschule Darmstadt, Petersenstrasse 20, 64287 Darmstadt, Germany* [91]

C.N.R. RAO *Solid State and Structural Chemistry Unit and CSIR Centre of Excellence in Chemistry, Indian Institute of Science, Bangalore 560012, India* [237]

G. RESCH *Institute of Inorganic Chemistry, Technical University of Vienna, Getreidemarkt 9, A-1060 Vienna, Austria* [1]

R. ROY *Materials Research Laboratory, The Pennsylvania State University, University Park, Pennsylvania, USA* [253]

V.A. TOLKATCHEV *Institute of Chemical Kinetics and Combustion, Siberian Branch of the Russian Academy of Sciences, Institutskaya, 3 Novosibirsk-90, 630090, Russia* [185]

Preface

The concept of reactivity seems to have been formulated in chemistry at the close of the 19th–beginning of the 20th century. At that time, when speaking of the reactivity of a substance one meant the ability of *individual molecules* to take part in various chemical reactions, as well as the rate of these reactions. Indeed, chemical properties of gases and liquids are, to a large extent, determined by the characteristics of the individual molecules. However, when the same atoms and molecules form solids, the cooperative interactions may make the properties of a solid noticeably different from those of the individual molecules. Hence, *reactivity of solids* is essentially not the same as *reactivity of molecules.*

Our understanding of reactivity has undergone an evolution with time. Not a very long time ago, historically speaking, reactions of solids such as decomposition or interaction with gases were well known. However, it did not seem to be obvious that reactions between solids could be at all possible. A statement: 'Corpora non agunt nisi fluida' is ascribed to Aristotle. It was not until the beginning of this century that it became evident that solid state reactions do take place. This has stimulated the first special studies of solid state reactions and the first attempts to understand the mechanisms of these processes and the factors determining the reactivity of solids. It is important to remember that at the time of these first studies of solid state reactions, most of the presently used tools of studying solids were not available, for example the X-ray diffraction techniques were not developed and even the crystal structures of solids were not known! Understanding of the physics of the solid state at the beginning of this century was also very different from what it is now.

More and more experimental data were accumulated, which showed clearly that chemical reactions of solids are different in many respects as compared with reactions of liquids and gases. Reactivity of a solid has proved to depend on the 'biography' of the solid, i.e. on the method of preparation, storage conditions, preliminary treatment, etc. It turned out that kinetics of a solid state reaction cannot be described adequately only by indicating the degree of transformation as a function of time. Solid state reactions tend to be heterogeneous, the reaction rate usually depends on the site in a solid sample and, therefore, not only time, but also spatial coordinates are required for an adequate kinetic description.

Studies of chemical reactivity of solids were greatly influenced by the achievements in the developing of the physical theories of solids. As understanding of solid state physics was rapidly progressing, the theories of chemical reactivity of solids were also revised.

The first theories of solid state reactivity were based on considering the idealized crystal lattice. Then it turned out that real solids are never perfect (just like people!), and every solid contains deviations from its idealized structure, termed *defects*. The defects can play an important role in solid state reactivity. First, defects can serve as potential centres at which the reaction starts. This phenomenon is usually termed as the *localization* of the process. Second, defects are important for the transport processes in solids, such as diffusion of ions and electron transfer through a layer of product. Many of the first detailed studies of the reactivity of solids were aimed at elucidating the role of defects in solid state reactions, and even to suggest some semi-quantitative theories. Many of the ideas proposed in the 1930s–1950s have not lost their value up to now. As examples, I can mention: A. Hedvall's model interpreting the changes in the reactivity of a solid at the point of phase transition; the idea of C. Wagner to treat the thermodynamics of point defects in solids in terms of the theory of the electrolyte solutions; the model of N. Mott and P. Gurney describing the processes in photographic systems; Bernard's model of oxidation of metals and of the reduction of metal oxides; the shear structures proposed by J. Anderson, etc. These highly original studies can now be considered as classical. They impress by their clarity, coherence and the breadth of generalization. The authors of these studies did not have the experimental tools that we have now; they did not even know what a computer was. However, they could formulate the problem very clearly, choose the 'key points' and will have come closer to the solution than many of us today. Unfortunately, many of our contemporaries have lost these abilities in our time of vanity and profit.

By the late 1940s it became clear that not only the chemical composition and the idealized crystal structure of a solid are essential for its reactivity, but also the types and concentration of the defects. At first, defects, especially the non-equilibrium ones, were expected to increase the reactivity in all the cases, since each defect is a distortion of crystal structure and accumulation of the defects results in an increase in the energy of the solid. However, the interrelationship between the reactivity of a solid and the concentration of the defects in it turned out to be not that simple. More and more experiments showed that some defects did not affect the reactivity at all, whilst others could even inhibit the reaction. The effect of the defects on a particular solid state reaction is interrelated with the participation (direct or indirect) of these defects in this reaction. The experimental study of the effect of various types of defects on a solid state reaction proved, therefore, to be helpful in elucidating the reaction mechanism and finding the methods of the control of the reactivity. The theories used to describe thermodynamics and kinetics of solid state reactions were also to be modified in such a way, so as to take into account the role of the defects. As just one of the numerous examples I can cite the concept of an 'active state' of a solid (referring to a solid with a non-equilibrium concentration of defects), used when considering the thermodynamics of solid state processes.

One of the most characteristic features of solid state reactions (as compared with reactions in gases or in liquids) is that they are usually heterogeneous. Since the reactivity at different sites in a solid is often different, the reaction rarely starts simultaneously throughout the bulk or over the entire surface of a solid. Although this fact was mentioned as early as, for example, Faraday's working diaries, it seems to be Langmuir who first paid proper attention to the fact that a solid state reaction tends to be the reaction at the *interface* between the starting reactant and the product. The interface is often also referred to as the *reaction zone.* Understanding that heterogeneous solid state reactions are reactions at the interface was essential for the development of the kinetic theories of these processes. The 'topokinetics' aimed to take into account the variation of the size of the 'reaction zone' with time when calculating the reaction rate constants. The theory has borrowed some concepts from the theory of crystal growth and provided a set of the equations, which are now considered to be 'classical' (and, unfortunately, as often happens with 'classical equations', are often applied without a proper justification). Topokinetics flourished at the middle of this century, but at the present time the interest in it has considerably decreased. This may be, at least partly, because some of the basic postulates of topokinetics turned out to be inconsistent with the experimental data. A real solid state reaction turned out to be even more complicated than a process of crystal growth. Also, the forms of the feedback in the interface processes can be more diverse.

The first studies of solid state reactions paid most attention to the striking differences between homogeneous reactions in gases or liquids on one side and heterogeneous reactions in solids on the other side. This seems to be quite natural. Nevertheless, this trend can no longer form the basis for the progress in solid state chemistry. Critical analysis should be followed by a positive synthesis, and the time to throw stones must give way to the time to collect them. A theory of the reactivity of solids is still to be worked out. This theory will be of value not only for solid state chemists but also for specialists in physical, inorganic and organic chemistry. Chemistry is only artificially subdivided into separate 'subdisciplines'. It seems to be high time to start searching for links between the commonly accepted concepts of 'classical molecular chemistry' and the ideas of solid state chemistry.

Another important point in developing the theory of reactivity of solids is related to possible industrial applications. Solid state reactions are already used in technologies, and it is important to be able to control their course. The demand of new technologies that would be ecologically 'cleaner' and also advantageous economically is permanently growing. In many cases this means the necessity to develop 'dry technologies', based primarily on solid state reactions without using liquids at all. It is a commonplace that progress in fundamental science is usually related to the demands of industry, but this obviously is true for solid state science also. I am sure that although up to now attempts of technological applications of solid state reactions were only sporadic, an increasing interest in these reactions from industry will also stimulate the developments of the fundamental aspects of solid sate science.

It is also a commonplace that each science is progressing owing to the joint efforts of many scientists. Therefore, it is important that scientists have an opportunity to exchange ideas, to review the results and to outline the prospects for the future. As far as reactivity of solids is concerned, this opportunity is provided by organizing international symposia, which have been held regularly every 4 years since 1948 (Paris 1948, Gothenburg 1952, Madrid 1956, Amsterdam 1960, Munich 1964, Schenectady 1968, Bristol 1972, Gothenburg 1976, Cracow 1980, Dijon 1984, Princeton 1988, Madrid 1992). The next symposium is due to be held in Hamburg in 1996. Another way to review previous studies is to publish books, monographs and collective monographs. I would like to mention the 'classical' books of the 1950s, which were the 'milestones' in the development of the theory of reactivity of solids, i.e. the books by J.A. Hedvall, K. Hauffe, W.E. Garner and P. Budnikov. In the 1960s–1980s some more books appeared (for example, those by P. Barret, V. Boldyrev, M. Bulens, B. Delmon, K. Meyer, D. Young, H. Schmalzried, M. Brown, D. Dolimore and A. Galwey). Special issues of journals were also devoted to the problems related to reactivity of solids, for example a special issue of the *Journal de chimie physique* edited by P. Barret and B. Delmon (1986). The present book—issued in the IUPAC series '*Chemistry in the 21st*

Century'—pursues the same goal. This is a collection of articles written by experts from different countries. The authors have tried to provide a wide overview on the progress in the fields, in which they are recognized to be leaders, to evaluate the present state of research and the prospects for the future. The book covers the problems related to the hierarchy of the defects in the solids, the kinetics of chemical reactions in solids and the reactions in solid matrices. It also considers in more detail some particular types of heterogeneous reactions, for example thermal decomposition and dissolution of solids, interaction of solids with gases, as well as solid–solid reactions. Further on, homogeneous solid state reactions and chemical reactions in organic solids are considered. Special chapters are devoted to the problems related to the application of solid state reactions in inorganic synthesis and in the manufacture of new materials.

Of course, it is impossible within the scope of one book to cover the wide range of problems related to the reactivity of solids. Thus, for example, the book does not have contributions related to the applications of the theory of the reactivity of solids in experimental geology and mineralogy, despite an obvious progress achieved in this field. Regrettably, because of the sudden death of Professor M. Figlarz, the problems related to 'soft chemistry' are also not covered in the book. Also, a chapter devoted to the applications of the theory of reactivity of solids in catalysis and in the industry of pharmaceuticals has not been included, although interesting results have been obtained recently in these fields. Being the editor/compiler of the book, I am well aware of all these shortcomings. At the same time, I hope the reader will understand that the editor of this kind of book is not the God, and cannot do everything absolutely to his liking. As the ancient Romans said: 'Feci quod potui, faciant meliora potentes'. Anyway, I tried to do the best I could, and I would like to believe that this book will be useful to chemistry students and to researchers in the field of reactivity of solids, as well as to those who are already engaged in technological applications of the reactivity of solids, or who are about to do this.

I am grateful to I. Konstanchuk and T. Shakhtshneider for their help during preparation of the manuscript and to A. Polyakova for technical assistance.

V. Boldyrev

1 Hierarchy of Defects in Crystals and Reactivities of Solids

V. GUTMANN and G. RESCH
Institute of Inorganic Chemistry, Technical University of Vienna, Getreidemarkt 9, A-1060 Vienna, Austria

1 Introduction

Changes in material properties due to actions of mechanical forces are related to changes in the so-called 'defect structure'. According to this terminology, the actual differentiation of a crystalline material is described by 'defects' or by 'imperfections' of the idealized crystal structure.*

We shall use this terminology, although the presence of defects is necessary for the structural differentiation, and this is a requirement for the existence and properties of an object.

2 Aspects of defect structure as related to material properties

For the development of materials science, a distinction between various groups of defects and the so-called 'regular' building units proved most useful. Because in each of the different groups the building units contribute in highly specific ways to different macroscopic properties, it was found possible to consider only one of these groups for the approximate description of a given property.

1 For the description of chemical reactivity, catalysis, adsorption, adhesion or corrosion phenomena, the specific properties at the interface area are taken into consideration. 'Stacking errors' and 'surface defects' have been described; bond contractions at and near the interface established (Gutmann and Mayer 1976); and modern techniques for detailed structural information have been developed. Centres of 'excellence', such as peaks, craters and grooves have been found.

2 In order to account for changes in mechanical properties in the course of elastic or plastic deformations, 70 years ago, the gliding of lattice planes had to be assumed, and dislocation theories have been developed. They are successfully applied, although the dislocations themselves could not be directly observed. Decades later, this idea was triumphantly proved correct by electron microscopy, which makes the dislocation network directly visible (Amelinkx 1964). The electron micrographs show a bewildering variety of patterns, which alter with every slight change in the environment. The changes in the dislocation network are carefully considered in studies of elastic and plastic deformation, of fatigue/failure of materials, of annealing and quenching procedures and of changes in mechanical properties, such as strength, elasticity, toughness, fracture resistance, etc.

* The ideal crystal structure is the result obtained by the statistical elimination of all differentiations within the real crystal. It provides a suitable illustration of the lattice-geometrical arrangement and an excellent background for the description of the actual differentiation of a real crystal. Because the state of 'perfect order' has been ascribed to the fiction of the ideal crystal (at absolute zero temperature), the differentiation of a real crystal is considered as a manifestation of 'disorder'.

3 In order to describe changes in properties due to changes in chemical composition, pressure, temperature, as well as diffusion, thermal and ionic conductivity, another kind of imperfection is considered, the so-called 'point defects'. Even nearly stoichiometrically composed crystals are bound to contain both 'vacancies' and 'interstitial' positions. In 'regularly grown' crystals about 10^{17} point defects per cubic centimetre are present. Point defects are known to change places (displacement reactions), and 'point defect thermodynamics' has been worked out, although the direct observation of the behaviour of point defects is not possible.

4 Thermodynamic and idealized structural features are determined by the vast majority of building units, which are not in the immediate neighbourhood of any of the above mentioned imperfections. From the statistical point of view they are considered as 'regular building units'. Actually, they do contribute to the actual differentiation, as they cannot be separated from the influence of the above mentioned imperfections.

It proved of enormous advantage for materials science that most questions can be mastered with reasonable success simply by assuming that for the consideration of a given property only one of these main groups is taken into account. Most solid state scientists are pleased about this situation and rather reluctant to accept an extension of this concept, which is the subject of the present chapter.

3 Relationships and changes within the defect structure

3.1 *General considerations*

It is well known that the accepted approach is a simplification, because an object is bound to react as a whole. In order to illustrate this point, we may be reminded of the effects which are caused by mechanical deformation of a metal. Even in the elastic and more clearly in the plastic range of deformation all building units undergo changes, although to a different extent. The changes in the surface area are considered as long as changes in chemical reactivity and catalytic activity are under investigation. For the changes in mechanical properties the changes in the dislocation network are of great influence and, hence, dislocation theories are successfully applied. Changes in point defects are always taking place, because their density is known to be regulated by the dislocations (dislocation climb). As the number and mobilities of point defects are altered, even the normal building units cannot remain completely unaffected; after mechanical deformation of a copper rod, its vibration spectrum is changed, but slightly, in the same direction as by an increase in temperature of the undeformed rod.

Boldyrev (1986) has emphasized that the increase in reactivity due to mechanical treatment cannot be explained in terms of increased surface area, and that all changes in defect structure must be considered; new structural features are developed and metastable states produced.

3.2 *Conservative and dissipative aspects*

A structure that is developing new structural features is no longer a 'conservative structure', but a 'non-linear system' or a 'dissipative structure'. In fact, a conservative structure would not be observable in the total absence of dissipative features, and a dissipative structure could not develop in the complete absence of conservative boundary conditions. This may be illustrated by considering a river (Resch and Gutmann 1987). Conservative boundary conditions are provided by embankment, river bed and gravitational forces. River bed and embankment are slightly, but continuously altered, in particular by changes in the rate of flow. By its decrease the river bed may be silted up, whilst its increase will cause the river to overflow the bed, to deepen, and the banks to change their shape. This means that any change in dissipative features (rate of flow) causes changes in conservative features (river bed), and these in turn have an influence on the dissipative aspects. Thus, there are always mutual independencies between dissipative and conservative features; the former follow the latter, and these provide boundary conditions for the former.

It is characteristic for dissipative features that they have a direction of flow influenced by conservative boundary conditions. River regulations, elimination of cascades, damming up, etc., have a decisive influence on the dissipative features of the river. It appears, therefore, necessary for knowledge of a conservative structure to study in great detail its dissipative aspects, and for an understanding of a dissipative structure to study the conservative aspects.

The great influence of the conservative boundary conditions on the development of dissipative aspects shows that a dissipative structure cannot develop on its own, and that the 'building plan' cannot be designed by the developing system itself. It is therefore misleading to use the term 'self-organization', and it might be more appropriate to talk of 'self-performance' according to the plan. Organization is always serving a certain purpose to which the parts yield and, hence, organization cannot be produced by the actions of the parts themselves, much less by chance (Resch and Gutmann 1987).

According to the present strategy, solid state science is mainly concerned with the conservative features. In order to include the dissipative aspects: (i) all phenomena must be carefully considered, in particular those which cannot be explained in terms of existing practice and theories, and (ii) the roles of the different parts within the whole must be considered.

3.3 *The role of craters and grooves (three-dimensional imperfections)*

An interface is highly differentiated. When a metal surface is etched, the electron micrographs reveal the so-called 'etch pits'. These provide a kind of crater at the interface with high local surface energies and well-developed surface curvatures, and they are known to act as active centres in catalytic phenomena (Yamashina and Watanabe 1967). Their highly developed dynamic properties are supported by high vacancy densities around them. The etch pits are the points of emergence of dislocation lines at the interface and, in some cases, dislocations, partially parallel and close to the interface, produce grooves along their length (Amelinkx 1964). When grooves or craters are produced, for example by polishing the interface, pulsations take place and emission of sound waves can be registered (Natsik 1968).

The development of dissipative structures is evident from the observations on vapour deposition. Islands are formed, which grow by coalescence of the smaller ones, and this rapid process is further accelerated as the temperature of deposition is increased. This is unexplainable by random walk on statistical grounds (Chopra 1969) and shows that the interface becomes increasingly differentiated in its morphology by the development of dissipative aspects as the temperature is raised.

3.4 *The role of the flat interface areas (two-dimensional imperfections)*

An interface is in continuous interaction with its environment, with exchange of information, matter and energy. It cannot be clearly defined on a molecular level, because of the continuous changes in local composition and local surface structure. The enormous influence of such interactions is seen from the behaviour of sodium chloride crystals in hot water. The crystals are brittle in air, in hot oil and in a hot saturated sodium chloride solution, but they can be easily plastically deformed in hot water of the same temperature. This shows the development of new structural features initiated by the interactions with hot water. Other newly developing structures are known as the Rehbinder effect (Rehbinder *et al.* 1964). When germanium specimens, after cutting and polishing, are etched, the properties of this newly-formed interface are determined not only by the properties of the bulk material, but also by the cutting and polishing procedure as well as by the etchant composition and the etching technique.

Adsorption phenomena cannot be fully understood by considering only the local interactions with the adsorbed molecules. Far-reaching effects have been observed for the adsorption of carbon monoxide on a tungsten surface (Yates and King 1972). According to the first bond length variation rule (Gutmann 1978), the bonds within the adsorbed C≡O molecules are lengthened, and this requires an electron shift from the tungsten surface to the oxygen atoms of the C≡O molecules. As the number of adsorbed molecules increases, the heat of adsorption is decreased, and so is the bond lengthening of the newly-adsorbed molecules, even though the adsorbed molecules are still separated from each other by a great number of tungsten atoms. The C≡O bond lengths of the initially adsorbed molecules appear to be decreased (Yates and King 1972). This means that the metal interface no longer acts as a source for electrons, but rather as a regulating unit, that redistributes the electron densities nearly unchanged. Because the interface responds specifically to every change at every point, it must have all information both from the crystal and its environment, i.e. it belongs equally to both phases.

Because of the lattice contraction at and near the interface (Gutmann and Mayer 1976), and because of the highly developed dynamic properties of the constituents in the boundary area, the interface may be characterized as a region of high density and of 'high temperature' (Resch and Gutmann 1981). Solids, when exposed to high irradiance produced by an arc furnace, appear hot much sooner than expected from heat theory (Harrington 1966). The interface is the first area to melt on heating and the last area to solidify on cooling. Below the melting point of the bulk solid the boundaries show a liquid-like character, and this is used in sinter techniques, including powder metallurgy. It is evidenced by the melting properties of powdered binary mixtures of eutectic composition, which melt at a lower temperature (the eutectic temperature) than either of the components. This requires that the crystals obtain information from each other well below the melting point.

3.5 *The role of the dislocations (one-dimensional imperfections)*

The uniqueness of each crystal is well demonstrated by the uniqueness of its dislocation network, which is readily changed by actions from outside, such as irradiation or mechanical forces. Most of these changes involve the development of dissipative features, although near the equilibrium state.

The transportation of matter by means of the dislocations is called 'pipe diffusion'. At each dislocation node (a 'two-dimensional defect'), a decision is made about the direction of its further transportation. A dislocation line tries to attain a lower state of energy by shortening its length. This process may cause additional strain in other parts of the dislocation network, which may again be released by creating strains in other areas, and so on. Thus, a mode of 'pulsations' is taking place within the dislocation network, which seems to be characteristic for the actual state of the solid material under consideration. Such pulsations are also necessary in order to perform the process of 'segregation', i.e. the removal of material particles which are poorly integrated in the system. The segregation process is supported as the dynamic properties of the system are improved, for example by increase in temperature (dispersion hardening).

The decisive roles of the three-dimensional, two-dimensional and one-dimensional imperfections are also documented by their abilities to act as sinks and sources for point defects. All of the said imperfections are not in thermal equilibrium, dynamically active and subject to the development of dissipative structures, although not far from thermodynamic equilibrium. They bear information about the history of the material under consideration. Since these features are statistically insignificant, the said imperfections almost have no influence on the statistical results of thermodynamic, kinetic and crystal-structural properties. They are known as 'structure insensitive properties' (Smekal 1933).

3.6 *The role of point defects (zero-dimensional imperfections)*

Unlike the building units discussed so far, point defects (zero-dimensional defects) and regular building units are in thermal equilibrium. The formation of a vacancy in copper requires about 1 eV, and formation of an interstitial position about 4 eV. As the temperature is raised, the total energy of the material is increased and the number of point defects is also increased, which is particularly high near the melting point (Fine 1976). It is increased by the actions of mechanical forces or by irradiation. As mentioned before, the point defect density is regulated by the defects of higher dimensionality, notably by the dislocations which are known to act as sinks and sources of point defects by the so-called 'dislocation climb' (Thomson 1958, Seidmann and Baluffi 1965). Each point defect is under a certain tension by the surrounding lattice, which is modified. For these reasons, the term structure modifying and modified (SMM)-centre has been proposed (Resch and Gutmann 1980).

Apart from these structural aspects, point defects are known to migrate within the lattice by changing places with other building units. The driving forces for these 'displacement reactions' are gradients in chemical potential, structurally represented by structural differentiations. A building unit that jumps from one position into another one causes structural changes in the respective areas, i.e. the surrounding lattice will be 'relaxed' and 'strained', respectively. The superposition of all relaxations and strains leads statistically to the vibration pattern of the building units within the lattice.

Due to the motions of the point defects, a 'vacancy-lattice', i.e. a superstructure of point defects may be established. Evans (1977) has shown that for molybdenum, enriched in vacancies by neutron irradiation, the superlattice shows the characteristics of a body-centred structure, also known for the molybdenum structure. Because of the greater mobilities of the voids, the superlattice must have a greater influence on the crystal structure than vice versa, and hence the crystal structure appears subordinated to the superlattice structure of the voids.

3.7 *The role of the so-called 'regular building units'*

All motions and all changes in the 'defect' areas discussed so far influence both the arrangement and the vibrations of the vast majority of the remaining building units. It has been mentioned that the vibrational pattern of a given metal is not only a function of its temperature, but it is also influenced by mechanical actions or by irradiation. After such actions, the vibration spectrum corresponds to that of the untreated material at a higher temperature. Despite such small modifications, the structural framework provides essentially conservative boundary conditions for the dynamic features mentioned above.

The thermodynamic or macroscopic state of a crystal represents a great number of dynamically coexisting states, i.e. fluctuating microscopic states. What remains constant is neither the local analytical composition nor the positions of the building units, but rather its characteristic motion pattern.

4 The hierarchy of defects and the concept of system organization

4.1 *General considerations about hierarchy*

All of these features cannot be accidental, but must be intentional with respect to the overall situation. The local and temporal differences in properties of the parts require the existence of superordinations and subordinations, and these are characteristic for a so-called 'hierarchic order' (Resch and Gutmann 1980, 1987).

The term 'hierarchy' is frequently met with scepticism, because of the shortcomings of hierarchies in everyday life. Such hierarchies are invented by humans and executed by them. The faults of these hierarchical orders are, however, not due to hierarchy itself, but rather due to the consequences of faultily designed and badly executed orders.

In order to group the gradual transitions and continuous changes by means of abstractions, artificial discontinuities are introduced, which provide a simplified division of the system under consideration into more clearly arranged categories, called hierarchic levels. The criterion for their choice is neither the simplicity nor the precision of a model assumption, but rather the possibility to integrate all observations. Each hierarchic level always remains an integral part of the whole system and not a separate entity in its own right. Because the levels have no natural boundaries, they cannot be clearly defined, but they may serve as tools in order to illustrate certain aspects of the real material.

4.2 *Hierarchic levels in crystals and amorphous bodies*

In order to choose hierarchic levels in such ways as to express differences in dominance between them, we started from common grounds, i.e. from the differences between the imperfections of 'different dimensionality'. We tried to find out in what ways their building units differ in their abilities to develop dissipative features, in the energy content per part, in flexibility, in the ability to stand stress and to bear strain, and in their significance for the whole system in maintaining its chief characteristics under different conditions. On the grounds of such considerations, we appropriated the highest hierarchic level that can be observed to the three-dimensional imperfections, to which the two-dimensional imperfections are subordinated. The one-dimensional imperfections are subordinated to the latter, but superordinated to the zero-dimensional imperfections. The lowest hierarchic level is assigned to the regular building units, with provide the conservative framework for the dissipative features of the higher levels. We may refer to this level also as 'thermodynamic level' or 'structural level'.

In order to provide a first guide to such considerations, the actual differences in a system organization have been illustrated in Fig. 1.1 in a plot of the logarithm of the number of building units in each of the hierarchic levels versus the mean energies per building unit in each level. Neglecting the steps in the diagram, a broken line has been drawn from the upper edge to the lowest edge of the truncated pyramid. The slope of this line will depend on the number of building units in the various levels and hence on the 'defect structure', which is associated with the differentiation of the system. The smaller the differentiation of the system, i.e. the greater the purity and regularity of the crystal, the smaller is the slope of the said line. As the differentiation of the system becomes greater, as well as the number of building units in the higher levels, the slope of the line becomes greater too (Fig. 1.1). For an amorphous solid, the truncated pyramid shows a greater slope than for a crystalline material. Thus, differences in this

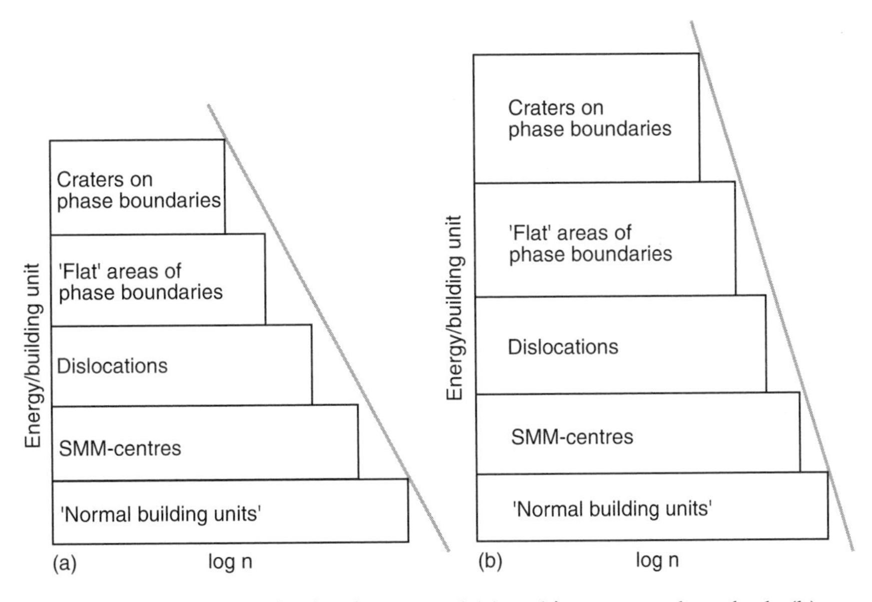

Figure 1.1 System organization in a crystal (a) and in an amorphous body (b).

slope express differences in system differentiation and in system organization. No peak is shown for the pyramid because the highest hierarchic level of a system is not sensually perceptible. It is impossible to attempt a scientific investigation of the highest hierarchic domain, which dominates and controls all observable aspects of a system.

As the external conditions are changed, the highest hierarchic levels 'decide' about the reorganizations in the subordinated levels, in order to allow the solid to respond properly through their concerted interactions for the maintenance of its chief characteristics. The higher levels seem to be mainly dynamically active, resistant against pressure and load and responsible for the structural and thermodynamic properties (Resch and Gutmann 1980, 1981, 1987). In no way is the lowest level 'enslaved' or 'exploited', as stated in synergetics (Haken 1983), but rather given all necessary help in order to provide its task within the whole of the system. We cannot say that the introduction of hierarchic levels is merely a matter of a new terminology. The main difference between the established and the new presentation is that none of the levels can be considered individually, but rather in their continuous relationships to each other. Although certain properties may be described approximately by referring only to one of these levels, it is impossible to account for all observable changes, unless all changes in all levels—in their mutual relationships—are adequately considered (Resch and Gutmann 1987).

5 Effects of mechanical actions on system properties

5.1 *General aspects*

Mechanical actions on a solid lead to strengthening of the higher levels: the number of their building units is increased; energy is preferably stored by them; and new features are developing. Both the complicated motion pattern and its structural boundary conditions cannot be understood on structural grounds alone. The properties of the building units in the complex relationship are lost by the physical or mental dismemberment of the system under consideration (Gutmann 1990), because the dominating influence of the 'continuum' has been eliminated (Gutmann and Resch 1982, 1990, 1992). The continuum of matter is described within the framework of quantum mechanics as a 'vacuum field that is not empty', and this is subject to quantum fluctuations and responsible for the 'zero-point energy' (Puthoff 1987, 1989). It is a requirement both for the observability and for the existence of atoms and molecules (Resch and Gutmann 1987; Puthoff 1989). Thus, isolated atoms represent a first-order correction to vacuum physics, but they cannot provide the basic starting point for the description of nature.

As all that is observable is bound to be subject to the continuum properties and to a system organization (Gutmann and Resch 1982, 1990), all of the interacting phases are organized (Resch and Gutmann 1987; Gutmann 1991). One of the consequences for an understanding of mechanical activation is the fact that both of the coexisting phases are mutually interacting by means of their highest hierarchic levels. These must contain all information from both of the phases, and any interaction must lead to reorganizations of both of the interacting phases.

5.2 *Mechanochemical effects on phase transitions*

Mechanochemical activation is related to changes in the interface area, and this is dominating for the reorganization of the whole system under consideration. This is illustrated by the mechanochemical activation of phase transitions, i.e. the lowering of the transition temperature. Crystalline zinc blende is transformed into wurtzite at about 970°C, but a vibro-milled sample was found to undergo the phase transition at 750°C (Imamura and Senna 1982). The transition at a lower temperature of the vibro-milled product shows that the latter is dynamically better developed than the crystalline product.

The well-developed dynamic features in the course of a transition are reflected in maximum values for the heat capacity and for the entropy changes, as well as in the unexpectedly high chemical reactivity within this range. The latter is known as the 'Hedvall effect'; for example, barium oxide and silver iodide show greatest reactivity at the temperature where a phase transition is taking place. (Hedvall 1938). During the transition, both the differentiation and the dynamic features of the interface are drastically increased, although the size of the interface area remains unchanged!

Precise magnetic and Mössbauer-spectroscopic data are available about the phase transitions in so-called 'spin–crossover complexes' (Gütlich 1981). In Fig. 1.2, curve A shows the sharp transition from the low-spin into the high-spin phase for the crystalline complex; curve B for a less crystalline complex; and curve C for the ground complex (Haddad *et al.* 1981). The temperature range in which both phases are coexisting is much wider after grinding. The greater differentiation of the ground material at any temperature is seen from the presence of certain amounts of the high-temperature phase at low temperatures and of the low-temperature phase at high temperatures (the so-called 'incompleteness' of the phase transition; Fig. 1.2, curve C).

The point of intersection indicates the greatest differentiation due to the mutual penetrations of both phases. Only at this point are the properties virtually independent

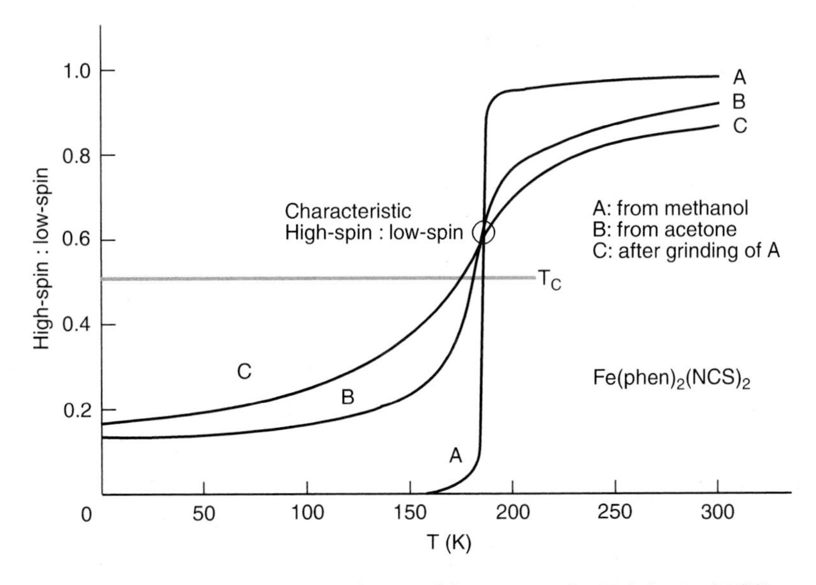

Figure 1.2 Low-spin $\rightleftharpoons$ high-spin transition curves for $Fe(phen)_2(NCS)_2$.

from the mechanical pretreatment, and the thermodynamic equilibrium of the system is fully defined by the variables of state. ΔH and $T\Delta S$ have the same numerical values (compensation effect), and the entropy is independent from the previous history and hence from the structural aspects of the defect structure! The system appears to be in a state of optimal system organization (Gutmann and Resch 1983a), and it is most adaptable towards changing conditions under optimal conservation of its chief characteristics.

5.3 *Consequences for the production of materials with special properties*

In the state of optimal organization, the system shows greatest 'openness' to the environment, i.e. greatest ability to integrate new information, thereby modifying itself. On the other hand, the well-developed system organization allows the system to provide information to the environment and, hence, to lose information. Their preservation is possible by rapid removal of the system from this dynamically active state into a state of better developed static boundary conditions. Such techniques have been empirically developed, for example the application of the Martensitic transformation, quenching techniques or the production of shape memory alloys. It may be suggested to produce materials of high strength by application of mechanical stress and radiation-resistant materials by exposure to radiation in the transition range, followed by fast removal from the latter. In these ways, highly metastable states may be produced and dynamically maintained within more rigid boundary conditions, for example at a lower temperature. Whereas many efforts are undertaken in the field of catalysis* to decrease the activation parameters, material scientists should be interested in their increase. This may be achieved simply by the improvement of its system organization!

5.4 *Interface properties*

The uniqueness of each interface is illustrated by the fact that between two copper electrodes, taken from the same material in a copper sulphate solution, differences between 0.1 and 10 mV are observed (Linert *et al.* 1984). Oxidic superstructures seem to exist on copper surfaces, which cannot be completely removed by acids, and appear to provide a semiconducting selvage layer (Gutmann *et al.* 1984). Its static aspects have been described as 'reconstructions' and considered as manifestations of disorder, as illustrated by Christmann's (1987) statement: 'The worst case for a metal surface is represented by the interactions with oxygen.' This might be the worst case for a simplified description and not for the system, the organization of which is definitely improved!

The system organization is also improved by mechanical actions. Many examples for the mechanical activation of solids and its application to technology are known (Boldyrev 1979, 1986, 1987). The mechanical activation of a metal in a liquid is reflected in the shift to more negative potential values. The potential shifts are smaller

* A catalyst is highly organized as it has a decisive influence on its environment, without losing its integral structure and functionality.

as the number of previous deformations is increased: the system reacts to mechanical deformations in such ways as to improve its resistance against further actions.

An interface is bound to be in a state of 'tension': a kind of 'existential conflict' between the phases under consideration, which are in immediate contact by means of the highest hierarchic levels. The existential conflict may lead either to the annihilation of the less organized phase or to the establishment of an equilibrium between both phases. The better organized system may be strengthened because of its better developed ability to integrate additional information without losing its integral configuration, whereas the system of lower differentiation may be weakened. At pH less than 2, passivated copper is even less resistant to acids than non-passivated copper. This is because the liquid medium becomes more active by decrease in pH value below 2. Gutmann *et al.* (1988, 1993) have suggested consideration of these possibilities in corrosion research.

The properties of thin films are known to be influenced not only by the ratio of surface to volume, but also by its previous history. It has been mentioned that by using quenching techniques some of the dynamic properties, developed at a higher temperature, are 'frozen in' at a lower temperature, and that the mechanical treatment may also be of importance.

Lamelles of gold expand on heating, but they begin to shrink above a certain temperature. A gold film of 0.4-μm thickness, prepared by electrolysis, begins to shrink at 350°C, but a gold film, prepared by hammering, does not shrink below 590°C (Tammann and Boehme 1932). The curves for films of 0.77-μm thickness in a plot load versus contraction and expansion for different temperatures show a 'point' of intersecting curves (Fig. 1.3), where the system appears to have greatest ability to maintain its state. At 650°C this state is maintained at a load up to 2 g (Tammann and Boehme 1932), and is considered as a state of optimal system organization (Gutmann and Resch

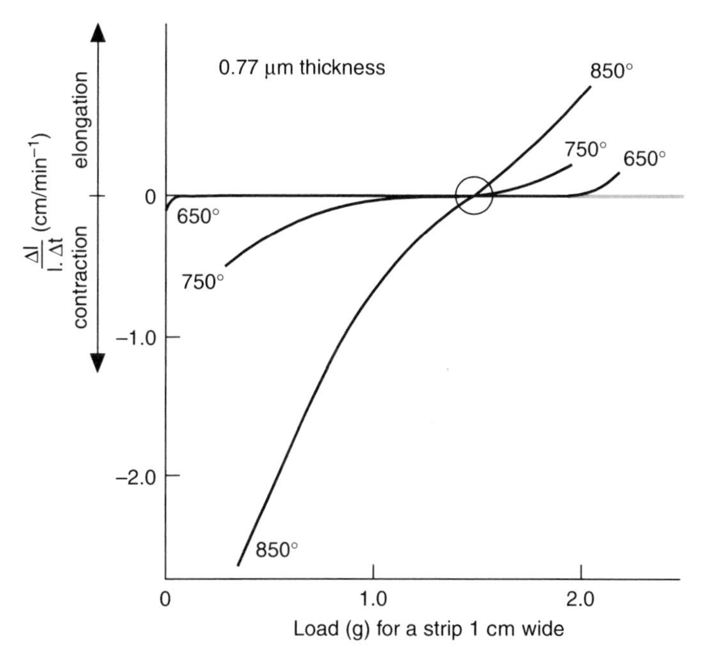

Figure 1.3 Rates of contraction and elongation, respectively, versus load for gold films of 0.77-μm thickness at temperatures above 650°C.

1983b). In this state the system is in the transition range between the solid state and a 'melt-like' state.

One may further expect that membranes should be best organized (Gutmann and Resch 1996) and hence most versatile, as a phase transition is taking place in them. Holzwarth (1986) points out that during a phase transition in a unilamellar vesicle formed by bilayers, the bilayer is transformed into a more fluid state, without losing its identity.

6 Conclusions

Each material system acts as a unity, which is differentiated in itself. The differentiation of a solid material is described by the so-called 'imperfections of different dimensionality'. They appear to be subject to hierarchically-ordered relationships. The cooperativities between all parts of the system appear to be governed by the actual conditions, i.e. by developing forces which minimize the effects of the actions of the forces from outside. The greater the system differentiation, the better is its resistance against forces and the greater its ability to influence less differentiated systems. The system approach is not a new theory, a mathematical construct nor an attempt to replace well-established theories, which remain unchallenged within their ranges of applications. Instead, it provides a wide framework in order to obtain a better understanding of qualities and their changes under different conditions. It is hoped that this concept may serve as a tool for new questions and for new answers, not only in mechanochemistry, but also in other fields of science.

7 References

Amelinkx, S. (1964) *The Direct Observation of Dislocations.* Academic Press, New York.
Boldyrev, V.V. (1979) *Ann Rev Mat Sci* **9**, 455–469.
Boldyrev, V.V. (1986) *J Chim Phys* **83**, 821–829.
Boldyrev, V.V. (1987) *Thermochim Acta* **110**, 303–317.
Chopra, K.L. (1969) *Thin Film Phenomena.* McGraw-Hill Co., New York.
Christmann, K. (1987) *Z Physik Chem* (Frankfurt) **154**, 145–178.
Evans, J.M. (1977) *Nature* **229**, 403–405.
Fine, H.A. (1976) In: *Treatise on Solid State Chemistry*, Vol. 1, (ed. N.B. Hannay), p. 283. Plenum Press, New York.
Gütlich, P. (1981) *Struct Bond* **44**, 84–183.
Gutmann, V. (1978) *The Donor–Acceptor Approach to Molecular Interactions.* Plenum Press, New York.
Gutmann, V. (1990) *Fresenius Z Anal Chem* **337**, 166–167.
Gutmann, V. (1991) *Pure Appl Chem* **63**, 1715–1724.
Gutmann, V. & Mayer, H. (1976) *Struct Bond* **31**, 49–66.
Gutmann, V. & Resch, G. (1982) *Comments Inorg Chem* **1**, 265–278.
Gutmann, V. & Resch, G. (1983a) *Inorg Chim Acta* **72**, 269–275.
Gutmann, V. & Resch, G. (1983b) *Monatshefte Chem* **114**, 839–850.
Gutmann, V. & Resch, G. (1988) In: *Innovation in Zeolite Materials Science.* (eds Grobet, P.J., Mortier, W., Vansant, E.F. & Schultz-Ekloff, G.), pp. 239–251. Elsevier Science B.V. Amsterdam.
Gutmann, V. & Resch, G. (1990) *Chem Ztg* **114**, 85–92.

Gutmann, V. & Resch, G. (1992) *Chimica Oggi Nr* **4**, 9–12, and *Chimica Oggi Nr* **5**, 15–19.
Gutmann, V. & Resch, G. (1996) *Lecture Notes on Solution Chemistry.* World Scientific Publishing, Singapore.
Gutmann, V., Resch, G. & Linert, W. (1993) *Proc. 1st Int. Conf. Mechanochemistry, Kosice.* **1**, pp. 71–5, Cambridge Interscience Publishing, Cambridge.
Gutmann, V., Resch, G., Kantner, W. & Linert, W. (1989) *Mh Chem* **120**, 11–20.
Gutmann, V., Resch, G., Kratz, R. & Schauer, H. (1984) *Mh Chem* **115**, 551–559.
Haddad, M.S., Lynch, M.W., Federer, W.D. & Hendrickson, D.N. (1981) *Inorg Chem* **20**, 123–131.
Haken, H. (1983) *Synergetics, an Introduction*, 3rd edn. Springer-Verlag, Berlin.
Harrington, R.E. (1966) *J Appl Phys* **37**, 2028–2034.
Hedvall, R.E. (1938) *Reaktionsfähigkeit Fester Stoffe.* Verlag Joh.A. Barth, Leipzig.
Holzwarth, J.F. (1986) *Dis Chem Soc* **81**, 353–358.
Imamura, K. & Senna, M. (1982) *J Chem Soc Farad Trans 1*, **78**, 1131–1140.
Linert, W., Stiglbrunner, K. & Gutmann, V. (1984) *Mh Chem* **115**, 905–920.
Natsik, V.D. (1968) *JETP Lett* **8**, 198–202 (English translation of Zhurnal Eksperimental noi i Theoreticheskoi Fitiki).
Puthoff, H.E. (1987) *Phys Revs* **D35**, 3266–3269.
Puthoff, H.E. (1989) *Phys Revs* **D40**, 4857–4862.
Rehbinder, P.A., Likhtnan, V.I. & Karpfenko, G.V. (1964) *Der Einfluß Grenzflächenaktiver Stoffe auf die Deformation von Metallen.* Akademie Verlag, Berlin.
Resch, G. & Gutmann, V. (1980) *Z Phys Chem* (Frankfurt) **121**, 211–235.
Resch, G. & Gutmann, V. (1981) *Z Phys Chem* (Frankfurt) **126**, 223–241.
Resch, G. & Gutmann, V. (1987) *Scientific Foundations of Homoeopathy.* Barthel and Barthel Publ., Germany.
Seidmann, D.N. & Baluffi, R.W. (1965) *Phys Revs* **139A**, 1824–1840.
Smekal, A. (1933) *Handbuch der Physik*, 2. Aufl., Band XXIV, 2. Teil (eds H. Geiger & K. Scheel), p. 795. Springer-Verlag, Berlin.
Tammann, G. & Boehme, W. (1932) *Ann Phys* **12**, 829–834.
Thomson, R. (1958) *Acta Metall* **6**, 23–28.
Yamashina, T. & Watanabe, K. (1967) *Bull Soc Chem (Jpn)* **46**, 2558–2563.
Yates, J.T. & King, D.A. (1972) *Surface Sci* **30**, 601–616.

2 The Reactivity of Solids in Thermal Decomposition (Crystolysis) Reactions

A.K. GALWEY

School of Chemistry, The Queen's University of Belfast, Belfast BT9 5AG, Northern Ireland (UK)

1 Introduction

Thermal decompositions of solid reactants have occupied an important, perhaps central, position in solid state reactivity by providing many of the model rate processes that have been widely accepted as most suitable for fundamental mechanistic investigations. Experimental observations, obtained from measurements for these relatively simple reactions, have notably contributed to the establishment of the theoretical concepts that are now routinely used in descriptions of a wider and more diverse range of chemical changes involving solids. This theory has been developed to explain aspects of the interactions of solids with other reactants: gaseous, liquid or solid. Reaction models developed from investigations of the simplest systems available can be valuable in contributing to the interpretation of more complicated phenomena. A review of solid state reactivity, developed in the context of the theory of the subject, was given by Boldyrev (1979).

Thermal decompositions of solids have been regarded as an identifiable subject area for several decades, and extensive reviews of the topic have been published (Garner 1955; Jacobs and Tompkins 1955; Young 1966; Delmon 1969; Brown *et al.* 1980; Sestak 1984). The term *crystolysis* was suggested (Galwey and Mohamed 1985) as a descriptive label for this subdiscipline because reports of the topic tend to be dispersed rather widely in the literature (Carr and Galwey 1984). The use of a keyword could be most valuable in providing coherence to the subject, because research article titles, and even the abstracts, do not always state explicitly that the report is concerned with a solid state process. A probable reason for this omission is the difficulty of demonstrating conclusively, from the experimental evidence usually available, that the reaction takes place in the solid. A second general feature of the subject is the apparent absence of systematic chemical correlations between decomposition characteristics and reactant compositions. The kinetic behaviour patterns for thermal reactions of solids containing common constituents are often different, whereas chemically dissimilar substances sometimes exhibit comparable features (Young 1966).

A characteristic feature, typical of many (but not all) crystolysis reactions, is that chemical transformations occur within an active zone at a reactant–product contact interface that progressively advances into unreacted material. These zones of locally heightened chemical reactivity are initiated at *nuclei* (Fig. 2.1), and their systematic development is termed *growth* (Fig. 2.2). Typical isothermal *nucleation and growth reactions* exhibit characteristic sigmoid-shaped yield–time curves, representing the aggregate product formation resulting from interface advance. This chapter identifies the reaction interface as a central, unifying characteristic in mechanistic discussions of solid state reactions (although other types of behaviour are also considered below). Aspects of the development and the current status of the theory concerned with the

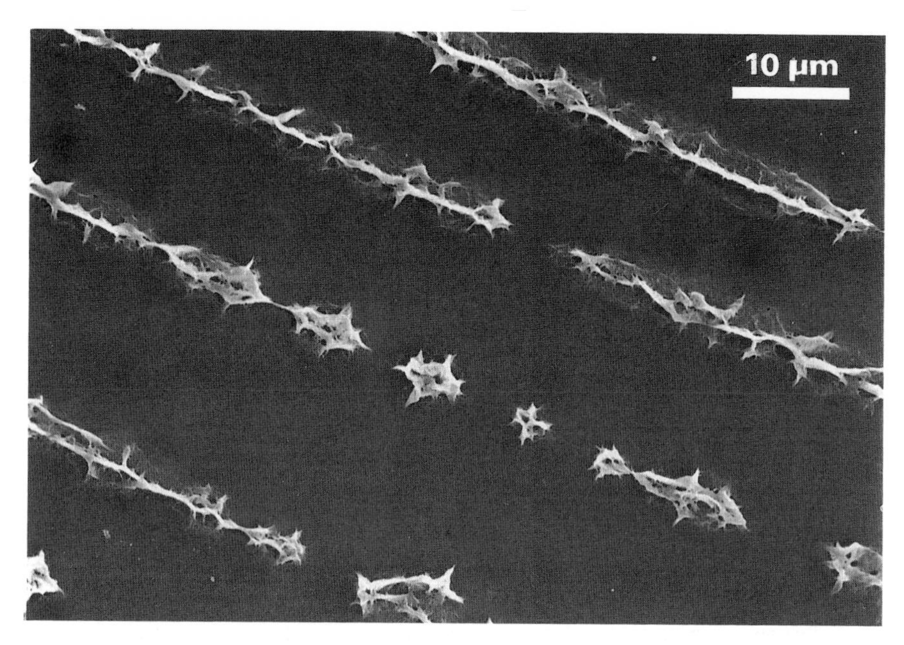

Figure 2.1 Replica of α-$NiSO_4.6H_2O$ surface after initiation of dehydration by the appearance of small nuclei, scanning electron micrograph (SEM). Nuclei are distributed along lines of imperfection, probably including surface terminations of dislocations, and crack systems coalesce. (A.K. Galwey, unpublished work.)

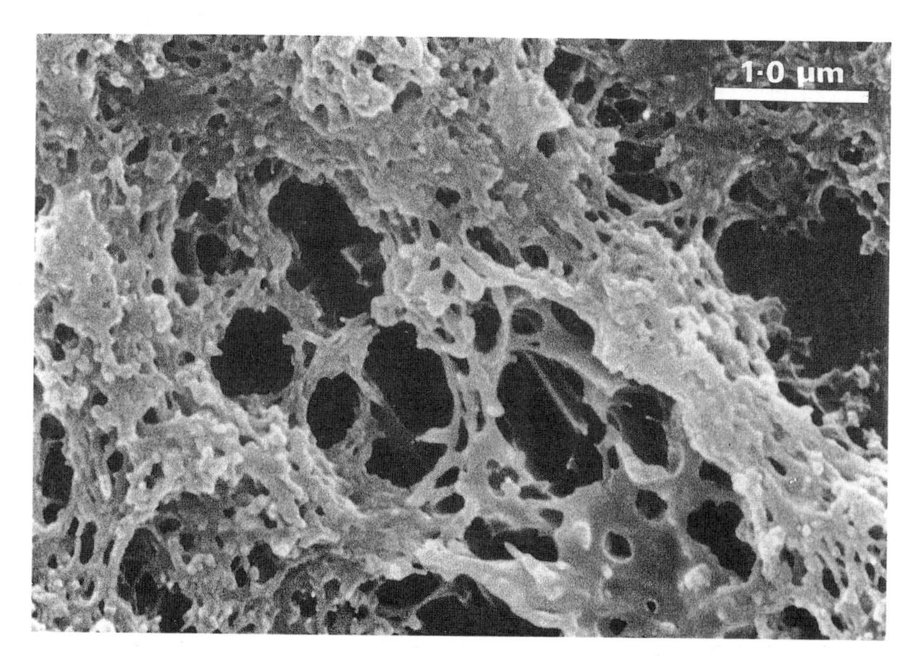

Figure 2.2 Replica of intranuclear crack structure within product Li_2SO_4 crystallites during the early stages of $Li_2SO_4.H_2O$ dehydration. (A.K. Galwey, unpublished work.)

types, functions and reactivities of interfaces are discussed, together with the essential precursor step, nucleation. Progress in the field is illustrated through references to a small number of crystolysis reactions, which are considered below in an historical perspective. These selected solids are amongst those that have attracted the greatest and also the most sustained attention. Towards the end of the chapter I suggest, from an essentially personal perspective, some areas that I believe (with my reasons) to be of promising potential for future research. Work in these areas might be expected to yield useful contributions towards the advancement of the subject.

2 The reaction interface

The forms and functions of reaction interfaces are considered first because these structures can be regarded as the most characteristic features participating, perhaps controlling, reactions of solids. The interface may be defined as the contact zone within which chemical changes preferentially occur, and it is between juxtaposed crystalline reactant and product phases. This zone of active reaction progressively advances into unchanged reactant, without substantial movements of material, although volatile products may escape. After decomposition, the residual material tends to occupy the space formerly filled by the reactant from which it was derived, although there may be displacements and cracking during product recrystallization and reorganization. The advancing interface may be portrayed as a *wave* of chemical change wherein the substance of the medium through which it moves undergoes relatively minor displacement, but does not travel with the disturbance. The majority of the constituent groups comprising the reactant do not undergo chemical transformation until reached by the active front (interface, wave of reactivity), and thereafter reaction is relatively rapid.

Specific interface structures are described below for particular reactants. Essential background material and more detailed descriptions of interfaces are to be found in the general references to the subject already mentioned. Nucleation, the essential precursor to interface development, is discussed in a subsequent section and other paragraphs are concerned with rate processes that do not proceed through interface advance.

Enhanced reactivity within an interface requires local specialized conditions capable of promoting chemical change. The effects most usually identified as enhancing reaction rates are strain and catalysis. Confirmations of the precise modes of action are difficult because the active reaction zone may be very thin, perhaps even of molecular thickness. Moreover, interface textures may undergo irreversible changes on removal of the stabilizing effects of the contiguous reactant and/or product crystals, between which it is located.

• *Strain.* A coherent contact between the different structures of reactant and product crystals is a zone of locally enhanced strain, an extended (usually two-dimensional) crystal imperfection. Bonds across the plane of mismatch differ from those within the contiguous crystals. This can be expected to facilitate the electron redistribution steps required to transform the reactant components into products. Strain is identified as diminishing the energy barrier (the activation energy) for the rate determining process.

• *Catalysis by-product.* A solid product capable of catalysing a controlling chemical change can account for the enhanced interface reactivity. The reactant constitutes a source of the unreacted species that break down after transfer onto the active product surface. This type of mechanism is familiar from heterogeneous catalysis where the changes occurring involve chemisorbed intermediates. Crystolysis reactions of this type include the decompositions of organic anions, such as carboxylates, in salts of the less electropositive elements including the transition metals, which are reduced to the elements. These metals may also be catalysts for the decompositions of the appropriate organic acids (Bond 1962).

Kinetic and mechanistic investigations of any crystolysis reaction require the consideration of both *reaction geometry* and *interface chemistry*, described in greater detail below. These complementary aspects of behaviour are often studied individually but the formulation of a meaningful reaction model requires information of both types which must, of course, be synthesized into a consistent overall representation of the reaction under consideration.

• *Reaction geometry.* It is usually accepted that all areas of reaction interface are identical (Brown *et al.* 1980). Consequently, the rate of an isothermal decomposition, proceeding within one (or more) crystals and involving one (or more) nucleus, is directly proportional to the total reactant–product contact area. The kinetic characteristics, expressed by an equation of the form: $f(\alpha) = kt$ (where α is the fractional reaction at time t), is, therefore, that rate is directly proportional to the systematic changes of area of active interface as reaction proceeds from nucleation to completion. This treatment assumes that the linear velocity of interface advance is constant, as demonstrated in representative observational studies, some of which are cited in the above general references. Accordingly, the rate equations accepted as being applicable to solids are based on *geometric* models and, unlike homogeneous processes, do not include concentration terms. Kinetic interpretations should always be supported by complementary microscopic observations to confirm the geometric conclusions.

• *Interface chemistry.* Identification of the bond-breaking and bond-making steps occurring within the reaction zone, together with the detection of any participating intermediates, must often be interpreted from indirect evidence. Relevant information may usefully include:

(a) consideration of all essential bond reorganizations (deduced from the overall reaction stoichiometry);

(b) the magnitude of the calculated activation energy (often regarded as a measure of the energy investment required for the controlling bond-rupture step);

(c) any information that can be obtained concerning the interface structure (including topotaxy detected by diffraction studies and textures identified from microscopic observations).

2.1 *Interface types*

Although there have been many discussions of the precise roles of interfaces in crystolysis reactions, less progress has been made in characterizing and classifying the different types. To contribute towards the development of this central feature of solid state reaction theory, the present author proposed a classification scheme at the *Conference on Contemporary Problems on the Reactivity of Solids* held in Novosibirsk in 1988. Essential features of the types distinguished are summarized here, together with references to mechanistic discussions concerned with the roles of these interfaces in specific reactions. Diagrammatic representations of structures and discussions of functions have already been given (Galwey 1990).

2.1.1 FUNCTIONAL NUCLEI

At the active interface, the product solid functions by promoting chemical change through a catalytic-type mechanism (Bond 1962), as described above in Section 2, *Catalysis by-product.* This reaction model explains satisfactorily kinetic and microscopic observations for the decompositions of nickel squarate (Galwey and Brown 1982) and of silver malonate (Galwey and Mohamed 1985).

2.1.2 FUSION NUCLEI

At reaction, temperature, the reactant decomposes to yield an intermediate or product that is either molten or forms an eutectic with the reactant. There is evidence that many rate processes occur more rapidly in the liquid phase than in the solid, because the intracrystalline attractive (stabilizing) forces are relaxed (Galwey 1994a). There is also greater stereochemical freedom to adopt the least-energy configuration in the precursor step to chemical change. More rapid reaction within an expanding liquid phase results in autocatalytic behaviour, as reported for the decompositions of ammonium dichromate (Galwey *et al.* 1983), ammonium perchlorate (Galwey and Mohamed 1984) and copper(II) malonate (Carr and Galwey 1986) (Fig. 2.3).

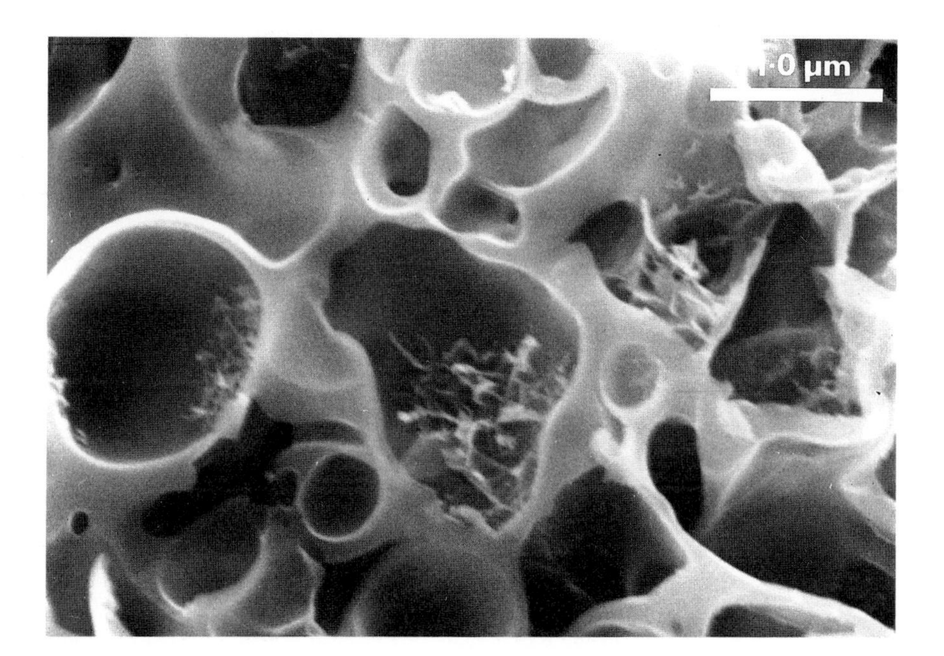

Figure 2.3 Thermal decomposition of copper(II) malonate (Carr and Galwey 1986). Partly reacted solid fractured to reveal internal froth and bubble texture indicating control by surface tension forces for reaction proceeding in a melt containing Cu^{2+}, Cu^{+}, acetate and malonate.

2.1.3 FLUID-FLUX NUCLEI

It is suggested that nucleus texture may be developed specifically to retain temporarily a proportion of the fluid product that promotes recrystallization, regarded as the difficult step in the reaction. It is believed that nuclei of this type occur in dehydrations where a proportion of the water of hydration released may be retained within nuclei, perhaps through multilayer adsorption or even condensed as liquid in microcracks and pores. This explanation of interface activity has been proposed in discussions of alum dehydrations (Galwey *et al.* 1981; Galwey and Mohamed 1987; Galwey and Guarini 1993), and also in the $KBr + \frac{1}{2}Cl_2$ reaction (Galwey and Pöppl 1984) which is a nucleation and growth process (but is neither a dehydration nor a crystolysis reaction) (Fig. 2.4).

2.1.4 FLUX-FILIGREE NUCLEI

During the dehydration of $Li_2SO_4.H_2O$, water is lost from the zone in advance of the sharp recrystallization front (Boldyrev *et al.* 1987), at which the hydrate is transformed into an open but coherent assemblage of crystallites. This classification label describes the flux of water followed by the generation of product having an open, or filigree, texture (Galwey *et al.* 1990). (see Fig. 2.2).

Figure 2.4 Interface KCl product (irregular crystals)–KBr reactant (flat surface) revealed by fracture across nucleus during $KBr + \frac{1}{2}Cl_2 \rightarrow KCl + \frac{1}{2}Br_2$ reaction (Galwey and Pöppl 1984). The interface is irregular and there is evidence that the intranuclear cracks are filled with liquid Br_2 during the reaction.

3 Nucleation

Nucleation refers to the sequence of chemical steps that lead to the establishment of an active interface (Galwey and Laverty 1990). The reactions contributing may, in principle, be different from those that participate in subsequent nucleus growth (interface advance). A problem inherent in any attempt to characterize these reactions in detail is that the establishment of a growth nucleus necessarily results in the irreversible destruction of its generation site (see Fig. 2.1).

3.1 *Role of dislocations*

Although there have been many expressions of the opinion that dislocations are involved in nucleation, positive proof of this participation and the identification of the precise mode of action remain elusive. Several careful studies directed towards characterizing any association between line imperfections and sites of initiation of reaction remain inconclusive. Some authors state that a link does exist, although, at present, its exact nature is uncertain. In any nucleation, some factor must locally heighten reactivity, and dislocations provide an attractive explanation because the numbers ending at surfaces are often comparable with the numbers of sites at which reaction commences. Moreover, the strain energy that exists within zones of structural distortion might be expected to contribute to the initiation of chemical change. An alternative explanation is that atypical alignments of crystal constituents at the surface may provide a stereochemical environment that facilitates the steps leading to nucleation. A difficulty that introduces uncertainties into experimental investigations of this type is that cold working and, more significantly, cleavage of crystals being studied can unpredictably alter the dislocation distribution. These topics are illustrated through the following experimental studies.

Comparisons between distributions of dislocations and of nuclei can be made on matched pairs of surfaces exposed by cleavage of a large relatively perfect crystal of reactant. One surface is subjected to conditions sufficient to initiate reaction and small nuclei are generated. The other surface is etched to reveal pits at dislocation–surface intersections. An exact correspondence between nuclei and etch pits on the two faces can then be accepted as evidence that nuclei develop at dislocations.

Exact correspondences are not always found, however, as reported by Thomas and Renshaw (1967, 1969) from observations on $CaCO_3$ and on α-$NiSO_4.6H_2O$. These authors concluded that dislocations were probably involved in nucleation but the existence of discrepancies indicated that other factors must also participate. From the $CaCO_3$ studies it was deduced that calcite decomposition was preferentially initiated at those dislocations that glide on the $\{100\}$ and $\{2\bar{1}\bar{1}\}$ planes. The relatively low activation energy for decomposition observed at very low α was ascribed to the possibility that the intrinsic energy of a dislocation can contribute to the activation of ions in transition complex formation. Fox and Soria-Ruiz (1970), also working with $CaCO_3$, came to very different conclusions. The role of dislocations in locally enhancing reactivity is identified as being largely stereochemical and the reaction is not associated with the strain energy present.

Hutchinson and Stein (1974) conclude that a single mechanism controls both the initiation and advance reactions during the decomposition of α-$Pb(N_3)_2$. The intentional introduction of controlled amounts of the impurities Fe^{2+} or $[FeN_3]^{2+}$ in α-$Pb(N_3)_2$, whilst accelerating reaction, did not apparently change the magnitude of the energy barrier in the rate controlling step.

Begg *et al.* (1983) studied the solid state decomposition of $NaClO_3$ single crystals, a reaction that is restricted to a small fraction of the early pyrolysis. They found some evidence that dislocations may participate in the surface reaction but the contribution of these imperfections towards initiation of the bulk reaction was less.

3.2 *Surface modifications during onset of reactions*

Microscopic studies of alum dehydrations (Galwey *et al.* 1981) have shown that, when subjected to reaction conditions, the total surfaces of these salts undergo changes. Subsequent exposure to water vapour may result in reorganization of a smooth crystal face through the generation of the 'orange peel texture' or there may be a large increase in nucleation rate on the reimposition of dehydration conditions (Garner 1955). It was concluded that these reactant hydrates underwent limited water losses from all surfaces. Continued reaction, by a nucleation and growth mechanism, is only possible after the essential, but difficult step requiring recrystallization of this strained boundary zone to the product structure. The growth nucleus formed was identified as a texture of this recrystallized material that is capable of retaining, temporarily, a proportion of the product water, by intranuclear adsorption. This water is envisaged (Galwey and Guarini 1993) as promoting the structural reorganization of partially dehydrated material to crystalline product at the advancing interface: an intranuclear autocatalytic mechanism.

Guarini and Dei (1983) have demonstrated the occurrence of similar behaviour in the dehydrations of a variety of other hydrates, so the model is shown to be of more general applicability (see also Dei *et al.* 1984). Further evidence recognizing recrystallization as the difficult step in nucleation was provided by Galwey *et al.* (1994). Added crystalline (anhydrous) product acted as seed crystals in markedly diminishing the induction period for the dehydration of lithium potassium-*d*-tartrate monohydrate (d-$LiKC_4H_4O_6.H_2O$). This is a nucleation and growth reaction after the initial (limited) deceleratory dehydrations of all crystal surfaces.

3.3 *Comment*

We conclude, therefore, that no general theory of nucleation has received universal support and, indeed, it may be that no single representation is generally applicable. The usual appearance of nuclei in boundary layers is consistent with the slight reduction of stability in this zone and the facile escape of the volatile product, particularly for reversible reactions. Whilst most workers in the field regard it as probable that dislocations participate in nucleation, the reactivities of such sites may vary with type and with the presence of incorporated impurities. The demonstration that many surfaces may undergo limited reaction before or during nucleation (Guarini and Dei 1983) means that strain premeates all boundary layers and this can be expected to modify the reactivities and perhaps the distributions of nuclei. This is one acceptable explanation for the inconsistencies noted above. We point out also that crystolysis reactions other than dehydrations give evidence of initial rapid superficial changes to yield a modified crystal boundary layer, see, for examples, the decompositions of copper(II) malonate (Carr and Galwey 1986) and $KMnO_4$ (Brown *et al.* 1994). Further investigations of nucleation processes are essential. These may include improved microscopic examinations to characterize the textural changes participating in processes that are, at present, poorly understood.

4 Reaction kinetics

Academic interest in the kinetics of crystolysis reactions has often been concerned with characterizing the chemistry of the processes involved, elucidating the mechanisms and identifying the controls of reactivity. Much of this effort has been directed towards developing the theory of the subject. This knowledge can be of great value to industrial scientists, interested in exploiting reactivities in the application of efficient methods for manufacturing products, in preventing the deterioration of such preparations before use and in ensuring that the material functions correctly (within specification) as and when it is required.

A review of the technological applications of solid state chemistry is beyond the scope of this chapter, although some particular points of current interest will be mentioned. It is essential that physiologically active drugs and medicines must have a long shelf-life and that their rates of deterioration (and their products) should be known. Stability times, under expected storage conditions, can be estimated by the theory through extrapolation from deterioration rates measured at higher temperatures. Also, studies of reactivities have been concerned with the stabilities of detonators, explosives and combustible materials such as rocket fuels and pyrotechnic components (Ubbelohde 1955). Great interest has been directed towards the reactivity of highly exothermic solids and work has been concerned with chemical changes controlling energy release at high temperatures.

4.1 *Interface advance reactions*

4.1.1 THEORY

The essential processes that control the kinetics of interface reactions are the rate of interface creation (nucleation) followed by the rates and directions of interface advance (growth). Investigations of the rate characteristics of both processes require quantitative measurements of the changes in nucleus numbers and sizes (in three dimensions) during the reaction, but most particularly at low α. These observations are most easily obtained from microscopic examinations; see, for examples, Acock *et al.* 1947; Garner and Jennings 1954; Garner 1955; Lyakhov and Boldyrev 1972.

A systematic derivation of the various rate equations applicable to reactions of solids, based on the kinetic characteristics of nucleation and growth processes, was given by Jacobs and Tompkins (1955). This classic treatment of the topic remains a valuable and useful source. These authors identified four nucleation rate laws: instantaneous, linear, exponential (deceleratory) and power (acceleratory). (See also Allnatt and Jacobs (1968) for a more theoretical discussion of nucleation.) They also concluded that, for the most relevant reactions, the rate of linear interface advance was constant, although there may sometimes be an initial period of relatively slower growth of small nuclei. Many nuclei are hemispherical, although other shapes are sometimes observed (Garner 1955) due to preferred interface advance in particular crystallographic directions. Growth of nuclei can occur in three, two or one dimensions.

Integration of appropriate combinations of the above nucleation and growth laws, with due regard for the number of dimensions in which nuclei grow (and including necessary approximations and assumptions), has yielded the set of kinetic expressions now regarded as characteristic of crystolysis reactions (Jacobs and Tompkins 1955; Young 1966; Delmon 1969; Brown *et al.* 1980; Sestak 1984). The essential feature of this approach is that an isothermal α–time curve measured for a solid state decomposition is represented by a geometric model based on systematic changes in the spatial disposition of the reaction zone. Interpretations must be supported, whenever possible, by microscopic observations. It is important also to consider reactant crystal shapes and, for powders, the range of shapes and distribution of crystallite sizes.

4.1.2 REVERSIBLE REACTIONS

Reaction reversibility may exert an important control on reaction rate. Accumulations of volatile product in the vicinity of the reaction interface (and the solid product) may inhibit the dissociation rate or even result in the cessation of reaction. Such slowing may arise indirectly when an inert gas effectively prevents volatile product escape and thereby diminishes its overall rate of release.

A further possible complication, most pronounced in the kinetic analyses of reversible rate processes, is the influence of heat flow on reactant temperature. The endothermic character of these reactions may result in reactant self-cooling, in contrast with self-heating observed for some exothermic reactions which, in extreme conditions, can sometimes lead to combustion or explosion (Ubbelohde 1955).

In a comment on the problems encountered in his interpretation of the kinetics of the endothermic dissociation of calcite, Draper (1970) stated that he could *design* the shape of the α–time curve through control of heat flow to the reactant. Either sigmoid or deceleratory yield–time plots could be obtained by suitable selection of reaction conditions. Here, the supply of heat, not CO_2 escape, was identified as the rate controlling factor.

4.1.3 TESTING OF RATE DATA FOR KINETIC FIT

Inspection of the overall shape of an isothermal α–time plot can be used as a qualitative guide to the mechanism of a crystolysis reaction. A sigmoid curve, having pronounced induction and acceleratory periods, suggests a difficult nucleation step followed by growth. A rate process that is deceleratory thoughout suggests facile initial nucleation, to establish rapidly a coherent superficial layer of product over all boundaries, and the interface advance is inwards thereafter. More precise methods are required for quantitative work. Approaches used to test the fit of kinetic data to rate expressions may be summarized as follows (Brown *et al.* 1980).

1 Linearity of a plot of $f(\alpha)$ against time. The slope of the linear region is a measure of the rate constant and the α-range of applicability can be readily interpolated from the graph.
2 Comparisons can be made to find the most precise fit between α–reduced time measured data and values calculated for the kinetic expressions, which have been tabulated (Sharp *et al.* 1966).
3 Comparisons are made between plots of measured rate ($d\alpha/dt$) values against α and the lines obtained from appropriate rate expressions on the same graphs (Delmon 1969).
4 Comparisons are made between plots of measured rate ($d\alpha/dt$) values against reduced time and the lines obtained from appropriate rate expressions on the same graphs (Selvaratnam and Garn 1976).
5 Plots are made of measured ($d\alpha/dt$) values against the differential form of the rate equation, $(d\alpha/dt) = f'(\alpha)$; a straight line is evidence of fit (Brown and Galwey 1995a).

Recent advances in instrumentation, exploiting pressure gauges or balances of increased accuracy, have enabled the collection of data that permit kinetic analyses to be based on rate ($d\alpha/dt$), t, α measurements in preference to α, time data. This is required by (3), (4) and (5) above. The increased availability of computers having flexible programs and large memories permits the wide use of statistical tests in kinetic analyses. Brown and Galwey (1979) discussed the distinguishability of different rate equations when deciding which of the possible alternatives gives the best fit to a set of data. An unusually careful kinetic analysis of the fit of α–time data for the decomposition of NH_4ClO_4 was given by Jacobs and Ng (1972). This article demonstrates the importance of precise and detailed investigation of rate behaviour during the earliest stages of reaction.

4.2 *Reaction mechanisms other than interface advance*

Whilst reaction geometry and effective surface areas remain an essential consideration in the kinetic analyses of all reactions of solids, other factors have been identified as exerting influence or control on rates of chemical change. The most important of these is *diffusion* and there are the less well-characterized *chain reactions* and *homogeneous mechanisms* discussed below.

4.2.1 DIFFUSION

Crystalline solids composed of two types of constituents: (i) extended coherent groups having strong internal bonding; and (ii) intercalated groups that are retained by weaker forces, may lose the latter components without recrystallization. In the absence of structural reorganization, no interface is generated, although there may be some change of lattice parameters. The rate of loss of the smaller and more mobile species is controlled either by the velocity of diffusive migration to the crystal edge or, less usually, the subsequent volatilization step. Under control by the former condition, a concentration gradient inwards from the reactant edge may be established. Reactions of this type include the dehydrations of some of the layer aluminosilicates, for example vermiculite, from which water coordinated with Mg^{2+} is lost from between the talc-like layers (Okhotnikov and Babicheva 1988; Okhotnikov *et al.* 1989a). Diffusion control has also been discussed for $CaSO_4.2H_2O$ dehydration where, again, the reactant has a layer structure (Okhotnikov *et al.* 1987). Ball and Norwood (1969) and Ball and Norwood (1970) find that loss of water from $CaSO_4.2H_2O$ is diffusion controlled above 383 K, but behaviour below this temperature is complicated where both product nucleation and boundary control are important.

At reaction interfaces, crack propagation resulting in disintegration of the product solid, arising through the volume changes and associated local strain accompanying recrystallization, provide the channels that facilitate volatile product escape. The absence of an interface in diffusion-controlled processes enables the bond structure characteristic of the reactant to be maintained across the reaction zone. This is of particular interest in mechanistic studies because it implies that reaction results in relatively small displacements and distortions of the reactant structure. The loss of order accompanying reaction is small. Reactions are said to be *topotactic* when the product bears a crystallographic relationship with the reactant and where this is maintained during the chemical change of the entire crystal. Topotactic reactions have been the subject of important reviews by Oswald (1980) and Boldyrev (1990), which place the phenomena in the perspective of the subject.

The dehydration of uranyl nitrate hexahydrate ($UO_2(NO_3)_2.6H_2O$) crystals is unusual in that it does not occur at an advancing interface and diffusion control is excluded on kinetic evidence. Water site vacancies were shown to be highly mobile in the reactant phase. It was concluded (Franklin and Flanagan 1972) that the slow step is water desorption from the crystal boundary where the water concentration is maintained effectively constant. This crystal boundary may be regarded as an immobile interface.

4.2.2 CHAIN REACTIONS

Reaction mechanisms in which the energy released during a specific decomposition step is transferred directly to activate another reactant species have been shown to be untenable (Brown *et al.* 1980). Energy dispersal is expected to be more effective than the transfer step required. Interruption of reactions exhibiting sigmoid-shaped α–time curves by cooling, which is expected to destroy energy chains, is usually followed on reheating by the resumption of the rate measured prior to the chilling. No further induction period is required, showing that the entities or structures, responsible for maintaining the reaction at its previously established rate, have not been dispersed or destroyed, see, for example, the decomposition of $KMnO_4$ (Prout and Tompkins 1944).

An alternative representation of a chain-type model identifies the factor propagating the change as strain-induced crack development within the active reaction zone. Reactant fragmentation exposes new surfaces upon which reaction is soon continued. This, in turn, generates further strain that, in time, causes new fractures, which exposes further reactive surfaces (Prout and Tompkins 1944). The decomposition of $KMnO_4$ was further considered by Hill *et al.* (1966) in the diffusion chain theory. The reaction model described may alternatively be regarded as an irregular and expanding reaction interface that can be treated as a nucleation and growth reaction (Brown *et al.* 1994). The recent literature contains relatively few references to chain reactions. It may be concluded, therefore, that this theoretical approach to solid state reactivity is currently regarded as having little value at temperatures below the onset of combustion or explosion.

4.2.3 HOMOGENEOUS REACTIONS

Reactions envisaged as proceeding *within* the crystalline reactant, without extensive distortion, have sometimes been referred to as homogeneous reactions. This descriptive label has been applied to the dehydrations of $Mg(OH)_2$ and of $Ca(OH)_2$ where proton transfer steps are involved (Brett *et al.* 1970). This reaction type is mentioned here for completeness, although few examples of such behaviour are to be found in the recent literature. Aspects of the mechanism of $Ca(OH)_2$ decomposition have been discussed by Chaix-Pluchery and Niepce (1988), who were concerned with the crystal modifications that precede water loss. Galwey and Laverty (1993) discuss a reaction mechanism involving water release followed by later product recrystallization.

4.3 *Composite reaction interface: chemical change followed by subsequent product recrystallization*

The interface reactions and diffusion control models described above are not to be regarded as mutually exclusive. There is increasing evidence, particularly from studies of dehydrations, that an interface may be a composite structure. Diffusive loss of water from the zone beyond the recrystallization boundary yields a layer of metastable, partly dehydrated salt that later recrystallizes to the product structure at what may be regarded as the advancing interface. Boldyrev *et al.* (1987) have demonstrated the existence of progressive loss of water from reactant in the zone immediately in advance of the progressing crystallographic change during the dehydrations of $Li_2SO_4.H_2O$ and of $CuSO_4.5H_2O$. Later, Okhotnikov *et al.* (1989b) showed that the rate of water loss from $Li_2SO_4.H_2O$, before the establishment of nuclei, was attributable to the diffusion-controlled removal of water molecules through a metastable intermediate phase.

This model has subsequently been developed in a detailed kinetic study of the dehydration of *d* $LiKC_4H_4O_6.H_2O$ (Galwey *et al.* 1994). The rate of water loss during the first reaction was shown to be diffusion controlled and limited in extent to a thin superficial layer of the crystal. The same process was identified as controlling the rate of interface advance during the second reaction; a nucleation and growth process. Essential features in this reaction model are shown in diagrammatic form in Fig. 2.5. The linear velocity of the (possibly intermittent, step-wise) advance of the interface is determined by the rate of diffusive water loss, which generates the partially dehydrated (metastable) material. Conversion to the anhydrous product phase is promoted by the crystalline anhydrous particles with which it is in contact (seed crystals). Initially, water loss is facile but progressively decreases in rate due to loss of order, and continuation necessitates the difficult recrystallization step to the product structure. This model is consistent with the observed changes in reactivity and texture which occur after exposure of partially dehydrated surfaces to water vapour (Guarini and Dei 1983); all surfaces undergo limited water losses in vacuum. An essential contribution from a diffusion step in dehydrations is probably more widespread than has been recognized in the past.

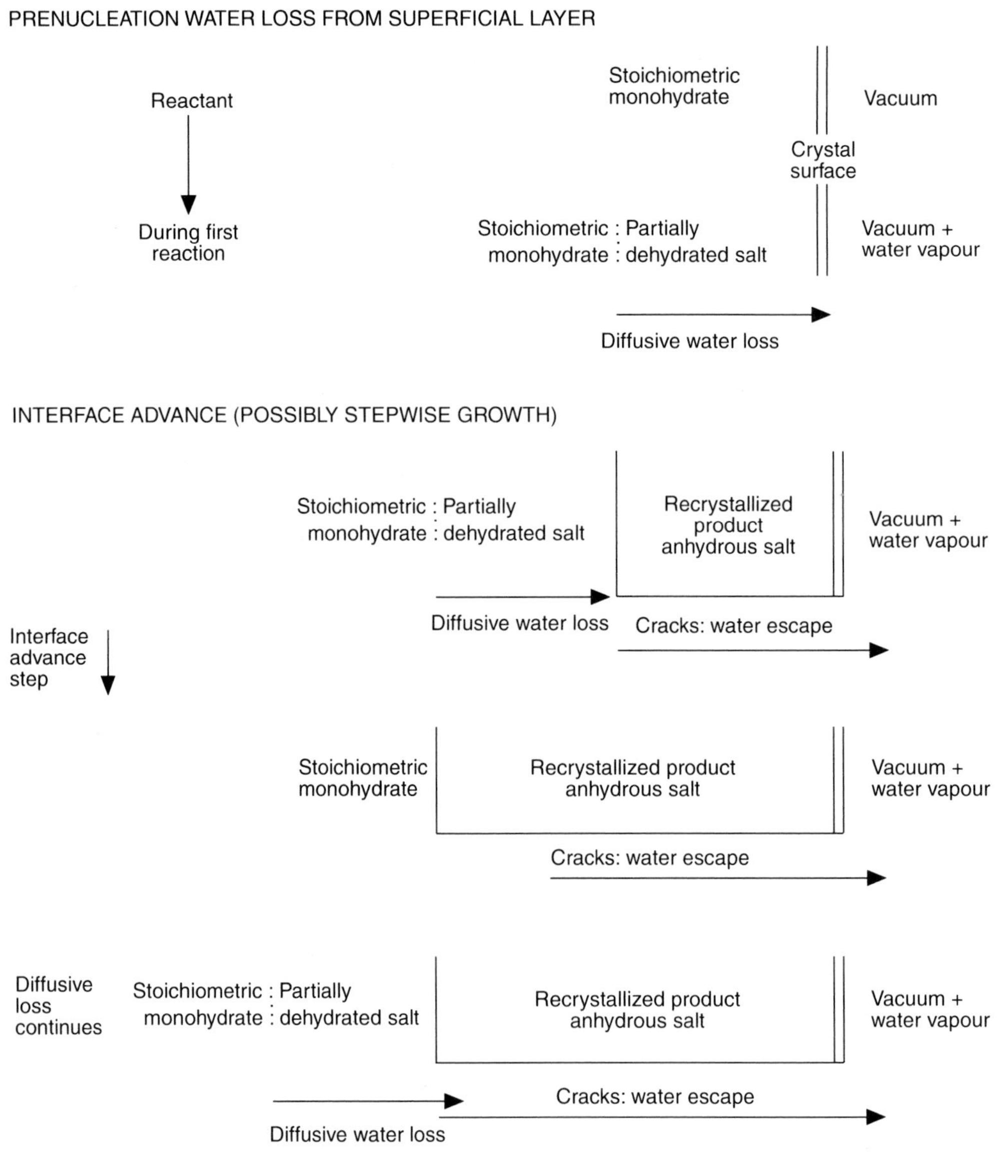

PRENUCLEATION WATER LOSS FROM SUPERFICIAL LAYER

Reactant

Stoichiometric monohydrate

Vacuum

Crystal surface

During first reaction

Stoichiometric monohydrate : Partially dehydrated salt

Vacuum + water vapour

Diffusive water loss

INTERFACE ADVANCE (POSSIBLY STEPWISE GROWTH)

Stoichiometric monohydrate : Partially dehydrated salt

Recrystallized product anhydrous salt

Vacuum + water vapour

Diffusive water loss

Cracks: water escape

Interface advance step

Stoichiometric monohydrate

Recrystallized product anhydrous salt

Vacuum + water vapour

Cracks: water escape

Diffusive loss continues

Stoichiometric monohydrate : Partially dehydrated salt

Recrystallized product anhydrous salt

Vacuum + water vapour

Cracks: water escape

Diffusive water loss

Figure 2.5 Diagrammatic representation of roles of diffusion and of recrystallization resulting in interface advance during the dehydration of *d*-$LiKC_4H_4O_6.H_2O$ (Galwey *et al.* 1994).

5 Dependence of reaction rate on temperature: the Arrhenius equation

The variation with temperature of rate constants for thermal decompositions of solids is almost invariably well represented by the Arrhenius equation (see Equation 2.2, p.31).

The precise fit of this relationship does not appear to have been frequently tested quantitatively in recent times. However, no alternative expressions relating k and T have been recommended for theoretical reasons, and the accuracy of fit of Equation (2.2) to most solid state reactions is probably as satisfactory as is, in general, found for most homogeneous rate processes. The empirical applicability of Equation (2.2) to crystolysis reactions appears to be universally accepted. Moreover, theoretical discussions of kinetic observations for solid state reactions are presented in the context of the theory developed most successfully in homogeneous kinetics. This was originally based on collisional activation and later refined by absolute reaction rate theory.

The temperature dependences of solid state reactions are conventionally reported as the calculated activation energy (E), usually implicitly regarded as the energy required to promote reactants to the transition state complex. The pre-exponential term (A) is usually identified as the frequency of occurrence of the reaction situation and includes the vibration term in the reaction coordinate. Often, the magnitude of A is regarded as providing a measure of the change of order during activation (entropy). An early mechanism of interface reactions was proposed by Polanyi and Wigner (1928), and subsequent theoretical discussions of reaction models, applicable to the reactant–product contact, have been given by Shannon (1964) and Cordes (1968).

Magnitudes of A and E measured for particular rate processes provide an important quantitative method of expressing and comparing reactivity. These parameters, together with a knowledge of the applicable rate equation, enable the extent of reaction to be calculated after heating at selected values of t and T. Numerous pairs of (A, E) values are to be found in the literature, reported for many and diverse crystolysis reactions (Brown *et al.* 1980); often, the reported E value might appear to be regarded as the most significant outcome of a particular study.

Despite these apparent successes of the Arrhenius model, there remains an important, general and fundamental difficulty in the direct application of homogeneous reaction rate theory to rate processes involving solids. The theory supporting Arrhenius behaviour is based on a quantitative consideration of the collisional interactions between freely moving entities; themselves possessing a Maxwell–Boltzmann distribution of energies. This is essentially different from the situation in a solid (Garn 1975, 1976, 1978), where all constituents of the reactant possess highly restricted mobilities and the energy distribution function is different.

The problem of extending our theoretical understanding of the factors determining rates, A and E for solid state reactions may be summarized as follows. Knowledge is required about the relevant properties of the participating species, including identification of the reacting entities, their interactions with their immediate environments and the energy distribution functions applicable to the reactive bonds. We have, however, little direct information concerning the reaction situations for most solids. Difficulties attend the characterization of the precise electron redistribution steps required to convert reactant into product. Investigations of interface structures on a subatomic scale are difficult and textures are liable to be distorted by the introduction of probes or exposure of the participating sites. The steps involved in an intercrystalline environment can be expected to be inherently more complicated than in homogeneous processes where the relatively simpler collisional transfer of energy can be modelled more acceptably.

A recent reconsideration of the activation step in solid state processes has been given by Galwey and Brown (1995b). The functions expressing the distribution of energy in solids are based on Fermi–Dirac statistics for electronic energy and Bose–Einstein statistics for phonon energy. In both of these treatments, when activation energy values are several times greater than kT (as observed for many solid state reactions), the distribution functions approximate to the same form as that given by the Maxwell–

Boltzmann function. If reaction follows promotion of energy (either electron or phonon) to an *interface level* (i.e. formally similar to an impurity level as represented by the band theory in considering semi-conductors), this accounts for the observed dependence of rate upon temperature for reactions of solids. This provides a theoretical explanation for the widely reported fit of data to Equation (2.2).

5.1 *Accurate measurement of activation energy values*

The accurate measurement of E from isothermal experiments requires precise control of temperature in each contributory experiment (see Brown *et al.* 1980, p. 83). Uncertainties in E values reported in the literature are difficult to estimate realistically and magnitudes of probable errors are important but are not invariably reported.

The determination of E values for reversible reactions is more difficult than is usually appreciated. In a limited number of exceptional studies, careful use of conditions designed to eliminate completely *any* contribution from the reverse process has demonstrated the inherent difficulties. Somewhat surprisingly, this type of approach has remained exceptional rather than a recognized route towards the advancement of the subject and the collection of meaningful values.

Flanagan *et al.* (1971) measured the dehydration rates of nickel oxalate dihydrate ($NiC_2O_4.2H_2O$) using a range of very small reactant masses, and extrapolated the results of zero mass. This eliminated the contribution from water vapour. The value of E determined (129 kJ mol^{-1}) was appreciably larger than previously reported results.

Beruto and Searcy (1974) measured the rates of CO_2 evolution during the decomposition of $CaCO_3$ at exceptionally low pressures ($< 10^{-4}$ Torr; 1.3×10^{-2} Pa), and showed that the E value thus measured (205 kJ mol^{-1}) was appreciably larger than the dissociation enthalpy (178 kJ mol^{-1}). These results show conclusively that it is essential to study reversible reactions under conditions where *only* the forward processes contribute, otherwise the activation energy will be comparable with the reaction enthalpy (Beruto and Searcy 1974). There must remain considerable doubt about the accuracy and, therefore, the value of many reported magnitudes of E in the literature for reversible reactions.

5.2 *The compensation effect*

For several groups of reactions, related through some common feature such as similar reactant constituents, it has been observed that the effect of a change in one Arrhenius parameter is compensated by a change in the other. This is most usually expressed by the relationship (Brown *et al.* 1980):

$$\mathrm{Ln}\, A = \mathrm{B} + \mathrm{e}E \tag{2.1}$$

An important property of sets of (A, E) values that exhibit this interdependency is that there exists an isokinetic temperature at which all the reactions of the group proceed at the same rate (Bond 1962). Compensation effects are not restricted to solid state decompositions and have been reported for a wide and diverse range of types of rate processes (Galwey 1977). There is, however, no generally accepted theoretical explanation of the phenomenon. Many kineticists regard at least some (perhaps most) examples of this type of behaviour as arising through unreliable rate data or through experimental artifacts. There are a variety of theoretical explanations (Galwey 1977). One possible reason for the existence of an isokinetic temperature is that there is a common temperature of onset within a set of reactions involving compounds of similar reactivities, but for which the individual temperature coefficients are appreciably different.

Two examples of reports describing compensation behaviour may be cited to emphasize the absolute importance of using reliable measurements for the interpretation of kinetic data. These examples identify systems in which it would appear that compensation effects have arisen through experimental artifacts. Zsako and Arz (1974) identify compensation behaviour within an exceptional range of variability for Arrhenius parameters measured for $CaCO_3$ decomposition ($108 < E < 1600$ kJ mol^{-1} and $10^2 < A < 10^{69}$). This range sharply contrasts with the precise measurement of E reported by Beruto and Searcy (1974), under the conditions described above, where dissociation only was investigated. Alvarez *et al.* (1981) studied reactions of $(CH_3NH_3)_2\,MnCl_4$ under isothermal conditions and four rates of temperature increase (2.5–20°C min^{-1}) (non-isothermal studies are discussed below). Data were analysed by several alternative theoretical methods for the calculations of A and E. Reported magnitudes of E, for the same reaction, were about 60–215 kJ mol^{-1} and A values were about 10^5–10^{18} min^{-1}. That these widely varying values exhibited compensation behaviour is hardly surprising because all the observations refer to the same process studied within comparable temperature intervals.

6 Non-isothermal reaction kinetic studies

In principal, the non-isothermal approach to kinetic measurements offers an exceptionally attractive method for the determination of the rate equation, $f(\alpha) = kt$, together with A and E, from a single experiment for the selected reactant by the use of the three applicable equations:

$$k = A \exp(-E/RT) \tag{2.2}$$

$$(d\alpha/dt) = k f'(\alpha) \tag{2.3}$$

$$T = T_0 + b(dT/dt) \tag{2.4}$$

This method has motivated many enthusiasts so that the literature now reports uncounted numbers of kinetic measurements of this type, concerned with an exceedingly diverse range of reactants. Instruments are available commercially that will record values of $(d\alpha/dt)$, α, t, T at programmed time intervals for specified heating rates. Such equipment includes, or may be connected to, computers that store the raw data and provide facilities for kinetic analyses.

This chapter cannot comprehensively survey the extensive literature devoted to non-isothermal investigations of solid state reactivity. It should be pointed out, however, that this field would benefit from authoritative and critical surveys to present the information available in a systematic accessible form and to place the reactivities reported in the context of related areas of chemistry. Also in need of comprehensive and critical comparisons are (in my opinion) the strengths and weaknesses of the mathematical methods used for non-isothermal kinetic analyses, the assumptions employed in the analyses of raw data and in the extraction of kinetic conclusions, including the identification of the rate equation, $f(\alpha) = kt$, together with the calculations of A and E. A meaningful account of the achievements of the past, presented in a form recognizing the weaknesses and gaps, together with signposts towards the directions of most profitable future progress, is well overdue in this subject. Whilst non-isothermal measurements have undoubtedly contributed much to our knowledge of chemical reactivity, it still appears that there are a number of shortcomings inherent in this general method which have received insufficient attention. Some of these are mentioned below. These comments are intended to be constructive in identifying the directions that would enable the methods to realize their greatest potential.

The non-isothermal method has considerable and obvious value in providing a rapid preliminary method for the identification of chemical changes occurring on heating a solid, and indicating the temperature intervals appropriate for use in kinetic studies. Thermal methods (differential scanning calorimetry (DSC) and differential thermal analysis (DTA)) detect both chemical and physical changes and enable reaction enthalpies to be measured. Weight loss determinations (thermogravimetric analysis (TGA)) by gravimetric techniques, can be concerned with chemical changes only and contribute towards establishing reaction stoichiometry.

The reliance that can be placed upon kinetic conclusions deduced from non-isothermal measurements is more controversial (Carr and Galwey 1986). Magnitudes of E from isothermal experiments are usually accepted as more trustworthy (see, for example, Thomas and Clarke 1968), because fewer variables are involved in the calculation. The following problems now appear to require further critical consideration to enhance the reliability of kinetic conclusions based on rising temperature measurements, and to enable the results to contribute to the interpretation of solid state reactivity and mechanistic problems.

6.1 *Kinetic fit*

Difficulties are experienced in finding the best fit for rate expressions, $f(\alpha) = kt$, for isothermal α–time data (Brown *et al.* 1980). Accordingly, the uncertainties inherent in determining the rate expression and concurrent measurement of A and E from non-isothermal data must be very much greater. The problem is not overcome by the use of the powerful computing techniques available because inherent inaccuracies are introduced in the assumptions required by the mathematical approaches used. A useful simplification, that has gained less widespread acceptance than might have been expected, is to establish the rate equation from an isothermal experiment for use in the calculation of A and E from a rising temperature experiment. Thomas and Clarke (1968) demonstrated the particular value of this approach in a study of $Mn(HCOO)_2.2H_2O$ dehydration, which exhibits the additional advantageous feature of proceeding at a constant rate between $0.15 < \alpha < 0.5$.

Geometric conclusions deduced from kinetic analyses usually benefit from confirmation through microscopic observations, although these are certainly not always reported in rising temperature studies. Many workers appear to prefer the automated instrumental approach to direct observational techniques. Perhaps future progress can be made more effective through the complementary use of both (and other) methods.

6.2 *Equations used in kinetic analyses*

The three complementary equations, (2.2), (2.3) and (2.4), given above, upon which non-isothermal kinetic analysis is based, cannot be integrated generally. A variety of approximate and special case solutions are to be found in the literature (see also Brown *et al.* 1980). Further methods and refinements (sometimes relatively minor) continue to be reported. However, the most widely employed calculations are possibly those that assume deceleratory behaviour, expressed as a reaction order. This approach clearly must introduce errors if applied uncritically to reactions that are represented by sigmoid α–time plots.

Many studies to be found in the literature report non-isothermal kinetic results, but the chemical significance to be attached to these observations is not always clear. Alvarez *et al.* (1981) showed that kinetic parameters calculated by different analytical methods give appreciably different A and E values for the same reaction. Zsako *et al.* (1981) report *apparent* values of A and E that show significant variations with reactant masses and heating rates. The chemical contribution from non-isothermal measurements of kinetic parameters will be substantially increased when reasons for these variations are understood, and it is possible to obtain reproducible Arrhenius parameters that agree with those obtained by isothermal methods. It would also be helpful if authors provided justifications for the analytical procedures selected for use in their kinetic analyses.

6.3 *Unusual kinetic behaviour*

Changes in kinetic characteristics with temperature, or perhaps the overlap of consecutive rate processes, may not be detected during non-isothermal studies, unless specifically sought. Automation of the mathematical analytical calculations reduces the ability of the researcher to detect unusual or untypical patterns of reactivity. This potential problem is most threatening when using sophisticated computers that are designed to be 'user-friendly'. Ability to inspect raw data and to vary methods of mathematical analysis avoids the possibility of unquestioningly accepting as absolutely reliable the data printed out by a 'black box'.

6.4 *Conditions within the reaction vessel*

Reaction conditions experienced by the (often rather small) reactant sample during a progressive heating regime are not always appropriate for the determination of the most accurate and reproducible kinetic data. Although minimized as far as practicable by instrument design, the sample temperature must lag behind that of its progressively heated container and there may be further variations within a layer of fine crystals.

The prevailing atmosphere may change during reaction, depending on the relative rates of volatile product release and escape (if possible) from the reactant container. This may not be important for studies of an irreversible rate process but, where a dissociation is reversible, the kinetic characteristics will at least be influenced and at worst dominated by interactions between the products. Melting, or a phase transition, cannot be distinguished from a chemical change with which it overlaps in a heat detection instrument, and this contribution must introduce an error into the calculated kinetic results. The presence of any gas capable of interacting with reactant, intermediate or product, even if present in small amount (for example, oxygen, water, CO_2, etc., impurity) in the gas flowing through the apparatus, may introduce an error into the heat output measured and, thus, for the kinetic analysis. The extent of secondary reactions between primary products, perhaps catalysed by containing vessel walls, is a further source of uncertainty. Conditions obtained within the small containers, characteristic of rising temperature experiments and in a gas flow, are probably more susceptible to disturbing influences than in most types of vacuum apparatus used in isothermal studies.

7 Selected crystolysis reactions: histories of mechanistic developments

The present status of the theory of thermal decomposition reactivity, to be used to provide pointers to the future, must be considered in the context of the lessons from history. It has been a feature of the field that the early studies focused attention on a relatively small number of reactant solids that became, perhaps, implicitly regarded as models representative of the whole subject. Some of these are discussed in the present section to illustrate the somewhat irregular and spasmodic progress that has characterized the field. No general theory of reactivity solids has, as yet, gained acceptance. It is not possible to predict reliably the kinetic behaviour for thermal decomposition of a hitherto untested solid reactant.

It is perhaps the (often unstated) hope, inspiring many mechanistic studies of crystolysis reactions, that the factors determining reactivity can be identified and will be shown to have wider application. However, a realistic historical appraisal of the field must lead us to conclude that recent advances tend to point towards increasing complexity of interface structures and that different reactants often exhibit individual behavioural characteristics. Early quantitative reaction models, for example the Polanyi–Wigner representation of $CuSO_4.5H_2O$ dehydration (Topley 1932), have not attained or maintained general acceptance. Now, more than 60 years later, a quantitative explanation of reactivity in interface phenomena remains an unrealized but still sought-after objective. The descriptions of reversible and irreversible reactions as 'normal' or as 'abnormal' (Garner 1955), on the basis of magnitudes of apparent A and E values, has been the subject of adverse comment (Galwey 1994b). Nevertheless, substantial progress has been made towards the more limited goal of formulating mechanistic representations of particular reactions that have been the subject of detailed examinations. The histories of some of these advances are outlined below.

7.1 *Dehydration reactions*

7.1.1 $CuSO_4.5H_2O$

This salt has been identified (Galwey and Laverty 1990) as one of the earliest to which the term nucleation was meaningfully applied (Cumming 1910). Subsequent work, reviewed by Young (1966) and Brown *et al.* (1980), was concerned with establishing and explaining the alternative reaction conditions under which this reactant yielded the monohydrate directly, or under which the trihydrate was formed as an intermediate. Investigations have been concerned with the effect of water vapour pressures on reaction rates, shapes of nuclei and whether the first-formed product is amorphous or a zeolite glass. An isothermal kinetic study of $CuSO_4.5H_2O$ dehydration in the atmosphere was reported in Ng *et al.* (1978). Plots of α against time were sigmoid shaped and E values were close to the reaction enthalpy. In a high-vacuum study, Okhotnikov and Lyakhov (1984) chose this reaction to test the reliability of their newly-developed quartz crystal microbalance. This was demonstrated by the 'substantial agreement' ($E = 73$ kJ mol^{-1}) they obtained with the earlier results by Smith and Topley (1931). Guarnini and Dei (1983) showed that there was a characteristic roughening of crystal faces after partial dehydration of $CuSO_4.5H_2O$ followed by exposure to water vapour. The behaviour observed was relatively complicated and there was surface modification prior to nucleus formation. Boldyrev *et al.* (1987) obtained evidence that there was at least partial loss of constituent water from salt immediately beyond the advancing recrystallization interface (see also Okhotnikov & Lyakhov 1984 and Okhotnikov *et al.* 1987).

7.1.2 ALUMS

Studies of the dehydrations of alums ($KAl(SO_4)_2.12H_2O$ and the NH_4^+ and Cr^{3+} salts) have contributed notably to the development of the nucleation and growth reaction model, particularly through the early careful microscopic observations (Cooper and Garner 1940; Acock *et al.* 1947; Garner and Jennings 1954; Garner 1955). The topic was also reviewed in detail by Young (1966). More recent studies, including the use of the scanning electron microscope to examine replicas of surfaces, have concluded that the function of the nucleus is to retain a proportion of the water product that participates in the difficult solid product recrystallization step (Galwey *et al.* 1981; Galwey and Guarini 1993).

7.1.3 $Li_2SO_4.H_2O$

Interest in the dehydration of $Li_2SO_4.H_2O$ developed later than the other hydrates mentioned. However, much recent attention has yielded an extensive literature (for example, Koga and Tanaka 1989, 1990, 1991). This reaction is appropriately mentioned here because it was proposed as a possible standard rate process to be used in comparative reactivity and kinetic studies by different laboratories. (Results were to have been critically compared in the same way as thermochemical and analytical standards are defined by observations by a selected group of acknowledged and independent experts.) This interesting enterprise appears, however, to have been abandoned before completion. The main reason may have been the unacceptably wide variations in behaviour between different studies of this possibly reversible reaction. Data reviewed by Brown *et al.* (1992, 1993) indicate that the accuracy of kinetic measurements for crystolysis reactions may be generally less reliable and reproducible than might have been hoped. The problems of irreproducibility and quantitative determinations of errors inherent in kinetic measurements for crystolysis reactions merits urgent and comprehensive consideration by kineticists.

7.2 *The dissociation of $CaCO_3$*

This is an interface reaction in which the nucleation step is believed (Thomas and Renshaw 1967) to occur mainly at the sites of surface intersections of dislocations that glide in the {100} and $\{2\bar{1}\bar{1}\}$ planes. There is some doubt as to whether this initiation arises from the strain energy associated with the dislocation or is due to local stereochemical distortion (Fox and Soria-Ruiz 1970). Kinetic behaviour is sensitive to reaction conditions, which may be varied to produce changes in rate characteristics that, in certain circumstances, are predictable (Draper 1970). Zsako and Arz (1974) discussed compensation behaviour for this dissociation, for which Arrhenius parameters probably show the greatest apparent range reported for any simple reaction (E from 108 to 1600 kJ mol^{-1} and A from 10^2 to 10^{69}).

Beruto and Searcy (1974) conclusively demonstrated the necessity for selecting reaction conditions capable of measuring only the reaction of interest by excluding here any contribution from the reverse process. At these very low CO_2 pressures the value of E (205 kJ mol^{-1}) was appreciably greater than the dissociation enthalpy, ΔH (178 kJ mol^{-1}); a result differing from much previous work that reported $E \approx \Delta H$. The importance of this result and reasons for differences from earlier studies are discussed in this significant paper by Beruto and Searcy (1974).

7.3 *The decomposition of $KMnO_4$*

The study of $KMnO_4$ decomposition by Prout and Tompkins (1944) was important because it provided a mechanistic explanation of autocatalytic behaviour that was of potentially wider applicability. The successful use of a chain-type reaction model to explain various homogeneous reactions was applied here to a solid state rate process, where continued crack disintegration of the crystal followed strain-inducing product formation on the surfaces so exposed. This model was later developed in the diffusion chain theory proposed by Hill *et al.* (1966).

The next significant advance was by Boldyrev (1969) who identified the controlling step as electron transfer within the anionic sublattice:

$$2MnO_4^- \rightleftharpoons MnO_4^{2-} + MnO_4 \rightarrow MnO_4^{2-} + MnO_2 + O_2 \qquad (2.5)$$

and $K_3(MnO_4)_2$ is now accepted (Herbstein *et al.* 1994) as a reaction intermediate. Recent microscopic studies (Brown *et al.* 1994) have encountered difficulties in attempting to characterize interface structures. The residual products may be ill-crystallized and amorphous and probably do not completely recrystallize to identifiable phases at reaction temperature.

7.4 *The decomposition of NH_4ClO_4*

The thermal decomposition of NH_4ClO_4 is certainly one of the most intensively studied crystolysis reactions, although interest has declined in recent years. The overall pattern of behaviour is explained in the comprehensive review by Jacobs and Whitehead (1969). The low-temperature reaction unusually (perhaps uniquely) results in the breakdown of only about 30% mass of the reactant salt, leaving a high-area residue with the same crystal structure and composition as the original NH_4ClO_4. Mechanistic studies have variously identified the controlling step in this reaction as:

1 breakdown of a transitory NH_4ClO_4 'molecule' (Galwey and Jacobs 1960);

2 proton transfer followed by decomposition of the $HClO_4$ so formed (Davies *et al.* 1967; Boldyrev *et al.* 1972);

3 electron transfer (Owen *et al.* 1972, 1974).

To resolve these inconsistencies and to explain the observed behaviour, Galwey and Mohamed (1984) and Galwey *et al.* (1988a,b) proposed an alternative mechanism in which decomposition proceeds through the molten intermediate NO_2ClO_4.

8 Classification of chemical characteristics of solid state decompositions

Classification of observed phenomena is an essential feature of the scientific method and progress. The grouping of like phenomena, thereby contrasting dissimilar patterns of behaviour, enables controlling principles to be recognized which may have value in predicting properties of hitherto untested systems. No useful classification criteria have, however, been identified for crystolysis reactions, except for the wide and relatively imprecise division into endothermic reactions, regarded as often being reversible, in contrast with many exothermic reactions that are irreversible. This basis for separation was used in the early general review by Garner (1955).

Other approaches to a systematic treatment of the subject that merit consideration are classification through the chemical constituents in the reactant or some aspect of the reaction mechanism, possibly the nature of the rate-controlling step (Boldyrev 1965, 1975a,b). Many of the crystolysis reactions that have attracted greatest interest have involved anion breakdown, and this has provided a useful method of presentation. Some patterns and trends of chemical similarities that have been discerned in the literature are mentioned below in the hope that these or similar criteria may be capable of extension and development in the future.

- *Oxalates.* Decompositions of many metal salts of oxalic acid have been studied (sometimes following a precursor dehydration step). The residual product, carbonate, oxide or metal, depends on the electropositive character of the cation (Dollimore and Griffiths 1970; Boldyrev *et al.* 1970). Mechanisms of interface reactions can be expected to vary with the chemical characteristics of the residual phase, for example some metals are more active in promoting oxalate anion breakdown than an oxide or carbonate product.
- *Formates.* General similarities with the oxalates can be expected and parallels may exist with the metal-catalysed decomposition of formic acid. Both reactions probably proceed through the intervention of a chemisorbed formate ion (Bond 1962; Inglis and Taylor 1969).
- *Malonates.* Decompositions of several malonates give significant yields of acetate as an intermediate or product. However, despite this chemical similarity, reaction mechanisms may be quite different. The solid state nucleation and growth decomposition of silver malonate (Galwey and Mohamed 1985) contrasts with the intervention of a melt (see Fig. 2.3), during the similar reaction of copper(II) malonate (Carr and Galwey 1986).

- *Copper(II) carboxylates.* A series of decompositions of copper(II) carboxylates (Galwey and Mohamed 1994) were shown to proceed with stepwise cation reduction, $Cu^{2+} \rightarrow Cu^{+} \rightarrow Cu^{0}$.

These relatively limited identifications of common mechanistic features serve to emphasize the *chemical* controls that may provide a route towards understanding the reactivities of solids. The known chemical properties of participating groups may include information essential in identifying the breakdown step at the interface or other rate-determining process. At the present time, however, there remain gaps in our knowledge concerning the detailed forms and functions of interfaces and how behaviour may differ between various, perhaps even apparently similar, rate processes. Many reports present a mechanistic discussion that is restricted to a single compound and no adequate basis for generalization to other crystolysis reactions has been recognized, as yet.

9 Thermal decomposition reactions: some comments on the future

The title of this book requires that some consideration should be given to the directions of expected future development of the subject. This can alternatively be regarded as the author's recommendations for profitable advance or speculations as to how the topic might develop hereafter. My input here includes both approaches. These extrapolations from the above brief and selective citations from the literature towards the future can at best be described (rather literally) as 'crystal gazing'. The opinions expressed are my personal 'best guess' and are based on a wide range of studies, together with inputs from many friends and colleagues who are active in the field. In front of each of the following sections, which include some overlaps of content, there is an implicit 'I believe'.

9.1 *Subject area*

The accepted convention of considering *thermal decompositions of solids* (or *crystolysis reactions*) as a distinct and separate subject area may be unduly restrictive. It could be more profitable to develop and to expand links with other thermal reactions, for example where there is partial or total melting during salt pyrolysis or with heterogeneous catalytic reactions that exhibit parallels with interface processes. Such an expansion of the range of mechanistic models for rate processes involving active solids may enrich the subject by finding parallels in wider chemical contexts.

9.2 *Structures of reaction zones (interfaces)*

Detailed characterization of the interfacial zone, including the textures of both reactant and product, together with their interrelationship, including topotaxy and the identification of intermediates, is essential in the establishment of reactivity controls including the rate-determining bond redistribution step. Developments in this field, particularly through the use of increased resolution microscopy to characterize local structures and their variations with distance by precise diffraction methods (Boldyrev *et al.* 1987), can be expected to increase our understanding of the chemistry of both nucleation and growth steps in interface advance reactions.

9.3 *Trends of chemical characteristics*

The establishment of chemical reasons for similarities of behaviour observed between related rate processes has, so far, been relatively limited. Some examples were mentioned in Section 8 above. Advances could result from the consideration of crystolysis reactions in a wider chemical context, for example with references to compounds that melt before or during decomposition, heterogeneous reactions, etc.

9.4 *Melting during reaction*

The intervention of melting, perhaps local and temporary, must be expected to influence, perhaps significantly change, reactivity. The participation of a liquid (reactant, product or intermediate, see Fig. 2.4) is an essential feature of any proposed reaction mechanism. Melting is not, however, invariably discussed explicitly in all literature reports concerned with the formulation of models for crystolysis reactions (Galwey 1994a). Some of the above references do mention rate processes that involve liquid participants. There is a need for kinetic studies of thermal decompositions that include melting to establish the rate characteristics of reactions in solvent-free melts. Comparisons of rate measurements would then enable the distinction to be reliably made as to whether the reaction was heterogeneous, and controlled by geometric factors, or was concentration dependent, and proceeded in a homogeneous phase.

9.5 *Accuracy of measured data*

A limited number of experimental studies have taken exceptional care to ensure that reaction conditions are demonstrably applicable to the process of interest. Examples of such work include the decomposition of $CaCO_3$ reported by Beruto and Searcy (1974) and the dehydration of $NiC_2O_4.2H_2O$ by Flanagan *et al.* (1971). These results demonstrate conclusively that meaningful conclusions are obtained here only when contributions from the reverse process have been effectively eliminated. This careful approach has not, however, received general acceptance, and reports of these and similar processes studied under much less well-defined conditions, even in the atmosphere, continue to appear. Literature reports must be read with reference to the limitations inherent in the experimental conditions used.

An obvious statement, not always implemented in practice, is that the most reliable mechanistic conclusions are those that are based on the widest possible variety of experimental evidence. To provide an acceptably complete description of any reaction, it is appropriate to obtain stoichiometric data, rate measurements, microscopic observations and all other relevant information that may be accessible (Carr and Galwey 1984). Studies of related reactions may be appropriate. The value of reported kinetic data is increased by realistic assessments of the accuracy of the raw measurements and the calculated values of A and E, together with the ranges of applicability of all reported kinetic fits.

9.6 *Literature*

The extensive literature (Carr and Galwey 1984) concerned with crystolysis reactions maintains growth through the continued publication of articles concerned with the reactivities of many diverse solids. A very approximate estimation indicated that over 100 articles appear in the latest 12 volumes of *Thermochimica Acta* (mid-1993–1994), which are identified from titles as probably being concerned with reactions of this type. Many further studies were reported at the most recent ICTA (1992) and ESTAC (1994) meetings. Crystolysis reactions continue to be studied as an important research topic and it is expected that interest will be maintained for the foreseeable future.

10 Acknowledgements

It is with the greatest pleasure that I offer my thanks for all the stimulation and help I have received from friends, colleagues, secretarial and technical staff as I now approach retirement after 40 years of activity in the field of crystolysis and other reactions.

11 References

Acock, G.P., Garner, W.E., Milsted, J. & Willavoys, H.J. (1947) *Proc Roy Soc Lond* **A189**, 508–526.
Allnatt, A.R. & Jacobs, P.W.M. (1968) *Can J Chem* **46**, 111–116.
Alvarez, M.R., Tello, M.J. & Bocanegra, E.H. (1981) *Thermochim Acta* **43**, 115–121.
Ball, M.C. & Norwood, L.S. (1969) *J Chem Soc A* 1633–1637.
Ball, M.C. & Norwood, L.S. (1970) *J Chem Soc A* 1476–1479.
Begg, I.D., Halfpenny, P.J., Hooper, R.M., Narang, R.S., Roberts, K.J. & Sherwood, J.N. (1983) *Proc Roy Soc Lond* **A386**, 431–442.
Beruto, D. & Searcy, A.W. (1974) *J Chem Soc Farad Trans I* **70**, 2145–2163.
Boldyrev, V.V. (1965) *Kinet Katal* **6**, 934–935.
Boldyrev, V.V. (1969) *J Phys Chem Solids* **30**, 1215–1223.
Boldyrev, V.V. (1975a) *J Thermal Anal* **7**, 685–694.
Boldyrev, V.V. (1975b) *J Thermal Anal* **8**, 175–194.
Boldyrev, V.V. (1979) *Ann Rev Mat Sci* **9**, 455–469.
Boldyrev, V.V. (1990) *React Solids* **8**, 231–246.
Boldyrev, V.V., Savintsev, Y.P. & Moolina, T.V. (1972) *Reactions of Solids, Proceedings of 7th International Symposium Bristol,* pp. 421–430. Chapman & Hall, London.
Boldyrev, V.V., Nev'yantsev, I.S., Mikhailov, Yu, I. & Khairetdinov, E.F. (1970) *Kinet Katal* **11**, 367–373.
Boldyrev, V.V., Gapanov, Y.A., Lyakhov, N.Z. *et al.* (1987) *Nucl Inst Meth Phys Res A* **261**, 192–199.
Bond, G.C. (1962) *Catalysis by Metals*, pp. 412–416. Academic Press, New York.
Brett, N.H., MacKenzie, K.J.D. & Sharp, J.H. (1970) *Chem Soc Quart Rev* **24**, 185–207.
Brown, M.E. & Galwey, A.K. (1979) *Thermochim Acta* **29**, 129–146.
Brown, M.E. & Galwey, A.K. *Thermochim Acta* (Submitted).
Brown, M.E., Dollimore, D. & Galwey, A.K. (1980) *Comprehensive Chemical Kinetics*, Vol. 22. Elsevier, Amsterdam.
Brown, M.E., Galwey, A.K. & Li Wan Po, A. (1992) *Thermochim Acta* **203**, 221–240.
Brown, M.E., Galwey, A.K. & Li Wan Po, A. (1993) *Thermochim Acta* **220**, 131–150.
Brown, M.E., Galwey, A.K., Mohamed, M.A. & Tanaka, H. (1994) *Thermochim Acta* **235**, 255–270.
Carr, N.J. & Galwey, A.K. (1984) *Thermochim Acta* **79**, 323–370.
Carr, N.J. & Galwey, A.K. (1986) *Proc Roy Soc Lond* **A404**, 101–126.
Chaix-Pluchery, O. & Niepce, J.C. (1988) *React Solids* **5**, 69–78.

Cooper, J.A. & Garner, W.E. (1940) *Proc Roy Soc Lond* **A174**, 487–503.
Cordes, H.F. (1968) *J Phys Chem* **72**, 2185–2189.
Cumming, A.C. (1910) *J Chem Soc* 593–603.
Davies, J.V., Jacobs, P.W.M. & Russell-Jones, A. (1967) *Trans Farad Soc* **63**, 1737–1748.
Dei, L., Guarini, G.G.T. & Piccini, S. (1984) *J Thermal Anal* **29**, 755–761.
Delmon, B. (1969) *Introduction á la Cinétique Hétérogène.* Technip, Paris.
Dollimore, D. & Griffiths, D.L. (1970) *J Thermal Anal* **2**, 229–250.
Draper, A.L. (1970) *Proceedings of the Robert A Welch Foundation Conferences on Chemical Research. XIV Solid State Chemistry*, pp. 214–219. Houston, Texas.
Fox, P.G. & Soria-Ruiz, J. (1970) *Proc Roy Soc Lond* **A314**, 429–441.
Flanagan, T.B., Simons, J.W. & Fichte, P.M. (1971) *Chem Commun* 370–371.
Franklin, M.L. & Flanagan, T.B. (1972) *J Chem Soc A* 192–196.
Galwey, A.K. (1977) *Adv Catal* **26**, 247–322.
Galwey, A.K. (1990) *React Solids* **8**, 211–230.
Galwey, A.K. (1994a) *J Thermal Anal* **41**, 267–286.
Galwey, A.K. (1994b) *Thermochim Acta* **242**, 259–264.
Galwey, A.K. & Brown, M.E. (1982) *J Chem Soc Farad Trans I* **78**, 411–424.
Galwey, A.K. & Brown, M.E. (1995a) *Thermochim Acta* **269/270**, 1–25.
Galwey, A.K. & Brown, M.E. (1995b) *Proc Roy Soc Lond* **A450**, 501–12.
Galwey, A.K. & Guarini, G.G.T. (1993) *Proc Roy Soc Lond* **A441**, 313–329.
Galwey, A.K. & Jacobs, P.W.M. (1960) *Proc Roy Soc Lond* **A254**, 455–469.
Galwey, A.K. & Laverty, G.M. (1990) *Solid State Ionics* **38**, 155–162.
Galwey, A.K. & Laverty, G.M. (1993) *Thermochim Acta* **228**, 359–378.
Galwey, A.K. & Mohamed, M.A. (1984) *Proc Roy Soc Lond* **A396**, 425–440.
Galwey, A.K. & Mohamed, M.A. (1985) *J Chem Soc Farad Trans I* **81**, 2503–2512.
Galwey, A.K. & Mohamed, M.A. (1987) *Thermochim Acta* **121**, 97–107.
Galwey, A.K. & Mohamed, M.A. (1994) *Thermochim Acta* **239**, 211–224.
Galwey, A.K. & Pöppl, L. (1984) *Phil Trans Roy Soc Lond* **A311**, 159–182.
Galwey, A.K., Herley, P.J. & Mohamed, M.A. (1988a) *Thermochim Acta* **132**, 205–215.
Galwey, A.K., Herley, P.J. & Mohamed, M.A. (1988b) *React Solids* **6**, 205–216.
Galwey, A.K., Koga, N. & Tanaka, H. (1990) *J Chem Soc Farad Trans I* **86**, 531–537.
Galwey, A.K., Pöppl, L. & Rajam, S. (1983) *J Chem Soc Farad Trans I* **79**, 2143–2151.
Galwey, A.K., Spinicci, R. & Guarini, G.G.T. (1981) *Proc Roy Soc Lond* **A378**, 477–505.
Galwey, A.K., Laverty, G.M., Baranov, N.A. & Okhotnikov, V.B. (1994) *Phil Trans Roy Soc Lond* **A347**, 139–156; 157–184.
Garn, P.D. (1975) *J Thermal Anal* **7**, 475–478.
Garn, P.D. (1976) *J Thermal Anal* **10**, 99–101.
Garn, P.D. (1978) *J Thermal Anal* **13**, 581–593.
Garner, W.E. (1955) *Chemistry of the Solid State*, Chs 8 & 9. Butterworth, London.
Garner, W.E. & Jennings, T.J. (1954) *Proc Roy Soc Lond* **A224**, 460–471.
Guarini, G.G.T. & Dei, L. (1983) *J Chem Soc Farad Trans I* **79**, 1599–1604.
Herbstein, F.H., Kapon, M. & Weissman, A. (1994) *J Thermal Anal* **41**, 303–322.
Hill, R.A.W., Richardson, R.T. & Rodger, B.W. (1966) *Proc Roy Soc Lond* **A291**, 208–223.
Hutchinson, R.W. & Stein, F.P. (1974) *J Phys Chem* **78**, 478–481.
Inglis, H.S. & Taylor, D. (1969) *J Chem Soc A* 2985–2987.
Jacobs, P.W.M. & Ng, W.L. (1972) *Reactivity of Solids, Proceedings of the 7th International Conference Bristol*, pp. 398–410. Chapman & Hall, London.
Jacobs, P.W.M. & Tompkins, F.C. (1955) *Chemistry of the Solid State*, Ch. 7. Butterworth, London.
Jacobs, P.W.M. & Whitehead, H.M. (1969) *Chem Rev* **69**, 551–590.
Koga, N. & Tanaka, H. (1989) *J Phys Chem* **93**, 7793–7798.
Koga, N. & Tanaka, H. (1990) *J Thermal Anal* **36**, 2601–2610.
Koga, N. & Tanaka, H. (1991) *Thermochim Acta* **185**, 135–140.

Lyakhov, N.Z. & Boldyrev, V.V. (1972) *Russ Chem Rev* **41**, 919–928.
Lyakhov, N.Z., Chupakhin, A.P., Isupov, V.P. & Boldyrev, V.V. (1974) *Kinet Katal* **15**, 1224–1229.
Lyakhov, N.Z., Chupakhin, A.P., Isupov, V.P. & Boldyrev, V.V. (1977) *Kinet Katal* **19**, 84–89.
Ng, W.L., Ho, C.C. & Ng, S.K. (1978) *J Inorg Nucl Chem* **34**, 459–462.
Okhotnikov, V.B. & Babicheva, I.P. (1988) *React Kinet Catal Lett* **37**, 417–422.
Okhotnikov, V.B. & Lyakhov, N.Z. (1984) *J Solid State Chem* **53**, 161–167.
Okhotnikov, V.B., Babicheva, I.P., Musicantov, A.V. & Aleksandrova, T.N. (1989a) *React Solids* **7**, 273–287.
Okhotnikov, V.B., Simakova, N.A. & Kidyarov, B.I. (1989b) *React Kinet Catal Lett* **39**, 345–350.
Okhotnikov, V.B., Petrov, S.E., Yakobson, B.I. & Lyakhov, N.Z. (1987) *React Solids* **2**, 359–372.
Oswald, H.R. (1980) *Thermal Anal Proc 6th Int Conf Bayreuth.* Birkhäusen, Basel **1**, 1–14.
Owen, G.P., Thomas, J.M. & Williams, J.O. (1972) *J Chem Soc Farad Trans I* **68**, 2356–2366.
Owen, G.P., Thomas, J.M. & Williams, J.O. (1974) *J Chem Soc Farad Trans I* **70**, 1934–1943.
Polanyi, M. & Wigner, E. (1928) *Z Phys Chem* **A139**, 439–52.
Prout, E.G. & Tompkins, F.C. (1944) *Trans Farad Soc* **40**, 488–498.
Selvaratnam, M. & Garn, P.D. (1976) *J Am Ceram Soc* **59**, 376.
Sestak, J. (1984) *Thermophysical Properties of Solids.* Elsevier, Amsterdam.
Shannon, R.D. (1964) *Trans Farad Soc* **60**, 1902–1913.
Sharp, J.H., Brindley, G.W. & Achar, B.N.N. (1966) *J Am Ceram Soc* **49**, 379–382.
Smith, M.L. & Topley, B. (1931) *Proc Roy Soc Lond* **A134**, 224–245.
Thomas, J.M. & Clarke, T.A. (1968) *Nature Lond* **219**, 1149–1151.
Thomas, J.M. & Renshaw, G.D. (1967) *J Chem Soc A* 2058–2061.
Thomas, J.M. & Renshaw, G.D. (1969) *J Chem Soc A* 2753–2755.
Topley, B. (1932) *Proc Roy Soc Lond* **A136**, 413–428.
Ubbelohde, A.R. (1955) *Chemistry of the Solid State*, Ch. 11. Butterworth, London.
Young, D.A. (1966) *Decomposition of Solids.* Pergamon, Oxford.
Zsako, J. & Arz, H.E. (1974) *J Thermal Anal* **6**, 651–656.
Zsako, J., Varhelyi, C., Csegedi, B. & Zsako, J. (1981) *Thermochim Acta* **45**, 11–21.

3 Problems of Reactivity in the Dissolution of Inorganic Solids

M.A. BLESA

Umidad de Actividad Química Comisión Nacional de Energía Atómica, Avenida del Libertador 8250, (1429) Buenos Aires, Argentina and Facultad de Ciencias Exactas y Naturales, Universidad de Buenos Aires, Buenos Aires, Argentina

1 Introduction

Thermodynamically, a pure solid substance in contact with a liquid solvent should dissolve until the chemical potentials in both phases become equal. This condition may be achieved in a wide variety of conditions regarding the activity of the dissolving substance in the fluid phase, a_l. In this chapter we shall be mainly addressing the case in which a_l is not vanishing, and, furthermore, the time required to reach solubility equilibrium is not very short. We shall discuss the dissolution of inorganic solids essentially in aqueous solvents and, therefore, the equilibrium condition is in many cases best described by the equality of electrochemical potentials:

$$\tilde{\mu}(s) = \tilde{\mu}(l) \tag{3.1}$$

$$\tilde{\mu}(s) = \tilde{\mu}(l) = \mu(l) + z\mathcal{F}E \tag{3.2}$$

In Equation (3.2), z is the ionic charge, $\mathcal{F}$ = 96 500 C and E is the potential.

Metal oxides constitute a large class of inorganic compounds that are especially important for this review. For metal oxides, application of Equation (3.2) merits some comments. For stoichiometric ionic metal oxides $MO_{(z/2)}$, $\mu(s)$ is a constant at constant P and T, and Equation (3.3) describes the usual pH-dependency of solubility, at constant potential:

$$\tilde{\mu}(l) = \tilde{\mu}^0(l) + \text{RTln}\,(a_M/a_H^Z) \tag{3.3}$$

For non-stoichiometric oxides, MO_n where n may span a range around $(z/2)$, the solubility product, $K_s = (a_M/a_M^Z)$, changes with oxygen partial pressure:

$$MO_n + zH^+ = M^{z+} + \frac{z}{2}H_2O + \frac{1}{2}\left(n - \frac{z}{2}\right)O_2 \tag{3.4}$$

or:

$$\frac{\partial \ln K_s}{\partial \ln P_{O_2}} = 0.25\,(z - 2n) \tag{3.5}$$

where K_s is the solubility product as defined by Equation (3.4). For equilibrated oxides, the value of n is defined by the oxide defect structure (Heusler 1983).

For transition metal oxides several well-defined stoichiometric or nearly stoichiometric phases may exist, corresponding to different z values. In these cases, the solubility product is also P_{O_2} dependent or, alternatively, potential dependent. The solubilities of the various phases, in the potential range in which they are stable, are governed mainly by the cation charge : radius ratio and by the hard- or soft-acid character of the metal cation (Baes and Mesmer 1976).

The analysis of the thermodynamics of solubility does not require a description of the interfacial region. In contrast, the kinetics of dissolution are heavily influenced by the details of the interface structure. The importance of the truncation of the solid framework is already present in the theories of dissolution morphology. A nice example is the description given by Angus and Dyble (1975) of the dissolution patterns on (111) diamond surfaces. On these surfaces, both tri- and tetra-coordinated atoms are present. The assumption that di-coordinated atoms are rapidly transferred to the (undersaturated) liquid phase, leads to the prediction that triangular features develop on the surface, limited by <110> steps. All theories of crystal growth or dissolution based on nearest-neighbour interactions in the truncated solid surface also stress the importance of low coordination surface atoms (Sangwal 1987). These theories, when applied to the ideal Kossel crystal, lead to the usual identification of kink atoms as semi-crystalline, *i.e.* as the crucial entities in growth or dissolution. Thus, surface diffusion from kinks is identified as the transition from solid to adsorbed state, and is believed to regulate the rate of dissolution, when this process is not fast enough to be limited by mass–transport phenomena in the fluid phase.

Any attempt to achieve a comprehensive description of a dissolving solid should provide with an explicit expression for the rate of movement of the interface. For a surface roughly parallel to the (x,y) plane, the movement is essentially along the z direction, but the rate is a function of all spatial coordinates, as well as of time:

$$R = \frac{\partial z}{\partial t} = f(x,y,t) \tag{3.6}$$

In the kinematic theories of crystal dissolution, the changes in z result from the motion of steps along the (x,y) plane, and the rate R is heavily influenced by the density of steps and step height (Cabrera 1960; Sangwal 1987). It has been shown that these two characteristics are time dependent (Mullins and Hirth 1963). Even for atomically flat surfaces, dissolution is anisotropic and etch pits may develop upon nucleation. The etch pit nucleation Gibbs energy is, however, appreciably small at a dislocation site.

The evolution of surface morphology in the course of dissolution shall depend on the ratios of at least three individual rates (Sangwal 1987): the rate of attack of the smooth surface, and the normal and lateral rates of etch pits advance (Fig. 3.1). These rates are essentially determined by solid state properties, and the morphology of attack is

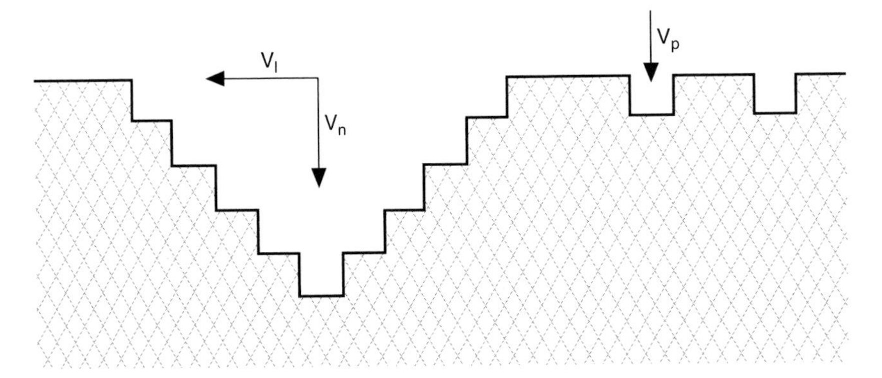

Figure 3.1 Generalized attack, characterized by the rate R_G, versus etch pit formation, characterized by the rates R_N and R_L. (From Sangwal 1987.)

therefore best addressed on the basis of crystallographic considerations (Angus and Dyble 1975; Wokulska 1978; Sangwal 1987).

The subject of this review is not the theory of etch pits formation; the interested reader may consult the thorough revision by Sangwal (1987). Here, we shall discuss the factors that define the aggressivity of a given solvent to an inorganic solid, essentially on the basis of the involved chemistry. The subject is of much practical importance: amongst other examples, one could mention the behaviour of ceramic materials in contact with water or even moist atmospheres; or the need to develop repositories for nuclear wastes capable of withstanding weathering for long (even geological) time spans. The outlook is heavily influenced by our previous monograph (Blesa *et al.* 1994a). Within inorganic solids, the dissolution behaviour can be sorted according to whether the solid is a metal, a metal oxide or calchogenide, or a salt. This chapter focuses on the second group. Dissolution of metals is essentially an electronic charge transfer phenomenon, and its description would take us into classical electrochemistry. Dissolution of salts is, on the other hand, a phenomenon essentially governed by the thermodynamics of solubility. Thus, we shall be dealing with metal oxides, using this label in a very broad sense, to encompass compounds like SiO_2 and $BaTiO_3$ that either involve non-metallic atoms, or that could be classified as oxometallates.

A large variety of chemical equilibria and irreversible reactions take place at the interface when these compounds contact water solutions: (i) hydration and hydroxylation; (ii) surface reconstruction; (iii) protolytic equilibria; (iv) other ionic adsorption; (v) anion and cation phase transfer (dissolution and precipitation); (vi) (electronic) charge transfer; or (vii) surface precipitation of new solid phases. Usually, these reactions take place sequentially; dissolution is mediated by hydration, hydroxylation and ionic transfer. Phase transformations in aqueous media often involve dissolution–reprecipitation. Thus, we shall describe these reactions in the next sections sorted in two groups: (i) equilibria at the solid–liquid interface, Section 2; and (ii) dissolution mechanisms, Section 3.

2 The solid–liquid interface

2.1 *Hydration and hydroxylation*

Freshly cleaved metal oxide surfaces are highly reactive towards water. At low degrees of coverage, dissociative chemisorption of water vapour takes place on low-coordination metal ions, and low-coordination oxide ions. Schematically:

$$\left|\begin{array}{l}\equiv \text{M} \\ \\ \equiv \text{M–O}\end{array}\right. + H_2O = \left|\begin{array}{l}\equiv \text{M–OH} \\ \\ \equiv \text{M–OH}\end{array}\right. \tag{3.7}$$

Henrich (1983, 1989) has studied the adsorption of water on freshly cleaved surfaces of rutile and other oxides. Surface defects in metal oxide surfaces interact strongly with water, which chemisorbs dissociatively, as evidenced by ultraviolet photoelectron spectroscopy; hydroxylated sites formed by water adsorption on very reactive sites do not dehydroxylate reversibly. At high temperatures, dehydration and surface reconstruction lead to passive hydrophobic surfaces (Pashley and Kitchener 1979), with very low densities of sites characterized by a high Lewis acidity or basicity. The surfaces of many minerals and ceramic materials are of this type.

Upon more prolonged exposure to water vapour or to liquid water, less reactive sites also hydroxylate:

$$\left|\begin{array}{l}\equiv M\diagdown \\ \quad\quad O + H_2O = \\ \equiv M\diagup\end{array}\right. \left|\begin{array}{l}\equiv M\text{–}OH \\ \\ \equiv M\text{–}OH\end{array}\right. \tag{3.8}$$

Taking the important case of TiO_2, on the basis of the analysis of the crystal structure of anatase, Boehm (1971) concludes that (001) faces should undergo extensive hydroxylation to minimize charge unbalance, in accordance with Pauling rules. Two types of OH^- surface ions are formed, with coordination numbers of one and two. Whilst the former are basic in principle, the latter are predicted to behave as acids in aqueous media.

In order to complete their coordination shell, low-coordination surface ions must also hydrate (Regazzoni *et al.* 1988). Figure 3.2 depicts electroneutral atoms on (111) and (100) faces of a rocksalt oxide, MO. The presence of both hydroxide groups and undissociated coordinated water molecules has been demonstrated experimentally by infrared (IR) spectroscopy (Lewis and Farmer 1986; Tejedor-Tejedor and Anderson 1986). In what follows, we shall describe surface species with the shorthand notation $\equiv$M–OH, etc., but it should be remembered that more precise would be $\equiv M(OH)_n(OH_2)_m$, bound water molecules being involved in the protolytic reactions described below.

The maximum achievable surface density N_s, of hydroxylated $\equiv$M–OH sites is a parameter of considerable importance in the description of the various types of chemical reactions involving metal oxides in water. In principle, N_s is determined by

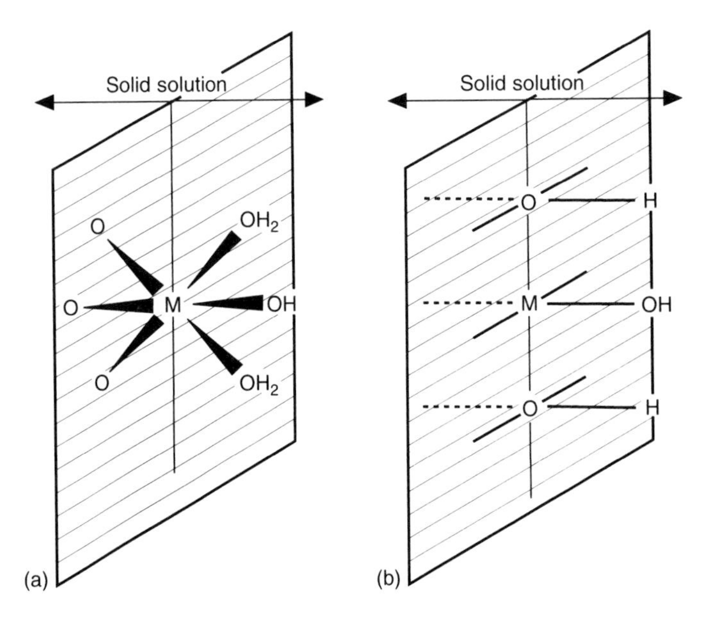

Figure 3.2 Hydration (a) and hydroxylation (b) of (111) and (100) faces of a rocksalt oxide. Reprinted with permission from Blesa *et al.* 1994.

the structure of each crystal face, and each type of site is characterized by its own maximum density, N_s. Although direct measurement techniques exist, 'experimental' N_s values are usually adjustable parameters used to fit ionic adsorption data according to a given interfacial model. Good agreement is found, however, in some cases between values derived from crystallographic considerations and ionic adsorption data. N_s is usually between 2 and 12 sites nm^{-2}.

As mentioned, hydroxylation is the first stage in the attack by aggressive aqueous solvents. Hydroxylation, especially of covalent oxides, may be a slow process that in some cases may limit the rate of dissolution.

2.2 *Protolytic reactions*

Upon adsorption of dissolved Brønsted acids or bases, metal oxide surfaces become charged, as evidenced by the electrokinetic mobility of the particles, and by the shape of the $(\Gamma_H - \Gamma_{OH})$ versus pH curves, where Γ denotes surface excess. The affinity of the surface for H^+ involves two contributions, chemical and electrostatic:

$$\Delta G_{ads} = \Delta G_{chem} + z\mathcal{F}\psi \tag{3.9}$$

For protons, ΔG_{chem} is dominated by bond formation: its value reflects the intrinsic acidity or basicity of surface sites. This term defines the pH value of the point of zero charge (pH_0) of each oxide, i.e. the solution pH at which the surface is uncharged. Table 3.1 collects some selected pH_0 values obtained essentially from potentiometric titrations or electrophoretic mobilities.

Surface protonation is commonly described by mass–law equilibria. In the simplest case, there is only one type of surface site (density N_s), and this site is assumed to be capable to exchanging protons with bulk water through an equilibrium described by *one* constant only (Hiemstra *et al.* 1989a,b):

$$\equiv M(OH)_n(H_2O)_m^{0.5+} = \equiv M(OH)_{n+1}(H_2O)_{m-1}^{0.5-} + H^+ \tag{3.10}$$

Table 3.1 pH_0 values for selected metal oxides at 25°C*

Oxide	pH_0
Al_2O_3	8.34–9.06
$Cr(OH)_3$	~ 8.5
α-Fe_2O_3	7.3–8.6
Fe_2O_3	9.0
Fe_3O_4	6.7
α-FeO(OH)	8.0
Co_3O_4	10.3–11.4
$Co(OH)_2$	8.3–11.5
NiO	9.8–11.3
SiO_2	2.5
TiO_2	5.8–6.0
ZrO_2	6.2

*For older data, see Parks (1965). Reprinted with permission from Blesa *et al.* (1994) (original source of data).

More elaborate models result by relaxing either the assumption of one single type of site, or the assumption of only one protolytic equilibrium, or both. The multisite complexation model (MUSIC) includes different types of sites actually seen, or inferred from crystal considerations, and attributes to each one a protolytic equilibrium constant (Hiemstra *et al.* 1989a,b). Table 3.2 collects the cases described by the MUSIC approach (Blesa *et al.* 1994).

Most usual models assume one type of site, characterized by *two* successive ionization constants (Blesa *et al.* 1994a):

$$\equiv M\text{–}OH_2^+ = \equiv M\text{–}OH + H^+ \tag{3.11}$$

$$\equiv M\text{–}OH = \equiv M\text{–}O^- + H^+ \tag{3.12}$$

$$pH_0 = ½(pK_{a1}^{int} + pK_{a2}^{int}) \tag{3.13}$$

The extent of H^+ adsorption at any $pH < pH_o$ is limited by surface charge and potential build-up. Bérubé and de Bruyn (1968) proposed that ψ_0 was a Nernstian function of pH: in thermodynamic equilibrium, both H^+ and M^{2n+} are potential determining ions:

$$\psi_0 = \frac{2.3RT}{2n\mathcal{F}} \log \frac{a_{M^{2n+}}}{a^0_{M^{2n+}}} = 2.3\frac{RT}{\mathcal{F}}(pH_0 - pH) \tag{3.14}$$

However, H^+ uptake is measured by the *fast titration technique*, with equilibration times very short compared to the time required to establish solubility equilibrium. For these 'insoluble' oxides, H^+ is the only pd ion.

Table 3.2 MUSIC description of metal oxides. Reprinted with permission from Blesa *et al.* 1994

Oxide	Acid/base surface pair	pK_a	N_s (sites nm^{-2})
α-$Al(OH)_3$	$\equiv Al\text{–}OH_2^{0.5+}/\equiv Al\text{–}OH^{0.5-}$	10	
	$\equiv Al_2\text{–}OH/\equiv Al_2\text{–}O^-$	12.3	9.6*
	$\equiv Al_2\text{–}OH_2^+/\equiv Al_2\text{–}OH$	- 1.5	4.8†; 13.8‡
α-FeO(OH)	$\equiv Fe\text{–}OH_2^{0.5+}/\equiv Fe\text{–}OH^{0.5-}$	10.7	3.3§; 7.1¶; 8.6**
	$\equiv Fe_2\text{–}OH/\equiv Fe_2\text{–}O^-$	13.7	
	$\equiv Fe_2\text{–}OH_2^+/\equiv Fe_2\text{–}OH$	- 0.1	3.3δ; 7.1¶; 8.6**
(111) Rutile	$\equiv Ti\text{–}OH_2^{0.66+}/\equiv Ti\text{–}OH^{0.33-}$	6.3	4.8††; 3.9‡‡
	$\equiv Ti_2\text{–}OH^{0.33+}/\equiv Ti_2O^{0.66-}$	5.3	4.8††; 3.9‡‡
	$\equiv Ti_3OH^+/\equiv Ti_3O$	- 7.5	
Silica	$\equiv SiOH/\equiv SiO^-$	11.9	4.6§§; 8¶¶
	$\equiv SiOH_2^+/\equiv SiOH$	- 1.9	
	$\equiv Si_2OH^+/\equiv Si_2O$	- 16.9	

*Present at crystal edges only.
†Edge faces of hexagonal crystals.
‡Planar faces.
§(100) face that also contains 3.3 triply coordinates sites nm^{-2}.
¶(010) face.
**(001) face.
††Dominant, (110) face.
‡‡Less important, (101) and (100) faces.
§§Well-annealed silica.
¶¶Hydrous silica.

The simple Nernstian equation is incompatible with the surface complexation model described by Equations (3.11–3.13) (Blesa and Kallay 1988). Modified expressions for ψ_0 must be used (Levine and Smith 1971); the influence of surface potential on H^+ adsorption is given by:

$$K_{a1}^{int} = K_{a1} \exp(-\mathcal{F}\psi_0/RT) \tag{3.15}$$

$$K_{a2}^{int} = K_{a2} \exp(\mathcal{F}\psi_0/RT) \tag{3.16}$$

The ratio $\{\equiv M\text{–}OH_2^+\}/\{\equiv M\text{–}O^-\}$, where brackets denote the surface concentrations, can be found only after an explicit interfacial model is introduced. The main models are the Gouy–Chapman model (or its statistical mechanical modifications) (Sposito 1990); the constant capacitance model, which equates the charged interface to a flat condenser formed by the solid surface and a plane of adsorbed counter-ions; the composite Gouy–Chapman–Stern–Grahame, or *triple-layer* model. Figure 3.3 shows the double-layer structures for each case (Blesa *et al.* 1994). From Equation 3.15 the constant capacity model derives a Frumkin–Fowler–Guggenheim isotherm for the adsorption of protons ($pH < pH_0$):

$$[H^+] = K_{a1}^{int} \frac{\{\equiv M\text{–}OH_2^+\}}{N_s - \{\equiv M\text{–}OH_2^+\}} \exp\left[\mathcal{F}^2\{\equiv M\text{–}OH_2^+\}/CRT\right] \tag{3.17}$$

As N_s and the capacitance, C, do not change much from oxide to oxide, the functional relationship between $\{\equiv M\text{–}OH_2^+\}$ and pH is determined essentially by K_{a1}^{int}, or the pH_0 (Furrer and Stumm 1986). A single master curve relates $\{\equiv M\text{–}OH_2^+\}$ to pH by shifting the origin of the pH axis by pH_0 as shown in Fig. 3.4 (Zinder *et al.* 1986; Wieland *et al.* 1988).

Ionic strength heavily influences surface charge. The triple-layer model uses mass–law equations to describe this effect also:

$$\equiv M\text{–}OH_2^+ + X^- = \equiv M\text{–}OH_2^+ \ldots X^- \tag{3.18}$$

$$\equiv MO^- + N^+ = \equiv MO^- \ldots N^+ \tag{3.19}$$

H^+ and OH^- adsorption is thus predicted to increase through the coadsorption of counter-ions, but the pH_0 does not change.

Pressure-jump relaxation measurements indicate that protolytic reactions are diffusionally controlled (Ikeda *et al.* 1985). Except at very low undersaturation, protonation rates should not influence dissolution kinetics.

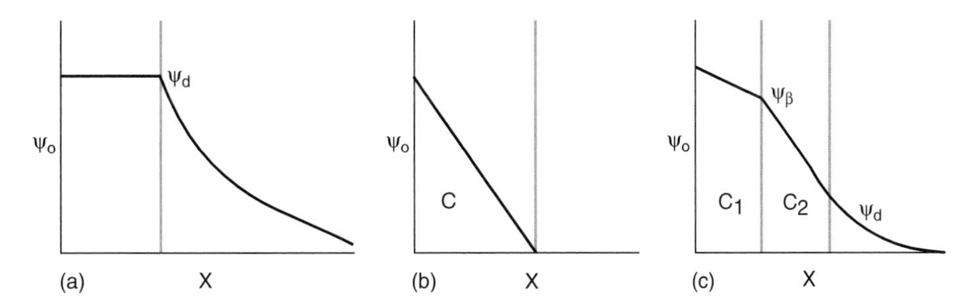

Figure 3.3 Most commonly accepted solution-side double-layer structures (a) Gouy–Chapman, (b) constant capacitance, (c) Gouy–Chapman–Stern–Grahame. Reprinted with permission from Blesa *et al.* 1994.

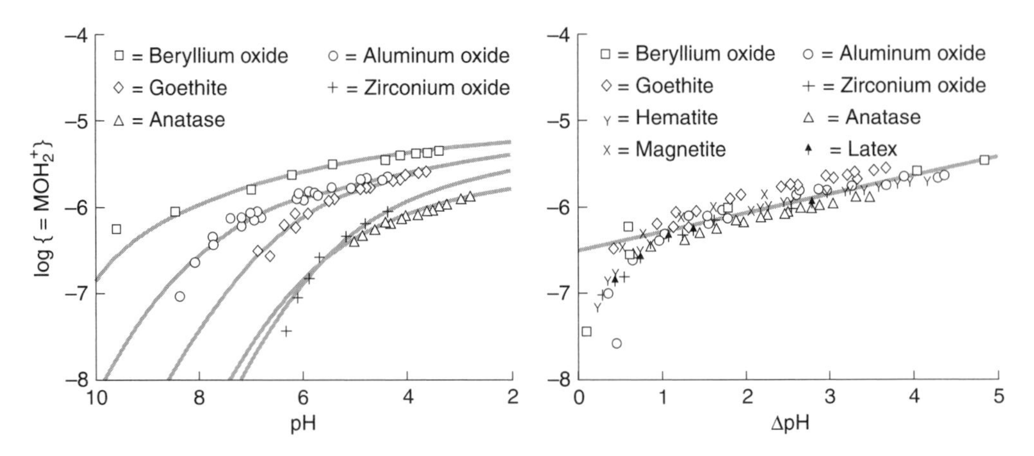

Figure 3.4 Master isotherm describing H^+ adsorption as a Frumkin–Fowler–Guggenheim (FFG)-type process. (From Wieland *et al.* 1988.)

2.3 *Anion and cation adsorption*

The surface complexation model describes adsorption of weak acids by:

$$\equiv \text{M–OH} + \text{HX} = \equiv \text{M–X} + \text{H}_2\text{O} \tag{3.20}$$

$$K_x = \frac{\{\equiv \text{M–X}\}}{\{\equiv \text{M–OH}\}}[\text{HX}]^{-1} \tag{3.21}$$

A large body of K_x values is available, but care should be taken to ensure consistency, because these constants are model dependent. Dzombak and Morel (1990), derived a self-consistent data set for adsorption on hydrous iron oxide using the generalized two-layer model.

Anion adsorption does not necessarily involve one type of site and one inner-sphere binding mode only. More elaborate models account for two or more adsorption sites, and/or more than one adsorption mode. Typical mass–law equilibria are those describing phosphate adsorption:

$$\equiv \text{M–OH} + \text{H}_2\text{PO}_4^- = \equiv\text{M–OPO}_2(\text{OH})^- + \text{H}_2\text{O} \tag{3.22}$$

$$\equiv \text{M–OH} + \text{HPO}_4^{2-} = \equiv \text{M–OPO}_3^{2-} + \text{H}_2\text{O} \tag{3.23}$$

$$\left|\begin{matrix}\equiv \text{M–OH} \\ \\ \equiv \text{M–OH}\end{matrix}\right. + \text{H}_2\text{PO}_4^- = \left|\begin{matrix}\equiv \text{M–O} & & \text{O}^- \\ & \text{P} & \\ \equiv \text{M–O} & & \text{O}\end{matrix}\right. + 2\text{H}_2\text{O} \tag{3.24}$$

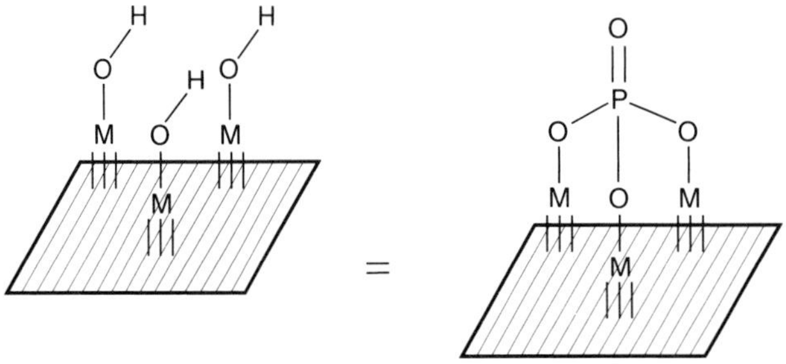

(3.25)

The multiplicity of adsorption modes has been documented by the adsorption isotherm of ethylenediaminetetra-acetic acid (EDTA) on magnetite (Fe_3O_4) at pH 4.7 (Blesa *et al.* 1984); at low EDTA concentrations the four carboxylic groups are coordinated to four surface iron ions, whereas at higher concentrations only two carboxylates bind to the surface. The shifts in the adsorption mode may alter the dissolution rates. In the case of EDTA/Fe_3O_4, dissolution rates are enhanced by the decrease in the number of anchoring positions (Blesa *et al.* 1984; Borggaard 1991).

Coadsorption of protons is implicit in all models, and more explicit in the dissociative chemisorption model of weak diprotic acids (Blesa *et al.* 1996a):

$$\left|\begin{matrix}\equiv M\text{–}OH \\ \equiv M\text{–}OH\end{matrix}\right. + H_2X = \left|\begin{matrix}\equiv MX^- \\ \equiv M\text{–}OH_2^+\end{matrix}\right. + H_2O \qquad (3.26)$$

Hydrolyzable cations adsorb through dehydration and formation of oxobridges. For a divalent metal ion:

$$\equiv M\text{–}OH + N^{2+} = \equiv M\text{–}O\text{–}N^+ + H^+ \qquad (\text{equilibrium constant } {}^*K_N) \qquad (3.27)$$

In this case, however, the single-site surface complexation approach does not describe adequately experimental data. It is necessary (Dzombak and Morel 1990) to use two types of sites characterized, respectively, by a strong affinity (large $^*K_N$) and low N_s, and by low $^*K_N$ and large N_s. Both sets of sites are assumed to have the same acidity, as described by the two pK_a^{int} values. A further mode of interaction is the (surface) precipitation of $N(OH)_2$ on top of the metal oxide.

The coadsorption of OH^-, implicit in cation adsorption, enhances the rate of attack on covalent oxides such as SiO_2. The changes in the surface charge operate in the same direction.

Dissolution frequently requires both acid conditions and another dissolving reagent, a complexant, reductant or oxidant, that must adsorb. Consequently, anionic reagents dissolve iron oxides smoothly acidic media, whereas cationic reagents (for example, reductants like V^{2+}, Cr^{2+}, Fe^{2+}) require further anionic complexants to alter the adsorption density/pH profiles. Dissolved metal complexes also adsorb. The general mass–law equilibrium for the *metal-like* (M) and *ligand-like* (L) adsorption are (Schindler 1990):

$$m \equiv M\text{–}OH + NL_n^{z+} = (\equiv M\text{–}O)_m NL_n^{(z-m)^+} + mH^+ \qquad (3.28)$$

$$\equiv M\text{–}OH_2^+ + LN^{z+} = \equiv M\text{–}L\text{–}N^{(z+1)^+} + H_2O \qquad (3.29)$$

The vast influence of this surface complexation mode on the dissolution kinetics is due to the easy possibility of internal redox reactions between the two bridged metal centres.

3. The dissolution of metal oxides

Dissolution of metal oxides is essentially an electrochemical phenomenon, involving the displacement of charges (ions) across an electrified interface. Furthermore, electronic charges are often involved in defining both surface potential and specific rates of ionic transfer. It is not surprising, therefore, to find dissolution models that follow closely the electrochemical formalisms. These models are described in Section 3.1.

The leaching of minerals, and open-circuit experiments in general, has led, on the other hand, to models that focus on the surface chemistry, in which surface potentials do not play the central role. Section 3.2 describes these models, including those that attempt to cover comprehensively both direct electrostatic and other chemical interactions.

Underlying both approaches, solid state properties are essential in defining reactivity, especially point and extended defects of all types. The influence of extended defects is in fact the subject of the kinetic theories of dissolution; as mentioned before, the reader is referred to the monograph by Sangwal (1987) for a comprehensive description of these aspects. Point defects influence has been much explored by Segall *et al.* (1988); point defects define both changes in reactivity from sample to sample, and the influence of external agents (light, redox reagents) on the reactivity. This latter aspect shall be dwelt upon throughout the following sections.

3.1 *Electrochemical models*

The first attempt to describe the kinetics of dissolution of metal oxides was made by Engell (1956a, b). According to Engell, the dissolution current density is given by Equation (3.30):

$$\vec{j} = (\vec{k}_{-}\vec{k}_{+})^{1/2} \exp [\mu \pm /2RT] \exp [(\alpha_{+} z_{+} + \alpha_{-} z_{-}) \mathcal{F} E/2RT] \tag{3.30}$$

$\vec{k}_{-}$ and $\vec{k}_{+}$ are the specific rate constants for dissolution (phase transfer) of anions and cations; the reverse (deposition) process is assumed to be negligible. The chemical potential of the oxide is $\mu_{\pm}$; α_{+} and α_{-} are the electrochemical transfer coefficients; and E is the electrode potential. Anion and cation transfer are independent processes. The exponential relationship between $\vec{j}$ and the chemical potential $\mu_{\pm}$ implies that at the oxide surface $\mu_{\pm}$ is constant and equal to the bulk oxide value $\mu_{\pm}^{b}$, and that the surface and bulk composition are almost identical. For metal oxides immersed in water, the charged surface species are *not* constituents of the solid lattice, and more recent models assume rather a constant total number of surface sites N_8:

$$2N_s = N_s^+ + N_s^- \tag{3.31}$$

Engell's experiments were performed on oxides of high electronic conductivity, and the applied potential E was used naturally. The extension of this magnitude to ionic oxides is, however, not straightforward. For the dissolution of semiconductors, E would be buffered by the space charge region, and a more detailed model for E would be necessary.

Vermilyea (1966) uses the Nernst's relationship between surface potential ψ_0 and the activity of the constituent ions (pd ions) in solution, to describe the dissolution of ionic compounds:

$$\psi_0 = \frac{RT}{z\mathcal{F}} \ln (a/a_0) \tag{3.32}$$

where a_0 is the activity at the point of zero charge. As discussed before, for metal oxides, the pd ions are H^+ and OH^-, which are not the constituent ions, whereas for

AgI, the pd ions are Ag^+ and I^-. For oxides dissolving in highly undersaturated media, the reverse (deposition) reaction may be ignored, and the current densities associated with cation and anion dissolution must by equal. A steady state surface is then achieved, characterized by a *freely-dissolving potential* ψ_f, given by Equation (3.33). The sign and actual value of this potential is determined mainly by the ratio $(N_s^+ \vec{k}_+ / N_s \vec{k}_-) \cong (\vec{k}_+ / \vec{k}_-)$:

$$\psi_f = \frac{RT}{(\alpha_+ z_+ - \alpha_- z_-)\mathscr{F}} \ln (N_s^- \vec{k}_- / N_s^+ \vec{k}_+) \tag{3.33}$$

On the other hand, in a saturated solution the rates of dissolution and deposition are equal, and Equation (3.34) describes the exchange of anions and cations without net chemical reaction:

$$\psi_{eq} = \frac{RT}{z_+ \mathscr{F}} \ln(\overleftarrow{k}_+ C_+ / N_s^+ \vec{k}_+) = \frac{RT}{z_- \mathscr{F}} \ln(\overleftarrow{k}_- C_- / N_s^- \vec{k}_-) \tag{3.34}$$

For metal oxides dissolving in acidic media, the transfer of anions is mediated by the attack of H^+ onto surface oxide ions; and the freely-dissolving potential becomes pH dependent:

$$\psi_f = \frac{RT}{(\alpha_+ z_+ - \alpha_- z_-)\mathscr{F}} \ln \frac{N^{-s}\vec{k}}{N^{+s}\vec{k}^+} C_{H^+} \tag{3.35}$$

The kinetic order on protons of the dissolution rate results in:

$$\mathscr{F}R = N_s^+ \vec{k}_+ (N_s^- \vec{k}_- / N_s^+ \vec{k}_+)^{\alpha^+ z^+ / (\alpha^+ z^+ - \alpha^- z^-)} C_H{}^{\alpha^+ z^+ (\alpha^+ z^+ - \alpha^- z^-)} \tag{3.36}$$

or

$$n = \frac{\partial \ln R}{\partial \ln C_{H^+}} = \frac{\vec{\alpha}_+ z_+}{\vec{\alpha}_+ z_+ - \vec{\alpha}_- z_-} \tag{3.37}$$

For anion transfer in the form of OH^-, and $\alpha_+ \cong \alpha_-$:

$$n = \frac{\partial \ln R}{\partial \ln C_{H^+}} = \frac{z_+}{z_+ + 1} \tag{3.38}$$

These equations describe the usual experimental fractional orders. Further fractional orders can also be accounted for by allowing $z_- = -2$ (corresponding to O^{2-} ions reacting with protons) coupled with either a first- or second-order dependency on C_{H^+}, or by allowing for a pre-equilibrium of H^+ adsorption previous to the transfer of OH^- as anions. Calling b the true kinetic order on H^+:

$$n = \frac{\partial \ln R}{\partial \ln C_{H^+}} = b \frac{z_+}{z_+ - z_-} \tag{3.39}$$

The results are summarized in Table 3.3.

If the rate of H^+ exchange is much faster than the rate of metal ion exchange, dissolution rates are limited only by the sluggishness of metal ion transfer, and the surface potential is solely determined by the activity of protons, at variance with the basic assumption made by Vermilyea (1966). However, for a dissolving oxide, the rates of transfer of metal ions are not negligible, and the applicability of Equation (3.32) or (3.35) depends on the absolute values of the rates of anion and cation transfer, and thus on the values of $pH_0 - pH$. Diggle (1973) identified H^+ as the pd ion, and the freely-dissolving potential becomes:

$$\psi_f = 2.3 \frac{RT}{\mathcal{F}} (pH_0 - pH) \tag{3.40}$$

Equation (3.40) implies that anion exchange (i.e. proton exchange) is fast, and $\vec{j}_-$ is essentially balanced by $\overleftarrow{j}_-$. Only the minor unbalance between $\vec{j}_-$ and $\overleftarrow{j}_-$ is compensated by $\vec{j}_+$, which measures the net rate of metal oxide dissolution. This condition is described as cation transfer control of the dissolution rate, and is experimentally realized in many cases. Under anion transfer control (cf. below), the assumption that H^+ is the only pd ion may not apply.

For metal oxides of high electrical conductivity, the oxide surface may be polarized by either an external applied voltage or by the influence of dissolved redox couples. High anodic potentials make cation transfer fast enough to reach rate limitations by anion transfer (potentials higher than ψ_f). Figure 3.5 depicts the conditions for cation and anion transfer, for the rather unlikely case of $b = 0$. The unbalance between anionic and cationic fluxes is compensated by redox reactions that change the oxide stoichiometry. By equating $\psi - \psi_f$ in Fig. 3.5 with an overpotential η, and calling j_0 the current density at $\psi = \psi_f$, the cationic and anionic dissolution fluxes are:

$$\vec{j}_+ = j_0 \exp(\alpha_+ z_+ \mathcal{F}\eta/RT) \tag{3.41}$$

$$\vec{j}_- = j_0 \exp(\alpha_- z_- \mathcal{F}\eta/RT) \tag{3.42}$$

The influence of applied potential depends on the sign of η: for $\eta > 0$, anion removal controls the rate and increasing η decreases the rate; for $\eta < 0$, rate control is by cation removal, and increasing the potential (decreasing $|\eta|$) increases the rate. The peak at -100 mV versus saturated calomel electrode in the polarization curve of Fe_3O_4, has been attributed to the shift in control from anion transfer ($E > -100$ mV) to cation transfer ($E > -100$ mV).

The order on protons, under cation and anion transfer control are, respectively:

$$n = \frac{\partial \ln R}{\partial \ln C_{H^+}} = \alpha_+ z_+ \tag{3.43}$$

Table 3.3 Relationship between dissolution mechanism and rate dependence on proton concentration according to Vermilyea (1966)

	Power dependence on H^+			
Dissolution mechanism	M_2O	MO	M_2O_3	MO_2
$\equiv M{-}O + H^+ \rightarrow \equiv M{-}OH^+$ (complete) $\equiv M{-}OH^+ + H^+ \rightarrow M^{2+} + H_2O$ rds	1/2	2/3	3/4	4/5
$\equiv M{-}O + H^+ \rightarrow M^{2+} + OH^-$ rds $OH^- + H^+ \rightarrow H_2O$ (fast)	1/3	1/2	3/5	2/3
$\equiv M{-}O + 2H^+ \rightarrow M^{2+} + H_2O$ rds	2/3	1	6/5	4/3
$\equiv M{-}O + H^+ \rightarrow \equiv M{-}OH^+$ (equilibrium) $\equiv M{-}OH^+ + H^+ \rightarrow M^{2+} + H_2O$ rds	Complex, depending on the value of the constant of the pre-equilibrium			

rds, rate-determining step.

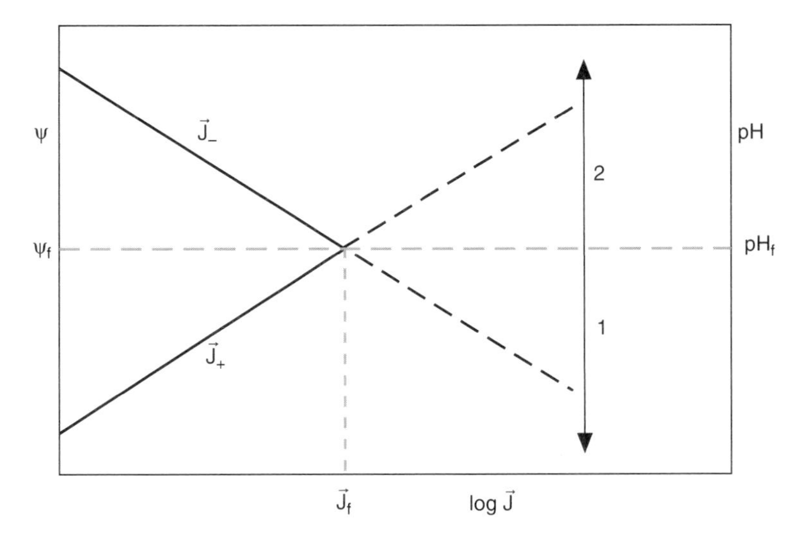

Figure 3.5 Schematic dependence of cationic and anionic fluxes on surface potential and solution pH, showing the freely-dissolving condition parameters. 1, cation transfer control; 2, anion transfer control. Reprinted with permission from Blesa *et al.* (1994).

$$n = \frac{\partial \ln \mathrm{R}}{\partial \ln C_{\mathrm{H}^+}} = b = \alpha_- z_- \tag{3.44}$$

In electrochemical experiments, conditions for anion transfer-limited rates can be achieved; in open-circuit conditions, the fast protolytic rates on the surface lead to cation transfer control, and the participation of aliovalent ions when ionic adsorption is important can be well described by surface complexation models.

Nicol *et al.* (1975) and Needes *et al.* (1975) describe the electrochemical mechanism of chemical redox dissolution, in which the same effect is produced by dissolved redox reagents or by an externally applied potential. The model has been successfully applied to the oxidative dissolution of UO_2. Net dissolution takes place when the potential at the oxide surface is higher than E_{eq} (UO_2). At high anodic polarization, the current densities are given by the modified Butler–Volmer equation:

$$j = 2\mathscr{F}\mathrm{k_a}(\mathrm{UO_2}) \exp\,[2.3E/b_\mathrm{a}(\mathrm{UO_2})] \tag{3.45}$$

According to Nicol's model, the chemical dissolution of UO_2 by a redox couple is also represented by the Butler–Volmer equation as applied to the reduction (or oxidation) of the redox couple:

$$j' = n\mathscr{F}\{\mathrm{k_{a(Red)}[Red]} \exp\,[2.3E/b_{\mathrm{a(Red)}}] - \mathrm{k_{c(Ox)}[Ox]} \exp\,[-2.3E/b_{\mathrm{c(Ox)}}]\} \tag{3.46}$$

The potential at the surface is now governed in some sites by the $UO^{2+}{}_2/UO_2$ couple [$E_{eq}(UO_2)$] and in others by the oxidation/reduction (Ox/Red) couple [E_{eq}(Ox/Red)]. Eventually, a steady state should be achieved, under which a unique rest or mixed potential E_M is attained, which equals the net rate of oxidation of UO_2 and the net rate of reduction of Ox, as depicted in Fig. 3.6. Under certain limiting conditions, the model leads to tractable expressions for the partial kinetic orders of reaction on [Red] and/or [Ox]; two cases are summarized in Table 3.4. Very seldom is a first-order dependency predicted, although it is not impossible; for example, if $2b_a(UO_2) \cong b_a(\mathrm{Red}) \cong b_c(\mathrm{Ox})$,

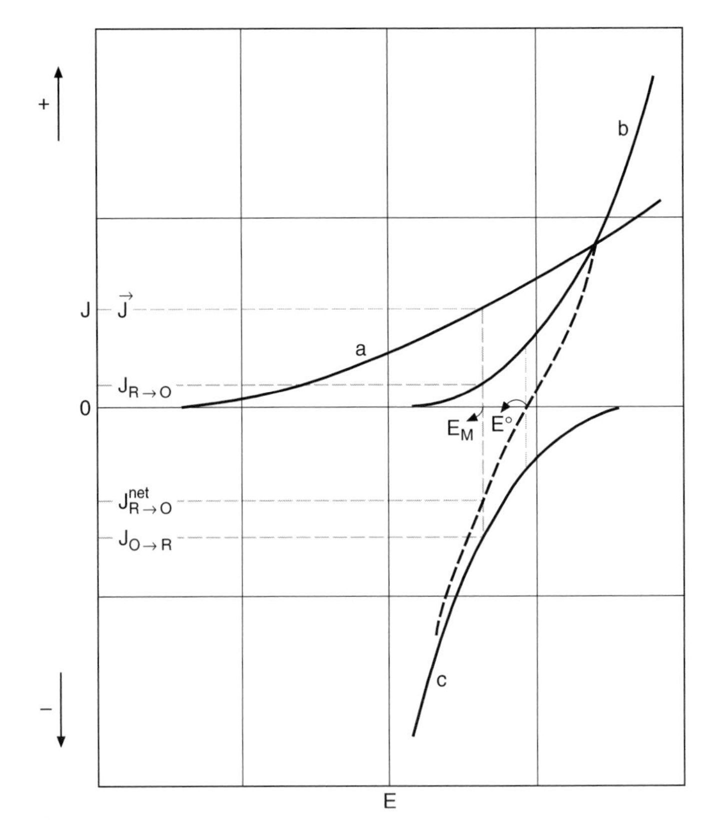

Figure 3.6 (a) Anionic current associated with U (VI) dissolution; (b) anionic current associated with the oxidation of dissolved reductant, R; (c) cationic current associated with the reduction of dissolved oxidant, Ox. Reversible (E^0) and mixed (E_M) potentials indicated. Reprinted with permission from Blesa *et al.* (1994).

$n_{Nicol} = N_{ox} = -n_{Red} = 1$. More often fractional orders close to 0.5 are predicted, which is in agreement with experiment.

Solution variables, such as pH or complexing agents, that affect E(Ox/Red), may also shift the anodic curve. The influence of pH on anodic and cathodic reactions can be measured separately and used to predict the affect of pH on chemical (open circuit) dissolution. The simplest pH effect, at low current densities, is a monotonous shift determined by the H^+ stoichiometric coefficient in the rate-determining step. Two cases are summarized in Table 3.5.

Table 3.4 Particular cases of the Nicol model

Assumptions	E_M	Order on Ox	Order on Red
One electron couples. All *b*s are equal	$0.5b_m \log \frac{k_{c(Ox)}[Ox]}{2\vec{k}_{a}(UO_2) + k_{a(Red)}[Red]}$	0.5	Variable, 0–0.5
$\vec{j}' \gg \vec{j}$	E(Red/Ox)	$\frac{b_{a(Red)}b_{c(Ox)}}{b_a(UO_2)[b_{a(Red)} + b_{c(Ox)}]}$	$-\frac{b_{a(Red)}b_{c(Ox)}}{b_a(UO_2)[b_{a(Red)} + b_{c(Ox)}]}$

Table 3.5 Particular cases of the Nicol model

Rate-determining step	E_M	Order on OH^-
$MO_x(s) + OH^- \rightarrow MO_xOH(s) + e^-$	$E_M = \frac{b_a(MO_x)b_{c(Ox)}}{b_a(MO_x) + b_{c(Ox)}} \log \frac{nk_{c(Ox)}[Ox]}{n'\vec{k}a(MO_x)[OH^-]}$	$\frac{b_a(UO_2)}{b_c[Ox] + b_a(UO_2)}$
$Ox + H^+ + e^- \rightarrow Red$		$-\frac{b_c[Ox]}{b_c[Ox] + b_a(UO_2)}$

At high values of j_{diss}, the rate becomes insensitive to the applied potential (closed-circuit experiments) or to E_M values (open-circuit experiments), due to the levelling off of the anodic dissolution branch. Saturation may arise when a step that does not involve charge transfer becomes rate determining.

Gorichev and Kipriyanov (1981) have presented a different model, that not only addresses the factors governing the specific rate for unit area R, but also the time evolution of the surface area $S(t)$. The total rate, v, is given by Equation (3.47), and $S(t)$ is modelled assuming that dissolution nuclei grow along dislocation lines, and that inward H^+ diffusion along these channels activates secondary nucleation sites:

$$v = S(t)R(t) \tag{3.47}$$

In modelling $R(t)$, electron-exchange currents are assumed to be faster than ionic-exchange currents, and the surface concentrations of reduced and oxidized metal ions are determined uniquely by the solution redox potential. The relevant redox couples may involve just dissolved ions (for example, M(III)/M(II)), two solid phases (for example, MO_x/MO_xH) or a solid phase and a dissolved ion (for example, M_2O_3/M(II)). The dissolution rate is considered to be first order on Γ_H. At constant potential, a Freundlich-type adsorption isotherm links Γ_H to $[H^+]$, the Freundlich constant K_F being an exponential function of the redox potential. Such a relationship implies a very complex interfacial structure (Blesa *et al.* 1994). For an oxide subjected to a polarizing potential E, the actual potential felt at the interface is assumed to be $\simeq 0.5E$, 0.5 being the electrochemical transfer coefficient (cf. with the mixed potential used by Nicol *et al.* (1975) to describe polarizable electrodes). Under these conditions, Equation (3.48) results for the case of reductive dissolution:

$$R = k_{unt}K^{0_F} \exp(0.5 \mathcal{F}E_{pzc}/RT)\, a^{n_H} \exp(-0.5 \mathcal{F}E/RT) \tag{3.48}$$

This equation identifies the partial order on $[H^+]$ with the Freundlich exponent. Due to the fast electron exchange, both members of the redox couple in solution influence the rate. For example, for the dissolution of Fe(III) oxides, if $\alpha_{\pm} \approx 0.5$, the partial orders on Fe^{2+} and Fe^{3+} are 0.5 and -0.5, respectively, values that have been realized in some experimental measurements. Complexing ions are assumed, essentially, to change the driving force; surface complexes are not important in this approach, in spite of the wealth of evidence that strongly supports their involvement.

The analysis of Gorichev and Kipriyanov (1981) is invalid when the redox reagent is not a reversible redox couple. It cannot be used for slow reductants or oxidants; for these, either the Nicol approach or the surface complexation models provide a better description.

3.2 *Adsorption models*

In the course of the study of the leaching of minerals, it was well established that not all mineral acids are equally effective in bringing about dissolution, and the anions influence the dissolution rates and the kinetic laws. Warren and Devuyst (1975), following the earlier work of Surana and Warren (1969), proposed that a sequence of interfacial chemical reactions describing ionic adsorption must be taken into account. These reactions include surface hydroxylation, proton and anion adsorption, and are considered as multistep Langmuirian equilibria. Thus, the metal oxide surface is described as a manifold of sites of different composition; each site may dissolve with a characteristic rate constant k_i, controlled either by the rate of its formation from its predecessor species, or by the detachment rate. The general rate expression is:

$$R = \frac{K_H \sum_{i=1}^{i=r} k_i a_{\pm}^{i} \prod_{m=0}^{m=i} K_m}{1 + K_H + \sum_{i=1}^{i=r} a_{\pm}^{i} \prod_{m=0}^{m=i} K_m} \tag{3.49}$$

where K_H is the hydroxylation equilibrium constant and K_m is the stepwise adsorption constant for species i. This complex expression simplifies when only one or two types of surface sites are considered.

In this approach the influence of interfacial potentials on adsorption and phase transfer rate constants is ignored. The first aspect is taken into account in the surface complexation models, originally derived by Stumm and coworkers (Furrer and Stumm 1986; Zinder *et al.* 1986; Wieland *et al.* 1988). This model converts empirical rate laws of the type:

$$R = k[H^+]^n \tag{3.50}$$

into ith-order processes of release of $\{\equiv MOH_2^+\}$ surface complexes:

$$R = k' \{\equiv MOH_2^+\}^i \tag{3.51}$$

The order i on surface complexes is usually identified with the charge z on M^{z+}, implying that full protonation takes place before phase transfer. In the Bragg–Williams approximation for two-dimensional square lattice (chessboard) statistics (Sposito 1990), the probability of finding (n/m) vicinal $\equiv M{-}OH_2^+$ sites for low coverages, is proportional to the (n/m) power of $\{\equiv M{-}OH_2^+\}$. The surface density of $\equiv MOH_2^+$ depends on $[H^+]$ as given by the Frumkin–Fowler–Guggenheim isotherm, which takes into account the influence of surface potential, Equation (3.17); it may be approximated by a power function (order m). The empirical order, n, is explained as $n = im$. In agreement, n/m was found to be 3 for the acid dissolution of δ-Al_2O_3 and α-FeO(OH), and 2 for the dissolution of BeO.

The values of k' are governed by the stability of the bonds linking the metal to the solid. For BeO, Al_2O_3 and SiO_2 a reasonably linear Gibbs energy relationship is found between $\log k$ and the average site energy, E_M, of the cation, which represents the resultant of the interaction with all its neighbours (Wieland *et al.* 1988).

Anion involvement in dissolution is well described by surface complexation models:

transfer of both $\{\equiv M{-}X^{-}\}$ and $\{\equiv M{-}OH_2^{+}\}$ contribute to the observed rate. For dissociatively chemisorbed HX_2, Equation (3.26) predicts that at high coverages, $\{\equiv M{-}X^{-}\} \cong \{\equiv M{-}OH_2^{+}\}$ and the total rate of dissolution is:

$$R_{tot} = R_L + R_H = 0.5\,(k_H + k_L)\,N_s \qquad (3.52)$$

where k_L and k_H are the first-order rate constants of phase transfer of each type of surface complex. If $k_L > k_H$, the observed rate shall be higher than the maximum rate achievable on a hypothetical totally protonated surface. Much of the discussion in the past on the 'ligand-assisted acid dissolution' has been based on attempts to compare k_L and k_H. Expectedly, k_L shall be larger for ligands that loosen the oxobonds linking the metal ion to the solid lattice. There is, therefore, a correlation between the electron-donation capacity of the ligand and its ability to promote acid dissolution (Regazzoni and Blesa 1991). In the case of transition metal oxides, this charge donation may imply a change in the oxidation state, and a new type of process, the reductive dissolution, may result. However, direct comparison between k_L and k_H is seldom possible. The rate of acid dissolution may increase substantially upon complexation, just because $\{\equiv M{-}OH_2^{+}\}$ has increased. Furthermore, ion transfer requires less protons when M is complexed by X^{2-}. Therefore, if Equation (3.51) is also applied to $\equiv M{-}X^{-}$, the value of i is expectedly lower. Lack of experimental data precludes an adequate modelling of i in this case (Blesa *et al.* 1996a). Finally, for reactions with a modest $-\Delta G$ value, the net rate may increase drastically because the solubility may increase by orders of magnitude (Blesa *et al.* 1996b).

Strong chemisorption may result in a decrease of the dissolution rate if $k_L < k_H$. Whether a ligand shall corrode or passivate an oxide surface shall depend on the structure and stability of the surface complex. Polydentate ligands such as EDTA may coordinate surface ions in a variety of ways: when several surface bonds are formed, the oxide may be protected from proton attack, whereas the attachment of partially protonated ligands $H_nX^{(m-n)-}$ to a single surface site may result in fast dissolution. The surface complexation model describes in a simple way not only systems with several adsorption modes, but also the effect of the presence of more than one potential ligand. If all the surface complexation constants are known, the competition for surface sites can be quantitatively described as a function of pH and total degree of coverage. For the simple case when an aggressive ligand A and a passivating one P are present:

$$\{\equiv MA\} = \frac{N_s K_A[A]}{1 + K_A[A] + K_P[P]} \qquad (3.53)$$

where K_A and K_P are the conditional (pH dependent) surface complexation constants. For iron oxides, [A] = oxalate and [P] = chromate. W. Fish (personal communication, 1991) has found an inhibiting effect of chromate when the degree of coverage is high (i.e. $\{K_A[A] + K_P[P]\} > 1$). If competitive effects compound with conformational changes in the surface, more complex behaviours may result. The enhancement by phosphate of iron oxides in EDTA solutions (Borggaard 1991) has already been described in terms of surface complexation.

These surface complexation models do not allow for the influence of surface potential on the rate constant. In contrast, Bruyère and Blesa (1985) assume that a simple first-order rate law suffices to describe the dissolution by mineral acids; the rate constant, k, being dependent on potential drop between surface and Stern planes:

$$R = k\{\equiv M{-}OH_2^+\} \quad (3.54)$$

$$k = k_0 \exp [\alpha_+ z_+ \mathcal{F}(\psi_0 - \psi_\beta)/RT] \quad (3.55)$$

The reactive surface sites are supposed to be a small fixed fraction β of the total protonated sites, and the triple-layer model is used to derive approximate expressions for $\psi_0 - \psi_\beta$ and for $\{\equiv M{-}OH_2^+\}$ as a function of pH. The pH dependence of rate is found to be:

$$-(\partial \log R/\partial pH) = \alpha_+ z_+ q + (pH_0 - pH)^{-1} \quad (3.56)$$

where q is typically about 0.5. This expression is appreciably more complex than Diggle's equation (Equation 3.43). In Vermilyea's approach, the case of adsorption of protons at a pre-equilibrium was not analysed quantitatively. The first term in Equation (3.56) is equivalent to Diggle's equation, modified for the influence of $\psi_\beta \neq 0$; the second term in Equation (3.56) arises from the pH dependency of $\{\equiv M{-}OH_2^+\}$.

The relevance of the potential drop attending ion transfer, may in fact depend on system parameters. On flat surfaces, characterized by a high α_J, Jackson's factor (Wehrli 1989; Blesa *et al.* 1994), all the relevant potentials may be meaningful. On rough surfaces, macropotentials may be irrelevant. Formally, Stumm's approach is equivalent to setting the electrochemical phase transfer coefficient, α, at a very low value, implying that the rate-determining step precedes phase transfer from the surface to the Stern plane (Blesa *et al.* 1994). The same assumption is made by Aagaard and Helgeson (1982), in their description of the dissolution of silicates (see below). The electrical work for ion transfer from the surface to the Stern plane, for reasonable values of z and $\Delta\psi$, is always below 30 kJ mol^{-1} (Blesa *et al.* 1994). Dissolution rates of sparingly soluble oxides are usually characterized by activation energies in the range $E_a \geqslant 70$ kJ mol^{-1}. It is, therefore, likely that steps previous to ion transfer may determine the rate. The exponential potential dependency of the transfer rate constant is perhaps most useful for oxides that dissolve with low activation energies from well-defined flat surfaces.

Transition metal oxides can often accommodate oxidized or reduced metal ions, with either trapped or mobile charge carriers. It has already been mentioned that electrochemical models describe well the influence of mobile charge carriers on dissolution. However, oxidized or reduced surface metal centres present at the interface drastically influence the dissolution behaviour in a more general way, irrespective of the characteristics of electronic conductivity. Surface complexation is adequate to describe this influence. For the case of a metal ion that can adopt two oxidation states, say II and III, each characterized by a different intrinsic reactivity under a given experimental condition, Valverde and Wagner (1976) wrote:

$$R = k^{II}{}_{o,uni}\Gamma_{II} \exp (\alpha_{II} z_{II} \mathcal{F}\Delta\psi/RT) + k^{III}{}_{o,uni}\Gamma_{III} \exp (\alpha_{III} z_{III} \mathcal{F}\Delta\psi/RT) \quad (3.57)$$

The two types of sites are interconvertible, and this characteristic defines much of the dissolution behaviour. Metallic oxides, such as the tungsten bronzes shall respond to both external applied potential and solution composition, whereas insulating oxides may achieve the interconversion only through reaction with dissolved redox couples.

Semiconducting oxides with high electrical conductivity (low band gap) also respond to externally applied potentials, to the action of light, and even to the thermal equilibrium value of electrons or holes concentration. For example, for $Ni_{1-x}O$ (*p*-type semiconductor); valence band holes h_{vb} and conduction band electrons e_{cb} mediate dissolution:

$$h_{vb} + {\equiv}Ni(II) \xrightarrow[\text{the depletion layer}]{\text{Charge transfer across}} {\equiv}Ni(III)$$

$$\downarrow \text{ Phase transfer}$$

$$Ni(II)\ (aq) \xleftarrow{+e_{cb}} Ni(III)\ (ads) \qquad (3.58)$$

The steps involved in redox dissolution are:

1 complexation of the surface metal ions;

2 Inner-sphere redox reaction within the surface complex, or eventually outer sphere electron exchange between the surface complex and the redox partner;

3 decomposition of the successor complex formed in the inner-sphere redox reaction, for example diffusion to solution of the;

4 phase transfer of the reduced or oxidized metal ion.

For this pathway to operate, it is required that *all* steps be faster than the direct transfer to solution.

In the electrochemical models described in Section 3.1, all these chemical processes are implicit (or hidden) in the current density potential slopes. The surface complexation models, on the other hand, focus on the relative reactivities of two surface ions of different oxidation state, say ≡ M(III) and ≡ M(II), and in the ways they may be interconverted by chemical reactions in which they participate as true chemical entities. In what follows, we shall derive steady state rate expressions for the reaction scheme in Equation (3.59), with $k_2 \gg k_3$ (for example, the reductive dissolution of iron oxides); k is rate constant.

$$\begin{array}{l} \equiv M(III) \xrightarrow{k_3} M(III)\ (aq) \\ k_{ET} \downarrow\uparrow k_{-ET} \\ \equiv M(II) \xrightarrow{k_2} M(II)\ (aq) \end{array} \qquad (3.59)$$

Although not explicitly indicated in Equation (3.59), all the steps indicated above must be considered.

The formation of outer-sphere precursors according to Equation (3.18) is fast, and can be safely described as a pre-equilibrium; $\log [K_x^{int} \exp (\mathcal{F}\psi_\beta/RT]$ plays a role equivalent to the work term for ion-pair formation in the Marcus–Hush theory of homogeneous electron exchange reactions. However, outer-sphere redox reactions, if real at all (see below) are usually preceded by an inner-sphere complexation by auxiliary ligands. Thus, outer-sphere redox reactions, except in very special cases, involve two adsorption processes:

$$\equiv M(III)\text{–}OH + L^- \rightleftarrows\ \equiv M\text{–}L(III) + OH^- \qquad (3.60)$$

$$\equiv M(III)\text{–}L + Red^- \rightleftarrows\ \equiv M(III)\text{–}L \ldots Red^- \qquad (3.61)$$

The reader is referred to the review by Stone and Morgan (1987) for the case of slow adsorption kinetics. We shall analyse the more usual case in which complexation may be treated as a pre-equilibrium. The surface concentration of precursor complexes dependency on the concentration of Red has been described both by Langmuir and Freundlich isotherms. The former predicts a dependency on Red with an exponent that decreases from 1, at very low degrees of coverage, to 0 at saturation. The Freundlich treatment yields an intermediate n_{Red} exponent, typically 0.4–0.6.

The precursor complexes formed in Equation (3.61) must evolve through electron exchange into the successor complexes:

$$M\equiv(III)\text{-}L\ldots Red^- \rightarrow \equiv M(II)\text{-}L\ldots Red \qquad (3.62)$$

For outer-sphere complexes, the process may be described by the formalism of outer-sphere electron transfer in solution. Far more common are inner-sphere reductants (or oxidants). In these cases, as a rule, electron transfer between bridged metallic centres is fast, whereas electron transfer from a reductant ligand bound to an oxidizing metal ion (or vice versa) can be appreciably slower. Low-oxidation metal ions are efficient reductants in the presence of adequate complexants. Reductive dissolution of metal oxides by Cr(II), V(II) and Fe(II) have been described (Segal and Sellers 1984; Blesa, *et al.* 1994).

There are two possible fates for the successor complex: either it reverts to the precursor, or it dissolves. The corresponding energy profiles along the idealized dissolution reaction coordinate are shown in Figs 3.7a and b, respectively (Fig. 3.7b also includes the case of concerted electronic charge transfer and ion phase transfer, to be discussed below). The case depicted in Fig. 3.7a leads to rates that are sensitive to the presence of conjugate oxidant in solution, whereas the rates corresponding to Fig. 3.7b should be altered only marginally by the oxidant. For low-oxidation metal ions both cases have been reported, although the former seems to be more common (Rueda and Blesa 1996). A different picture arises when the reductant is an organic molecule, such as a substituted phenol or a carboxylic acid.

Reductive dissolution by reducing complexing agents takes place systematically via inner-sphere pathways, and the rate data could in some cases be casted in the form $R = k\{\equiv M(III)\text{-}Red\}$ through the use of adsorption equilibria data (Torres *et al.* 1989; dos Santos *et al.* 1990; Regazzoni and Blesa 1991). Internal electron transfer within $\equiv$M(III)–Red may lead to reasonably stable one-electron oxidation products, Red; which may inhibit the course of reaction. Using the dissolution of iron oxides by thiocyanate as an example, the kinetic scheme is:

$$\begin{array}{l} \equiv Fe(III)\text{-}OH + SCN^- \rightleftharpoons \equiv Fe(III)\text{-}SCN + OH^- \\ \qquad\qquad\qquad\qquad\qquad \downarrow\uparrow \\ \quad Fe(III)SCN^-(aq) \leftarrow \equiv Fe(II)\text{-}SCN^{\bullet} \\ \quad SCN^{\bullet}\uparrow \qquad\qquad\qquad\quad \downarrow \text{scavenging} \\ \qquad\quad Fe(II)\,(aq) \leftarrow \equiv Fe(II) \end{array} \qquad (3.63)$$

The most important features that result from this scheme are:

1 the dissolution rate may be controlled by the rate of scavenging of $SCN^{\bullet}$ rather than by the rate of ion phase transfer or of charge transfer;

2 labilization of the surface ion may be achieved through a concerted partial charge transfer (see Fig. 3.7b);

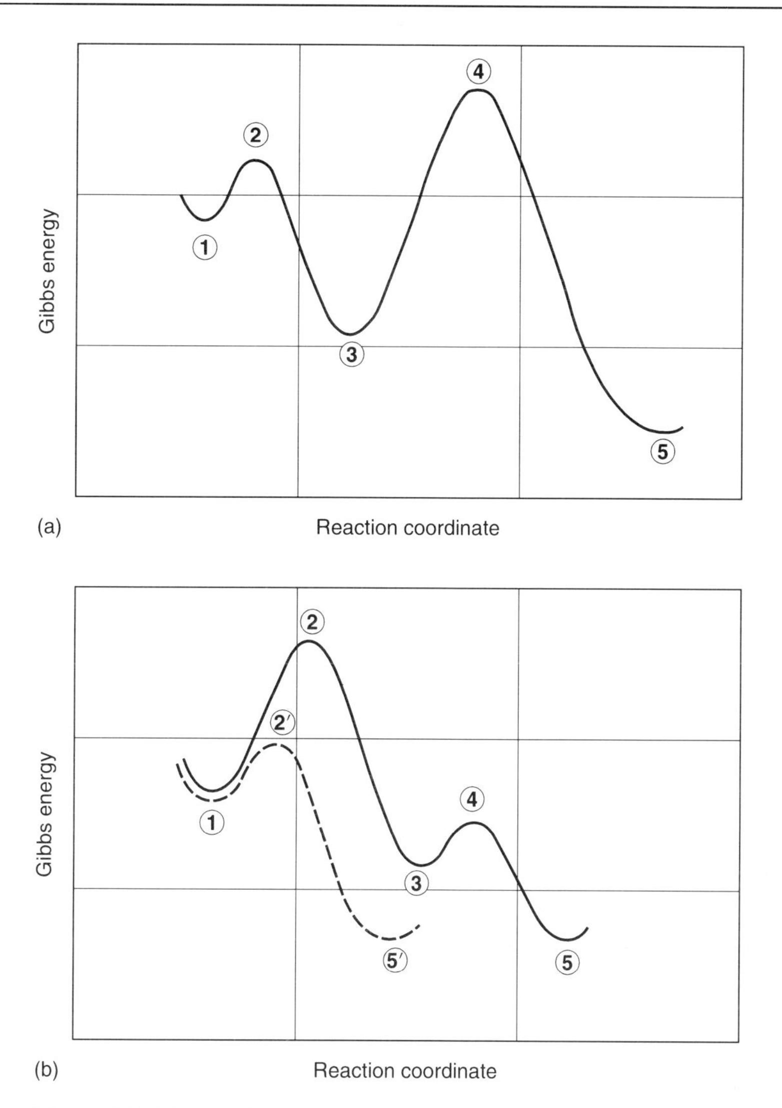

Figure 3.7 (a) Reaction coordinate profile according to Equation (3.62). 1; precursor complex; 2, electron-transfer activated state; 3, successor complex; 4, phase-transfer activated state; 5, dissolved species. (b) *Full line*: similar to (a), but under electron-transfer rate control. *Dashed line*: concerted electron transfer and dissolution. Reprinted with permission from Blesa *et al.* 1994.

3 either Fe(II) or Fe(III) may be put in solution, depending on the reversibility of the charge-transfer process and on the possibility to scavenge $SCN^{\bullet}$ (Regazzoni and Blesa 1991; Blesa *et al.* 1996b).

Thermal dissolution reactions can be accelerated by illumination, as in the case of dissolution of iron oxides by carboxylates (Litter and Blesa 1992, 1994) and mercapto-carboxylates (Waite and Torikov 1987). Photodissolution is a direct consequence of the involvement of electrons and holes in the dissolution mechanisms of semiconductors. Different mechanisms have been described.

1 Corrosion mediated by e^-_{cb} and h^+_{vb}. The rate of reaction may be controlled by availability of either type of charge carrier, or by the phase transfer of the dissolving metal ion. Segall *et al.* (1979) have shown that of both types of carriers, the minority carriers are not necessarily those less readily available for dissolution. Band bending at the surface poses a barrier for the majority of carriers, whereas minority carrier's diffusion is unhindered. Excess charge-carrier concentrations are created by light adsorption, the increase being more important for minority carriers. Dissolution rates controlled by the availability of these are, therefore, expected to increase by illumination. Trapped electrons or holes (surface complexes with energy in the band gap) are often involved. Photodissolution is best achieved in the presence of hole scavengers for *n*-type oxides, or electron scavengers for *p*-type oxides. Hole scavengers are organic complexing anions that also facilitate the phase transfer of the reduced metal ion. In the most simple cases the stoichiometry of these reactions is that of reductive dissolution with oxidation of the organic reagent. They are similar to the well-known homogeneous redox decomposition of dissolved complexes.

2 Photolysis of surface complexes. The ligand to metal charge transfer band of surface complexes may be excited directly, and a photochemical behaviour similar to that of aqueous complexes may lead to dissolution. However, photodissolution through this mechanism has not yet been documented, either because of the low affinity for inner-sphere surface complexation and the filtering effect of solution, or because of the lack of corrodibility of the aliovalent surface complex.

3 Fast thermal corrosion triggered by slow photochemical corrosion. The photochemical dissolution of iron oxides proceeds via a thermal redox reaction triggered by the photochemical production of Fe(II) (Litter *et al.* 1991). A thorough discussion of the rate laws can be found in the monograph by Blesa *et al.* (1994).

Surface complexation has also been used to describe dissolution of covalent oxides. In these systems, the salient features are: (i) very large band gap, as to make the oxides insulating; (ii) unavailable aliovalent metal levels; (iii) sluggish oxobond-breaking kinetics; and (iv) limited thermodynamic solubility. This group includes SiO_2, silicates and aluminosilicates and valve metal oxides. Chemically, dissolution involves electrophilic attack on oxygen atoms (by H^+, or Brønsted acids in general) and/or nucleophilic attack on the metal. The nucleophilic attack by OH^- or F^- on SiO_2 involves pentacoordinated silicon, and this reaction is slow enough as to become rate determining. Because of this feature, the preferred site of attack on mixed or substituted oxides, such as aluminosilicates, is the most labile metal ion. Various quantities have been used to describe this lability, such as the binding energy, the mean electrostatic potential, etc. (Blesa *et al.* 1994). All of these are descriptors that quantify the Pauling rules on the stability of complex oxides (Pauling 1960).

According to Aagaard and Helgeson (1982), the dissolution proceeds via several successive chemisorptive equilibria, which produce an ensemble of critical composition $\equiv S_c$. This species dissolves irreversibly through a critical activated complex $\equiv S_c^{+}$, at a rate given by the standard equation:

$$R = \frac{k_B T}{h} \{\equiv S_c^{\neq}\} \tag{3.64}$$

Solution variables influence the surface concentration of $\equiv S_c$, as well as the activation step. Aagaard and Helgeson's model also takes into account the reverse deposition process when the concentration of dissolved species approaches the solubility limit.

Hiemstra and van Riemsdijk (1990) have modelled the importance of various possible $\equiv S_c$ surface complexes in the dissolution of SiO_2 and silicates. The species

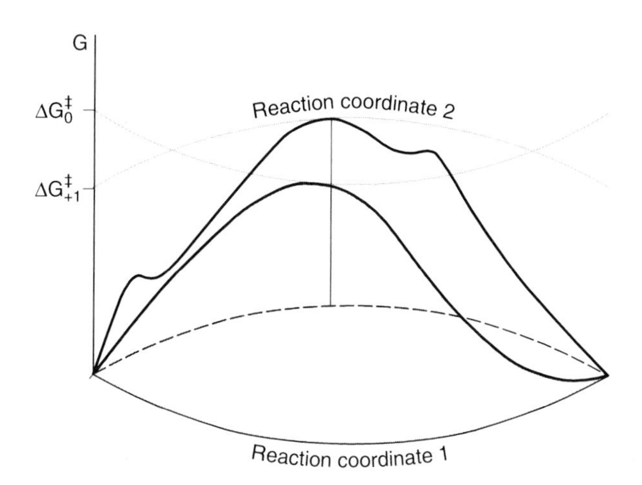

Figure 3.8 Comparison of two possible dissolution pathways. 1, no protonation pre-equilibrium involved; 2, faster dissolution pathway involving prior protonation. Reprinted with permission from Blesa *et al.* (1994).

considered are mononuclear $\equiv SiO_2H_{(2+z)}^{z+}$ complexes ($0 \leq z \leq 4$) formed through successive protolytic equilibria. The contribution of each species to the rate is assessed by assuming that their activation Gibbs energies are linked:

$$\Delta G^{\neq_z} = \Delta G^{\neq_0} - z\Delta H_a \tag{3.65}$$

where ΔH_a is the enthalpy of proton adsorption. Charge build-up thus leads to enhanced dissolution rates. Figure 3.8 depicts two possible reaction pathways.

4 Concluding remarks

Advanced materials for a large variety of uses are based on the metal oxides. Their extensive usage implies, in general, an acceptable chemical behaviour when exposed to water or moist atmospheres. It might not be wholly realized that very complex chemistry is involved in the current use of some of these materials. For example, CrO_2, or high-T_c superconducting oxides are readily corrodible by water and aqueous solutions (Blesa *et al.* 1994; Candal *et al.* 1996). It is remarkable that the simple ideas exposed here suffice to understand the behaviour of these materials, and also to design appropriate protection strategies.

5 Acknowledgements

The outlook presented here is the result of many years of research with a large number of associates and students, as exemplified in the Reference list. My special appreciation to A.E. Regazzoni, and also to P.J. Morando and A.J.G. Maroto for the fruitful exchange of ideas and unending support. The author is a member of Consejo Nacional de Investigaciones Cientificas y Técnicas.

6 References

Aagaard, P. & Helgeson, H.C. (1982) *Am J Sci* **282**, 237–285.

Angus, J.C. & T.J. Dyble (1975) *Surf Sci* **50**, 157–177.
Baes, C.F. & Mesmer, R.E. (1976) *The Hydrolysis of Cations.* Wiley Interscience, New York.
Bérubé, Y.G. & de Bruyn, P.L. *J Colloid Interface Sci* (1968) **27**, 305–318.
Blesa, M.A. & Kallay, N. (1988) *Adv Colloid Interface Sci* **28**, 111–134.
Blesa, M.A., Morando, P.J. & Regazzoni, A.E. (1994) *Chemical Dissolution of Metal Oxides.* CRC Press, Boca Raton, Florida.
Blesa, M.A., Regazzoni, A.E. & Stumm, W. (1996a) To be published.
Blesa, M.A., Alí, S.P., Morando, P.J. & Regazzoni, A.E. (1996b) To be published.
Blesa, M.A., Borghi, E.B., Maroto, A.J.G. & Regazzoni, A.E. (1984) *J Colloid Interface Sci* **98**, 295–305.
Boehm, H.P. (1971) *Disc Farad Soc* **52**, 264–275.
Borggaard, O.K. (1991) *Clays Clay Miner* **39**, 324–328.
Bruyère, V.I.E. & Blesa, M.A. (1985) *J Electroanal Chem* **182**, 141–156.
Cabrera, N. (1960) Kinematic theory of crystal etching and its application to etching. In: *Reactivity of Solids, Proceedings of the 4th International Symposium* (ed. J.H. de Boer), p. 345. North-Holland, Amsterdam.
Candal, R., Regazzoni, A.E. & Blesa, M.A. (1996) *J Mater Sci* **31**, 45–60.
Diggle, J.W. (1973). *Oxides and Oxide Films,* Vol. 2. (ed. J.W. Diggle), Ch. 4. Marcel Dekker, New York.
dos Santos, M., Morando, P.J., Blesa, M.A., Banwart, S. & Stumm, W. (1990) *J Colloid Interface Sci* **138**, 74–82.
Dzombak, D.K. & Morel, F.M.M. (1990) *Surface Complexation Modelling. Hydrous Ferric Oxide.* John Wiley & Sons, New York.
Engell, H.-J. (1956a) *Z. Phys Chem NF* **7**, 158–181.
Engell, H.-J. (1956b) *Z Elektrochem* **60**, 905–911.
Furrer, G. & Stumm, W. (1986) *Geochim Cosmochim Acta* **50**, 1847–1860.
Gorichev, I.G. & Kipriyanov, N.A. (1981) *Russ J Phys Chem* **55**, 1558–1568.
Henrich, V.E. (1983) *Prog Surf Sci* **14**, 175–199.
Henrich, V.E. (1989) *Surfaces and Interfaces of Ceramic Material* NATO ASI Series, Vol. E 173. (eds L.C. Dufor, C. Monty & G. Petot-Evans). Kluwer, Dordrecht.
Heusler, K.E. (1983) *Electrochim Acta* **28**, 439–449.
Hiemstra, T. & van Riemsdijk, W.H. (1990) *J Colloid Interface Sci* **136**, 132–150.
Hiemstra, T., van Riemsdijk, W.H. & Bolt, G.H. (1989a) *J Colloid Interface Sci* **133**, 91–104.
Hiemstra, T., de Wit, J.C.M. & van Riemsdijk, W.H. (1989b) *J Colloid Interface Sci* **133**, 105–117.
Ikeda, T., Sasaki, M., Hachiya, K., Astumian, R.D., Yasunaga, T. & Schelly, Z.A. (1985) *J Phys Chem* **86**, 3861–3866.
Levine, S. & Smith, A.L. (1971) *Disc Farad Soc* **52**, 290–301.
Lewis, D.G. & Farmer, V.C. (1986) *Clay Minerals* **21**, 93–100.
Litter, M.I. & Blesa M.A. (1994) *Canad J Chem* **72**, 2037–2043.
Litter, M.I. & Blesa, M.A. (1992) *Canad J Chem* **70**, 2502–2510.
Litter, M.I., Baumgartner, E.C., Urrutia, G.A. & Blesa, M.A. (1991) *Environ Sci Tech* **25**, 1907–1913.
Mullins, W.W. & Hirth, J.P. (1963) *J Phys Chem Solids* **24**, 1391–1404.
Needes, C.R.S., Nicol, M.J. & Finkelstein, N.P. (1975) In: *Leaching and Reduction in Hydrometallurgy* (ed. A.R. Burkin), pp. 12–19. The Institute of Mining and Metallurgy, London.
Nicol, M.J., Needes, C.R.S. & Finkelstein, N.P. (1975) In: *Leaching and Reduction in Hydrometallurgy,* (ed. A.R., Burkin), pp. 1–11. The Institute of Mining and Metallurgy, London.
Parks, G. (1965) *Chem Rev* **65**, 177–198.
Pashley, R.M. & Kitchener, J.A. (1979) *J Colloid Interface Sci* **71**, 491–500.
Pauling, L. (1960) *The Nature of the Chemical Bond,* 3rd edn. Cornell University Press, Ithaca, N.Y.
Regazzoni, A.E. & Blesa, M.A. (1991) *Langmuir* **7**, 473–478.

Regazzoni, A.E. & Blesa, M.A. & Maroto, A.J.G. (1988) *J Colloid Interface Sci* **122**, 315–325.
Rueda, E.H. & Blesa, M.A. (1996) To be published.
Sangwal, K. (1987) *Etching of Crystals. Theory, Experiment and Application. Defect in Solids*, Vol. 15 (eds S. Amelinckx & J. Nihoul). North Holland, Amsterdam.
Schindler, P.W. (1990) Mineral–water interface geochemistry. In: *Reviews in Mineralogy*, Vol. 23 (eds M.F. Jr. Hochella & A.F. White), Ch. 7. Mineralogical Society of America, Washington D.C.
Segal, M.G. & Sellers, R.M. (1984) *Advances in Inorganic and Bioinorganic Series, Mechanisms*, Vol. 3 (ed. A.G., Sykes), pp. 97–129. Academic Press, London.
Segall, R.L., Smart, R.St.C. & Turner, P.S. (1979) *Physics of Materials* (eds D.N. Borland & L.M. Clareborough), pp. 261–270. CSIRO Press, Melbourne.
Segall, R.L., Smart, R.St.C. & Turner, P.S. (1988) *Surface and Near-Surface Chemistry of Oxide Materials*, (eds J. Nowotny & L.-C. Dufour), Ch. 13. Elsevier, Amsterdam.
Sposito, G. (1990) Mineral–water interface geochemistry. In: *Reviews in Mineralogy*, Vol. 23 (eds M.F. Hochella, Jr. & A.F. White), Ch. 6. Mineralogy Society of America, Washington, D.C.
Stone, A.T. & Morgan, J.J. (1987) *Aquatic Surface Chemistry* (ed. W. Stumm), Ch. 9. Wiley-Interscience, New York.
Surana, V.C. & Warren, I.H. (1969) *Trans Inst Min Metall* **C78**, 133–139.
Tejedor-Tejedor, M.I. & Anderson, M.A. (1986) *Langmuir* **2**, 203–210.
Torres, R., Blesa, M.A. & Matijevic, E. (1989) *J Colloid Interface Sci* **131**, 567–579.
Valverde, N. & Wagner, C. (1976) *Ber Bunsenges Phys Chem* **80**, 330–333.
Vermilyea, D.A.A. (1966) *J Electrochem Soc* **133**, 1067–1070.
Waite, T.D. & Torikov, A. (1987) *J. Colloid Interface Sci* **119**, 228–235.
Warren, I.H. & Devuyst, E. (1975) *Inst Chem Eng Symp Series* **42**, 7.1–7.11.
Wehrli, B. (1989) *J Colloid Interface Sci* **132**, 230–242.
Wieland, E., Wehrli, B. & Stumm, W. (1988) *Geochim Cosmochim Acta* **52**, 1969–1981.
Wokulska, K.V. (1978) Doctoral Thesis, Silesian University, Poland, cited in Sangwal (1987).
Zinder, B., Furrer, G. & Stumm, W. (1986) *Geochim Cosmochim Acta* **50**, 1861–1869.

4 Some Data on Surface Properties and Reactivity of Metal Oxides

L.C. DUFOUR

Laboratoire de Recherches sur la Réactivité des Solides UMR 5613 CNRS, Université de Bourgogne, BP 138, 21004 Dijon cedex, France

1 Introduction

For several decades, many papers have been published on the properties and reactivity of metal oxides. Metal oxides are often considered as a special class of chemical compounds, which lead to materials having extremely varied chemical and physical properties. The recent development of research works on metal oxides usable, for example, as sensors or catalysts, or in energetics and electrical transportation, clearly evidences this interest. Figure 4.1 illustrates this activity with, as an example, the number of papers devoted in the last 40 years to reactivity and related properties of NiO, which can be considered as a typical oxide. It is noteworthy that, in the 1970s, a strong interest in bulk properties and reactivity of NiO was increasing, but a renewed interest has recently appeared and is orientated rather to surface properties and reactivity.

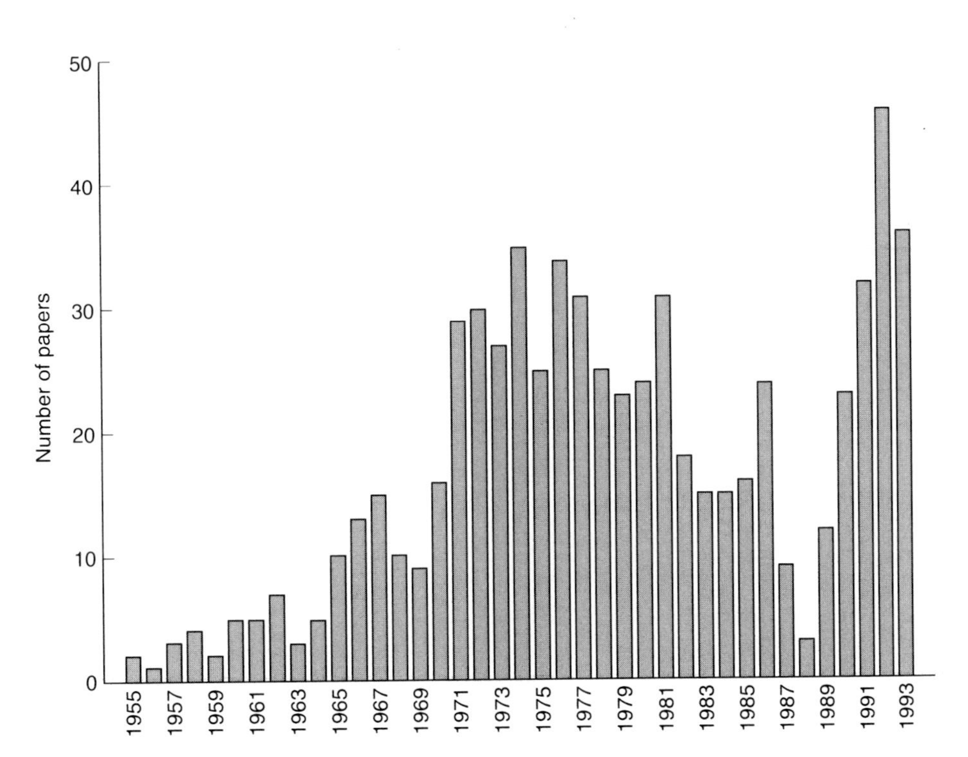

Figure 4.1 Evolution of the number of papers published on properties and reactivity of NiO in the last 40 years.

It is usually admitted that the properties of materials, and particularly of iono-covalent oxides, strongly depend on their mode of preparation and on their biographic story, before and during their use. This means that a great deal of attention must be turned to the properties of surfaces and interfaces, which often play a role important in the overall properties and behaviour of materials. This holds for the preparation, insofar as thermal treatment, sintering and shaping lead to possible changes in chemical composition or crystallographic structure of surfaces and interfaces. Unfortunately, for metal oxides these effects are generally poorly known and underexploited. However, when they are controlled they can be used to modify the properties of materials, for example of oxide ceramics. The same is true when the evolution of the properties of materials in use is considered: for example, ageing is a complex phenomenon which often is the consequence of solid state diffusion, thermal segregation or chemical reaction, which can modify specific properties localized at interfaces and also control the overall properties. This is the case, amongst other materials, for high-temperature oxide catalysts, oxide-based composites, oxide electrodes in fuel cells, oxide coatings in protective layers, gas oxide sensors or thin oxide layers for friction or adhesion.

With these observations in mind, it is obvious that a large gap exists between the absolute necessity to know better the interfacial properties of the oxide materials and the relatively low production of theoretical and experimental results in this field; relatively low when compared with metals or alloys and semiconductors of industrial use, for which a strong activity has been developing over the last 30 years in the field of the properties of surfaces. This increasing activity is related to the extraordinary development of experimental techniques enabling us to investigate crystallographic structure, nature of defects, local chemical composition, arrangement of adsorbates, chemical bonding, etc. In parallel to this interest for metallic or semiconducting compounds, results have recently begun to be published on surface properties of metal oxides. As pointed out above, the scientific world seems to have become aware of the lack in this field, and an increased flux of the published papers can be observed nowadays. Books recently published on these topics attest this growing interest (Pask and Evans 1981; Dufour and Nowotny 1988; Nowotny and Dufour 1988; Dufour *et al.* 1989; Kung 1991; Freund and Umbach 1993a; Henrich and Cox 1994; Noguera 1995).

The general feeling could be roughly summarized by repeating the terms used for the title of this book. Past, present and future in the field of the surface properties and reactivity of metal oxides can be expressed in a condensed way as follows.

- Distant past: nothing.
- Recent past and present: increasing interest.
- Future: much is to be done.

Keeping these schematic views in mind, we intend, in this chapter, to focus modestly on some of the problems, results and interests that make this scientific field so difficult to investigate and, nevertheless, so attractive. We present, in a not exhaustive and probably partial way, some of facets which seem to be significant today for fundamental and applied research. Nevertheless, these reflections are restricted to the oxide–gas interactions, excluding the field of the oxide–liquid reactions for which many of the ideas developed here apply. The exciting subject of the new high-temperature oxide superconductors and the field of non-crystalline oxides will not be considered in this chapter.

2 Bulk versus surface properties of metal oxides: nanocrystals and extended monocrystalline surfaces

In solid state physics and chemistry, it is usual to distinguish between large monocrystalline surfaces cut in a single crystal, polycrystalline surfaces with multi-orientated grains and grain boundaries, and more or less finely divided solids. Generally, the basic properties of the surface layers are studied from orientated planes presenting long-distance ordering on a large surface area. The properties of polycrystalline or micro- and now nanocrystalline materials are investigated in order to be directly transposable to industrial applications. For instance, work on the influence of grain boundaries for ceramics and on the size or orientation effect for catalysts will be favoured.

Surprisingly, the communities of scientists working on perfect large surfaces, polycrystalline materials or powders seem to be separated without real links, as if these various forms of a same solid were totally different. In fact, this attitude is often related to difficulties in adapting specific techniques of surface analysis to these various forms of solid materials. An original and, nowadays, underdeveloped, way compares the properties and reactivity of metal oxide surfaces as collected from perfect two-dimensional surfaces to those of very finely grained solids.

In reality, with nanocrystals, the surface influence is obviously important as a great proportion of atoms are located at surfaces. In all cases, it is noteworthy that the surface cannot be considered only as a two-dimensional boundary limiting the solid, but rather as a zone several atomic layers deep with specific properties different from the bulk. Then the question is to know how the influence of surface properties can be modified when the crystallite size decreases to the limit of stability of the nanocrystal. Even though this effect has been much studied for small metal particles deposited onto oxides (see, for example, Kern 1978; Rochefort and Le Peltier 1991; Poppa 1993; and literature therein), little has been done in the past on small metal oxide particles, since the famous works by Néel (1961, 1962) on the unusual magnetic properties of very finely divided particles. This lack of research is surprising as the technological demand for submicronic metal oxides is great, both for direct use and for low-temperature sintering or shaping of oxide ceramic-based materials. Mostly, these particles are elaborated by quite well-controlled techniques based on chemical precipitation, the sol-gel route, organic combustion or other more sophisticated processes of soft chemistry. The micro- or nanocrystallites obtained are often not in their equilibrium state even after long treatment at moderate temperature or after shaping in the form of ceramics. For these iono-covalent oxides, interface atoms will be in a compressive or tensile situation towards the bulk, according to size, nature of the interface curvature and environmental conditions.

This particular behaviour of very finely divided oxide grains in the form of powders or ceramics is difficult to predict perfectly and will be evidenced by subtle changes in physical properties (for example, the temperature of a phase transition) or in chemical reactivity. This constitutes a considerable field of investigation which is starting to be explored. As an example, one can quote works on the influence of grain size and the role of interfaces in the properties of dielectrics (Frey and Payne 1993; Fang *et al.* 1993; Takeuchi *et al.* 1993; Perriat *et al.* 1994; Schlag and Eike 1994; Zhong *et al.* 1994), or on the oxidation mechanisms of submicronic ferrites (Gillot and Rousset 1994).

3 Ideal and real surfaces of metal oxides

An ideal surface should be clean, free of defects, relaxed or polarized and at thermodynamical equilibrium for its structural and compositional features. This situation is never possible with metal oxides, and several interconnected levels should be considered in reality:

- surface stability and surface energy;
- surface defects;
- surface composition.

3.1 *Surface energy and surface tension*

If a solid metal oxide free of electric charges and not situated in a magnetic or electric field is considered, the specific surface work γ (specific = relative to the unit) is the reversible work done to create the surface unit at constant temperature, volume and total number of moles:

$$\gamma = f^{s} - \Sigma\mu_i^s\Gamma_i^s \tag{4.1}$$

where f^s is the Helmholtz surface free energy and μ_i^s and Γ_i^s, the chemical potential and the surface excess number of each component of the metal oxide and the surrounding phase, respectively. As observed above, the physicochemical properties of the constituents cannot be considered as strictly discontinuous at the surface, but it is useful to define a *Gibbs dividing surface* as a theoretical boundary of discontinuity between solid and vacuum or surrounding atmosphere. Then, the position of this 'dividing surface' can be selected in such a way that the second term of the above equation is equal to zero and γ can be considered as equal to the free surface energy. In fact, γ is rather an excess of free energy as it does not represent the total surface free energy but the excess of energy that the surface atoms possess as a result of their location at the surface.

This surface free energy is a scalar representing work. It should not be confused with the surface tension σ, induced by the possible surface stretching. Contrary to the case of simple liquids, the low mobility of atoms and the existence of long-range interactions in iono-covalent metal oxides make it impossible to preserve the microscopic surface configuration after deformation. Then, the surface atoms cannot reach their equilibrium state immediately when the surface is newly formed. Solid state diffusion and thermal activation are necessary and this point must be kept in mind when metal oxides are reacting and creating new surfaces. The surface tension can be considered as a two-dimensional pressure associated with tensile or compressive effects according to the sign of the change in the stress tensor describing the crystallographic planes (see Linford 1978; Tasker 1979; Stoneham 1981; Delannay *et al.* 1987; and literature quoted by these authors).

An original way to introduce these concepts (Linford 1978) consists of presenting the surface energy, γ^s, as being composed of two terms corresponding to plastic and elastic deformation, according to:

$$\gamma^s = \frac{d\varepsilon_p}{d\varepsilon_{tot}}\gamma + \frac{d\varepsilon_e}{d\varepsilon_{tot}}\sigma \tag{4.2}$$

where $d\varepsilon_p$ and $d\varepsilon_e$ represent the two surface strains, plastic and elastic, respectively, with $d\varepsilon_{tot} = d\varepsilon_p + d\varepsilon_e$. Surface free energy γ and surface stress σ have equivalent dimensions (MT^{-2}), but γ is expressed in millijoules per square metre and σ in millinewtons per metre.

Related to this surface tension, slight changes in the position of surface atoms, when compared to the three-dimensional crystal lattice, can be detected (Fig. 4.2).

- Surface relaxation that is the displacement of the surface planes leading to change in the interplanar distance.
- Rumpling that is the relative displacement of the atoms of the first surface planes with respect to their positions in the plane.

These various effects can lessen the heterogeneities in the atom position and energy of the surface sites, as developed below.

3.2 *Adsorption and segregation*

The equations of adsorption and segregation link together γ, temperature (T) and the surface concentrations of the species *i*. From the generalized surface Gibbs–Duhem equation:

$$s^s\, dT + \Sigma d\mu_i^s \Gamma_i^s + d\gamma + (\gamma - \sigma)\, d\varepsilon_e = 0 \tag{4.3}$$

where Γ_i^s is the excess of species *i* per surface unit and s^s is the surface entropy. The adsorption isotherm Gibbs equation is obtained, at equilibrium ($\mu_i^s = \mu_i$) and constant temperature ($dT = 0$), and neglecting the last term:

$$\partial\gamma = -\Sigma\Gamma_i^s \partial\mu_i \tag{4.4}$$

or:

$$\Gamma_i^s = -\left(\frac{\partial\gamma}{\partial\mu_i}\right)_{T,\varepsilon_e} \tag{4.5}$$

For a metal oxide having a very low vapour pressure and surrounded by a gas insoluble

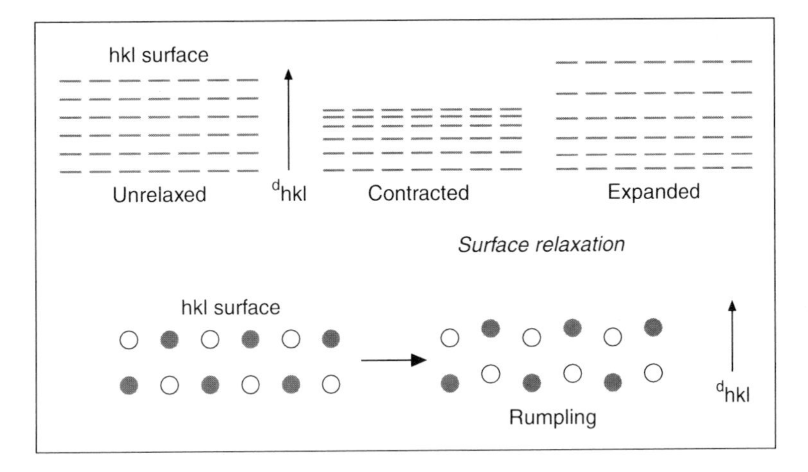

Figure 4.2 Relaxation and rumpling in the outer surface layers of metal oxides.

in the oxide, we have:

$$\mu_i = \mu_i^0 + RT \ln fg_i \tag{4.6}$$

where fg_i is the fugacity of the gas, i. That is:

$$\Gamma_i^s = -\frac{1}{RT}\left(\frac{\partial\gamma}{\partial \ln P_i}\right)_{T,\varepsilon_e} \tag{4.7}$$

where P_i is the pressure of the gas we suppose to be ideal.

Segregation can be considered as adsorption of species coming from the inside of metal oxide to the interface. Assuming no external adsorption, we have for an ideal solution:

$$\mu_i = \mu_i^0 + RT \ln a_i \tag{4.8}$$

where a_i is the activity of the component i, leading to:

$$\Gamma_i^s = -\frac{1}{RT}\left(\frac{\partial\gamma}{\partial \ln a_i}\right)_{T,\varepsilon_e} \tag{4.9}$$

For a diluted solution:

$$\Gamma_i^s = -\frac{1}{RT}\left(\frac{\partial\gamma}{\partial \ln X_i}\right)_{T,\varepsilon_e} \tag{4.10}$$

where X_i is the molar fraction of the species dissolved in metal oxide.

Equilibrium segregation (or adsorption) at surfaces (or at interfaces) corresponds to a distribution of the dissolved species (or in the external atmosphere) in such a way that the free energy of the system (volume and surface) is minimized. When these phenomena occur, they strongly modify the surface tension of the pure metal oxide. For example, for a compound 'sol' dissolved into the oxide 'ox', if $\Gamma_{sol}/\Gamma_{ox} > X_{sol}/X_{ox}$, segregation occurs at the surface or at the interface and the surface tension of pure metal oxide decreases.

Surface (or interface) faceting is a consequence of adsorption or segregation, leading to instability of the crystal planes in such a way as to minimize the product of the surface (or interface) energy by the surface (or interface) area of each of these planes. This phenomenon is reversible or irreversible and, for segregation, is often related to solid state diffusion.

3.3 *Polarity and charge distribution at surfaces of pure ionic metal oxides*

Ideal ionic crystals are terminated, theoretically, by three types of planes (Tasker 1979; Stoneham 1981) (Fig. 4.3). In the type I surface, the plane is electrically neutral as it is composed of the same number of cations and anions of opposite charge. In the type II surface, the successive planes are charged differently, but in such a way that the resulting dipole moment will be null perpendicular to the surface; each repeat sequence has a symmetry plane. In the type III surface, the successive planes are charged differently, but with a resulting dipole moment perpendicular to the surface, given that the repeat sequence has no symmetry plane. This type III surface is unstable, theoretically, but can be stabilized by segregated impurities (e.g. Floquet and Dufour 1983).

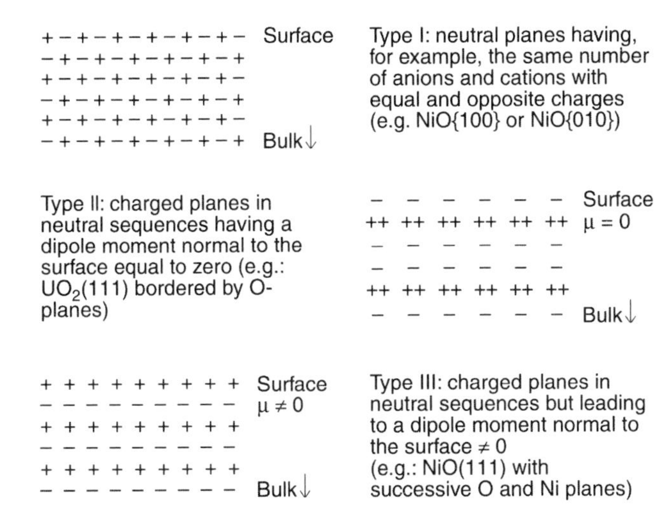

Figure 4.3 Distribution of charges in the surface planes for three sequences of stackings parallel to the surface. (From Tasker 1979; Stoneham 1981.)

Moreover, pure polar surfaces, theoretically unstable, can be reconstructed and stabilized in the form of ultrathin films, if these films have grown in perfect epitactic relationships on an appropriate surface; a recent example is given for the NiO{111} surface on Au{111} (Ventrice *et al.* 1994).

3.4 *Real surfaces and morphological features of metal oxides*

At the micrometre scale, a large variety of morphologies and related surface properties can exist for the same metal oxide according to the conditions of formation of this oxide from a precursor, in the rough sense of this term (metal, hydroxide, organic compounds, etc.). Amongst the main factors, involved before, during and after the reaction of formation and often interconnected, there are:

- the nature, purity and morphology of the parent;
- the experimental conditions and kinetics of the transformation parent–metal oxide;
- the crystallographic relationships and the mechanical features of the solids involved;
- the biographic story of metal oxide after its formation;
- the possible influence of mobile foreign species or impurities leading to segregational effects at interfaces.

The finest control of these solid state transformations should enable the technological demand to be satisfied, which now, and more and more in the future, will request oxides with specific morphologies and specific surface properties for specific applications. This involves a good knowledge of the physicochemical processes in order to orientate and control the reactivity of the parents and, therefore, to elaborate, amongst other properties, precise morphology and grain size of oxide materials for the planned use (Boldyrev *et al.* 1979).

The initial oxidation of Ni provides an example of the variety of morphologies of thin oxide layers. According to the temperature of oxidation, nature of oxidizing gas (pure oxygen or water vapour–hydrogen mixtures), nature and pretreatment of starting Ni (purity, grain size, conditions of initial annealing, initial oxidizing–reducing pretreatment, conditions of polishing etc.) the following morphologies, amongst others, can be observed: uniform large grain scale, geometric microcrystals, faceted microcrystals, honeycombed and filamentous layers, semicircular platelets, etc. (Morin *et al.* 1992; Dufour and Morin 1993; and references therein). Some of these morphologies have been investigated for their interest in specific applications, for example to be used in advanced water electrolysis.

Many other examples can be drawn from literature on basic or applied research devoted to optimize the surface properties and morphology of metal oxides for various applications in high technology: amongst others, oxides for catalysis (Delmon 1986a,b and references therein), metals supported on oxides for catalysis (Che and Bennett 1989, Rochefort and Le Peltier 1991; and references therein), chemical sensors (Moseley *et al.* 1991, and references therein), oxide electrocatalysts (Trasatti 1990, 1994 and references therein) and oxide materials for electrochemistry (Murugesamoorthi *et al.* 1993, and references therein; Siebert 1994, and references therein) may be mentioned. In recent years, an important progress has also taken place in the control of uniform shapes and very narrow size distributions of oxide powders in the micron and submicron range, often by a technique of precipitation in colloid science. Works of the group of Matijevic (1981, 1988, 1991) and the review of Haruta and Delmon (1986) may be quoted in this field.

Let us note also that theoretical calculations and modelling have started to be developed in order to predict structure, stability and morphology of pure or doped ionic crystals, particularly of metal oxides (see, for example, Mackrodt 1989; Gillan 1991; Egdell and Parker 1991; Titiloye *et al.* 1991; Oliver *et al.* 1993).

This variety in surface morphology is also probably very pronounced at the nanometric and atomic scale but, as observed below, these experimental features are not easily evidenced directly. The classical model terrace, ledge and kink (TKL), derived from works by Kossel (1927) and Stranski (1928) on crystal growth, is still of great interest. It enables us to propose a hierarchic description of the surface sites and reactivity, as reported in the next section.

4 Hierarchic order in the energy levels of the surface sites and reactivity

Although the surfaces of metal oxides are generally heterogeneous and non-stoichiometric, it is necessary to consider three successive levels in the description of their reactivity by referring to ideal surfaces, to real surfaces with intrinsic defects and to real surfaces with foreign atoms.

4.1 *The ideal (hkl) surface: clean and perfectly ordered*

This surface cannot exist, except *in vacuo* and in thermodynamical conditions where the oxide crystal would be stable, i.e. amongst others, in an appropriate oxygen potential. These two conditions are conflicting, except perhaps at low temperatures. In fact, due to their iono-covalent nature, the surfaces of metal oxides and particularly of ceramic oxides are considered as less reactive when compared to metals. This is due to strong charge exchange between adsorbate and the surface of metals in comparison with moderate electrostatic-type forces for ionic surfaces. This is evidenced through theoretical calculations of the electronic states of metal oxide surfaces and gas–surface interactions. These works are increasingly performed thanks to the development of powerful computers, and many papers have already been published for simple cases, often modelled by using the *ab initio* Hartree–Fock or local density function methods. Amongst the recently published examples, papers by Colbourn (1992), Neyman and Rösch (1992, 1993), Pöhlchen and Staemmler (1992), Shluger *et al.* (1992), Mackrodt *et al.* (1993), Freitag and Staemmler (1994), Nygren and Pettersson (1994), Birkenheuer *et al.* (1994) and Pacchioni *et al.* (1994a,b) can be quoted.

4.2 *The real surface of pure metal oxide with its intrinsic defects*

In reality, the reactivity of iono-covalent oxide surfaces at moderate temperatures is made possible by the existence of defects that play a basic role in the adsorption of molecules, which often is the step required before the nucleation of new solid phases. Therefore, an efficient control of the oxide reactivity characterizes both nature and number of structural defects pre-existing at surfaces, or those forming or vanishing as a consequence of the solid state transformation. These defects have various origins and are difficult to identify and characterize in the bulk and at the surface; point and extended defects, ordered or disordered defects are classically considered with dislocations, internal surfaces, voids and grain boundaries. The metal oxide surface can be itself considered defective as its ions are in a lower coordination state than those of the bulk. A variety of intrinsic surface defects exist according to the coordination number of the site, the type of defect and the possible effect of clustering modifying their energy level. For instance, whilst the bulk ions of the rock salt structured oxides are sixfold coordinated, the surface ions of the stable (001) surface can be five-, four- and even threefold coordinated according to their situation on terraces, ledges, kinks or corners. With point vacancies and extended defects (emergence of dislocations and crystallographic shear planes, voids, etc.), there are various and discrete energy levels corresponding to specific adsorption sites (Dufour *et al.* 1985, and references therein). In fact, the theoretical differences between energy levels are attenuated by the lattice relaxation effects (Colbourn and Mackrodt 1983; Tasker and Duffy 1984). For many simple cases, one can calculate the values of binding energy between a molecule, an atom or an ion and a surface site in order to predict the structure–reactivity relationships (see, amongst the mast recent papers: Pacchioni and Minerva 1992; Pacchioni 1993; Jug *et al.* 1993; Pettersson and Pacchioni 1994; Duffy *et al.* 1994; Pacchioni *et al.*, 1994c). These results agree more or less with experience (see below). However, this theoretical approach must be developed even if the initial data for calculations are inaccurate sometimes. Finally, it is noteworthy that the existence of surface heterogeneities can help to develop collective effects leading, for example to surface aggregation as theoretically calculated for the water–metal oxide interface (Lajtar *et al.* 1993).

4.3 *The real surface of metal oxides with foreign atoms*

In addition to these intrinsic defects, foreign atoms are often present at surfaces as a consequence of their existence in bulk as additives or impurities, or combined or not with segregational effects. Foreign atoms can also be intentionally deposited onto surfaces for specific purposes. The thermodynamical stability of these atoms at surfaces and, more generally, at interfaces, can be calculated in some simple cases, which is very useful for obtaining or improving the suitable surface properties in specific applications such as catalysis, elaboration of sensors, sintering of ceramic oxides or electronics. Theoretical predictions appear to be more necessary in this case where experimental results are limited and extremely difficult to obtain, for example, in thermal segregation (see below) (see, for examples, Egdell and Mackrodt 1989; Mackrodt and Tasker 1989; Egdell and Parker 1991; Mackrodt 1992; D.C. Sayle *et al.* 1994; T.X.T. Sayle *et al.* 1994).

4.4 *Hierarchic order and reactivity*

The concept of a hierarchic order in solid state reactivity and, in particular, in the reactivity of the surface sites is well accepted nowadays, although it is difficult to be quantitatively applied. The first papers on this topic have considered the general case of the energy levels in the solid state (Resch and Gutman 1980, 1981) but, for a long time, these differences in the thermodynamical features of the surface sites were qualitatively evidenced by infrared spectroscopy or temperature-programmed desorption. The most recent approach for metal oxide surfaces takes into account the acidic properties of this class of compounds, as discussed later, and aims to evaluate the energy distribution of surface acid sites for catalysis (Carniti *et al.* 1994, and therein), to show parallels with coordination chemistry (Köhler and Schläpfer 1993), or to correlate Madelung field and reactivity of low-coordinated sites (Stefanovic and Truong 1995).

If parallels between hierarchic energy level of sites and reactivity are not perfectly identified and quantitatively observed for surfaces, in some favourable cases, experimental results are published on the relationships between the reactivity of the bulk crystallographic sites and the strength of the cation–oxygen bonding. As an example, the selective oxidation of submicronic spinels was studied by differential thermogravimetry and infrared spectroscopy: by using a technique of deconvolution, specific oxidation peaks could be evidenced according to the oxidation state and crystallographic position of the various cations (Domenichini *et al.* 1992; Gillot and Rousset 1994). Correlations between cation–oxygen distance, crystallographic site position and reactivity measured by the oxidation temperature were deduced (Gillot 1994).

5 Metal oxide–gas interaction and analysis in terms of the donor–acceptor or acid–base concept

Although the most interesting investigations on gas–metal oxide interactions have been performed from clean, well-ordered and well-characterized surfaces of binary metal oxides, various other situations have been observed. However, this diversity is beginning to be understood in terms of both electronic configuration and geometric structure of the surface sites. In other words, nature and charge of cations, oxygens and defects, coordination number and acido–basicity of the surface sites must be considered.

The metal oxides are selected either for their apparent simplicity and the possibility of their elaboration in form of pure single crystals or for their interest as models or for technological applications. Often, the surface is prepared *in situ* by cleavage or by cutting and successive ion cleaning and thermodynamically controlled annealing. In some cases, the surfaces are ultrathin layers grown by deposition on or oxidation of an appropriate metal surface selectively orientated in order to stabilize a crystallographic oxide plane (Ventrice *et al.* 1994; Winkelmann *et al.* 1994). The main oxides classically investigated have a well-known crystal structure: face-centred cubic NiO and MgO, rutile (TiO_2), corundum (α-Al_2O_3) and wurtzite (ZnO), and are used either pure or doped.

The interaction of solid oxides with gases has been discussed in several recent papers (see, for example, Barteau 1993; Henrich and Cox 1993; Lin and Arribart 1993; Busca *et al.* 1993; Rösch *et al.* 1993; Arnaud and Bertrais 1994; Winkelmann *et al.* 1994; and references therein). Analysis using the acid–base or donor–acceptor concept is often put forward: it consists of postulating that the surface ions or defects that are in an unsaturated coordination state can act as acidic or basic sites towards molecules striking this surface. In reality, the gas–surface interaction will be controlled by two types of additives forces: electrostatic and acido–basic according to the nature of the coupled adsorbate–oxide ion. As the degree of unsaturation increases or the coordination number decreases at steps, kinks or defects, the reactivity is greater at these sites compared to the normal terrace sites. Then, mechanisms such as hydrogen bonding, proton or electron transfer, dissociative adsorption or acid–base interaction are considered. From several papers and books, non-exhaustively selected (Dufour *et al.* 1985; Kung 1991; Kuhlenbeck *et al.* 1991, 1992; Pacchioni 1993; Henrich and Cox 1993, 1994; Cappus *et al.* 1993; Freund *et al.* 1993b; Langell *et al.* 1994; Maier *et al.* 1994; Winkelman *et al.* 1994; Noguera 1995) and literature cited therein, several cases are given here as examples.

- The surface oxygens are mainly basic and can give their electrons to acidic molecules such as CO_2, which is an acceptor, although sometimes a weak donor. This acid–base interaction is relatively easy and can lead to surface carbonatation of metal oxides.
- The hydroxyl groups can easily adsorb onto oxide surfaces in residual atmospheres. From the dipolar water molecule, which is a good donor, strong chemisorption commonly occurs at steps and defects, and is often dissociative. Water molecules can adsorb onto the surface cations which then are considered as Lewis acids or acceptors of electron pairs. Hydroxide can form immediately with the most basic metal oxides such as calcium oxide or barium oxide. The hydroxyl groups stabilize the polar surfaces such as the basal oxygen plane of zinc oxide, which is totally inactive when hydroxyl is removed, whilst the basal zinc plane can easily dissociate many acidic molecules. The same is true for the normal sites of NiO{111} which, if hydroxylated, are very efficient at dissociating, for example acetic acid. Hydroxyl groups can also form by interaction of the surface with protons, but no existence of hydroxyl groups is detected on normal sites (5_c coordinated) of NiO{100}.
- Hyrogen is a non-polar molecule which is neither donor nor acceptor. As the hydrogen bond is very strong (energy of dissociation of about 4.5 eV), the hydrogen molecules can dissociate only at defects having an energy level high enough, for example a cation vacancy on the NiO{100} surface.
- CO is a weak donor with a small dipole moment and its adsorption is difficult except on low-oxidation state cations. For instance, adsorption is possible on the cations of a

clean $Cr_2O_3\{111\}$ surface in the absence of residual oxygen traces, given that oxygen can strongly adsorb on the same cations and totally inhibit CO adsorption.

- Contrary to hydroxyl groups, nitrogen monoxide is found to adsorb only onto normal 5_c coordinated sites of NiO{100} but, surprisingly, not onto lower coordinated sites.

These experimental examples clearly show the complexity of estimating readily the real behaviour of any gas–metal oxide couple. Nevertheless, qualitative predictions are possible and it may be expected that, if results are obtained from very well-characterized surfaces, a semi-quantitative approach to these phenomena will be developed in the not too distant future. Finally, as observed above, due to this complexity and thanks to the extraordinary development of calculation methods, modelling of the gas–surface site interaction will remain for a long time an incomparable source of prediction.

6 Some scientific fields of growing interest for applications of surface properties of metal oxides

As the metal oxides present a large variety of physicochemical properties, they are or can be used in various applications either in the form of very thin layers deposited onto appropriate substrates or in a massive form. In both cases, the surface layers often play an important role in the property which is interesting to exploit. Below, we give some examples of such fields of interest, keeping in mind that, in a very short time, novel industrial interest can be rapidly growing, screening or annealing the present development for economical reasons. Multiple industrial aspects of the processes and applications of the so-called 'surface engineering', particularly laser or ion surface modification of ceramics and laser, plasma or chemical vapour deposition of coatings, are left out of this short and modest review.

6.1 *Segregational effects at oxide surfaces*

Iono-covalent oxide materials are often elaborated or used in thermal conditions where solid state diffusion is kinetically active. Here, possible changes in chemical composition or crystal structure of the surface or interface (grain boundary) layers can occur. Thermodynamic and kinetic aspects of these phenomena of surface segregation have been little studied for the oxide surfaces by comparison with metals and alloys. One of the major problems results from the difficulty in preparing and characterizing the surface in perfectly controlled conditions and in using the classical techniques to analyse its properties at equilibrium *in situ*. Quenching is never totally effective and does not allow the high-temperature equilibrium state to be preserved. Nevertheless, theoretical calculations and experimental results are expanding as this field of materials science is considered increasingly important (see, for example, McCune and Wynblatt 1983; Hofmann 1987; Mukhopadhyay 1988; Wynblatt and McCune 1988; Nowotny 1989, 1991, 1994; Egdell and Mackrodt 1989; Mackrodt and Tasker 1989; Dowben and Miller 1990; Egdell and Parker 1991; Cabané and Cabané 1992; Dufour *et al.* 1992; Grabke *et al.* 1995; and literature cited by the authors). These phenomena of thermal surface segregation may have significant influence by modifying physical (for example, in thermoionic or electronic emission, optics, wettability, adhesion) or chemical properties (for example, in catalysis or electrocatalysis, reactivity including sintering, high-temperature electrochemistry). They can be used also for improving high-temperature oxidation resistance of alloys. Therefore, it may be expected that the interest for these phenomena of interfacial segregation will be increasing in the future as far as they are a key factor of the life of oxide materials when any thermal treatment is involved. Finally, it is noteworthy that these segregational effects can also be used to control and stabilize morphological features of oxide crystals.

6.2 *Metallic particles on metal oxides*

For three decades, small particles or ultrathin films of metals on metal oxides have been studied for various applications such as catalysis, optics, microelectronics or sensors (see, for example, Bauer 1958, 1991; Dufour and Perdereau 1989; Rochefort and Le Peltier 1991; Poppa 1993; Marks 1994; Noguera 1995; and literature therein). Investigating the properties of these small particles of metals on oxides is of interest in three directions.

1 In heterogeneous catalysis, metals supported on oxides have specific properties related to particle size and type of metal–oxide interactions. At present, the difficulties in preparing metallic particles monodispersed in nanometric size and supported by specific oxides having large surface areas, are overcome in many cases but much needs to be done in order to definitively understand the relationships between, on the one hand, size and structural state of metal particles and metal–oxide properties and, on the other hand, catalytic activity and selectivity of the material.

2 Metallic aggregates of some tens of atoms are much studied theoretically particularly if they present differences in their properties with respect to the bulk properties of the same material. Let us note that differences in electronic properties appear for cluster sizes below 30–50 atoms, whilst differences in chemical properties (catalytic) are evidenced in a much wider particle size range.

3 Studying the interaction between metal and oxide during the growth of the metallic particle on the oxide substrate, highlights important elements for understanding the metal–ceramic oxide bonding and wettability of oxides by metals.

Nature, growth mode and crystallographic relationships between metal and oxides can differ according to whether the metallic films are physical (ultrahigh-vacuum deposited) or chemical (*in situ* reduced from the oxide). The growth mode can be bi-dimensional (layer by layer or Frank–Van der Merwe mode), three-dimensional (island or Volmer–Weber mode) or mixed (layer island or Stranski–Krastanov mode), but it is difficult to predict. Multilayer mode and growth with alloying or interdiffusion can also occur. The mixed growth mode is often observed but, for the same metal–oxide system, various growth modes can be found according to temperature, crystallographic orientation and purity of the substrate. Qualitatively, it can be predicted that wettability of oxides by metals is improved when: (i) the oxygen activity in the metal, or for impurities or defects, increases; and (ii) when oxide ionicity, free energy of oxide formation, stoichiometry of the oxide and cohesive energy of the metal decrease.

Finally, it is probable that both preparation and properties of the small particles and thin films of metals and alloys on oxides for various applications will give rise to many studies in the near future.

6.3 *Thin films of metal oxides*

The recent emergence of ultrathin films of metal oxides (some tens of nanometres) on metallic or other conducting substrates is due to the idea of profiting from the properties of the oxides without being disadvantaged by their insulating and brittle character in the massive form. They enable us to probe the surface properties of oxides in more comfortable conditions by the classical techniques of analysis using electrons and ions (see below). Moreover, structural and electronic properties could be significantly affected or surprisingly modified by coupling these new oxide surfaces or layered oxide materials with metallic substrates (Hofmann 1990; Vurens *et al.* 1989). The overlayers of metal oxides can be grown by controlled oxidation of a similar metal or by controlled deposition by chemical (for example, sol-gel), electrochemical or physical (for example, evaporation–condensation or sputtering) methods.

Thin and ultrathin films of metal oxides are most likely destined for large development in various industrial areas, for example:

- electrochromic and photochromic devices elaborated mainly with tungsten, molybdenum, nickel, cobalt, ruthenium, iridium oxides (see for instance, Seike and Nagai 1991; Passerini and Scrosati 1992; Granqvist 1994; Cantao *et al.* 1994);
- chemical sensors and electronic 'noses' (see below);
- semiconductor devices with, for example silicon dioxide (Mayer *et al.* 1992) and electronic guns thermally activated;
- protective layers against high-temperature oxidation of alloys by using thin coatings of rare earth-based oxides (Czerwinski and Smeltzer 1993; Bonnet *et al.* 1994; Roy *et al.* 1993);
- tribological coatings for improving wear resistance, friction and lubrication of advanced ceramic materials with alumina, chromia and zirconia which have unique mechanical, electrical and chemical properties (Kitsunai *et al.* 1991; Vijande-Diaz *et al.* 1991; Fischer and Mullins 1992; and literature therein).

6.4 *Solid oxide fuel cells (SOFC)*

The technology of SOFC may become very important in providing electrical power in not very small polluting conditions in the distant future (Kartha and Grimes 1994; Plzak *et al.* 1994). SOFC are schematically constituted by two electrodes linked by an oxide ionic conductor, the electrolyte. The cathode and the anode are continuously supplied with air or oxygen and fuel (hydrogen, hydrocarbons, CO), respectively. The reaction products are permanently removed. SOFC have the best efficiency and lead to a high current density when compared to other fuel cells (Appleby 1993; Minh 1993; Murugesamoorthi *et al.* 1993). They work at high temperature (800–1000°C) and use various oxides, as cathodes but mainly *p*-type perovskite oxides such as manganites or cobaltites of lanthanum, partially substituted by strontium. Changes in cationic composition and segregative effects of impurities can arise at the surface layers of these cathodes submitted at high temperature to many changes in oxygen potential. These phenomena of ageing are due particularly to the differences in the affinity of the various cations for oxygen. The cathode materials must have an excellent electrical conductivity and a perfect substructural state enabling a perfect oxygen diffusivity into the crystal lattice, and the quality of their surface layers must lead to rapid surface re-equilibration. In reality, the interfacial steps can be limiting and make it necessary to investigate carefully the surface properties of these mixed oxides in their working conditions. These studies are also of interest in high-temperature oxidation catalysis which uses similar oxide materials. Such work is now in progress in our group.

6.5 *Environmental chemistry and chemical sensors*

Sensors are materials capable of converting a non-electrical signal into an electrical signal. Gas and chemical sensors convert signals due to adsorption and chemical changes, respectively. They must combine several qualities to be interesting and exploitable: small sized, inexpensive, sensitive and reliable. Often, they are metal oxides or metals codeposited on oxides having high electrical conductivity, high density, large thermodynamical stability in the working conditions and catalytically very active and resistant to thermal shocks. Mostly, they are polycrystalline or in the form of thin films, and it is necessary to know the incidence of both grain boundaries and external surfaces in the overall signal. Amongst the metal oxide sensors possibly involving the surface effects, there are the following.

• The humidity sensors ($ZnO–Cr_2O_3$, $LiZnVO_4$, $MgCr_2O_4$), which can be amorphous to improve conductivity in their surface layers.

• The electronic and thermal sensors ($BaTiO_3$) with specific microstructure according to the voltage.

• The oxygen sensors (rare earth (lanthanum)-alkaline earth (strontium) manganites or cobaltites, stabilized zirconium oxide, etc.) to be used, for instance in the $CO–CO_2$ mixtures. Kinetic limitations by surface reactions and segregation of additives at high temperatures are often detected.

Recently, a new technology using ultrathin films of oxides such as tin (IV) oxide, zinc oxide, titanium dioxide, has been developed (see, for examples, Collins *et al.* 1993, and literature therein). It is well adapted to gas sensors and uses the changes of surface conductivity with chemisorption related to the differences between the local work function at the surface site and the ionization potential or electron affinity of gases. Selectivity can be controlled by doping or creation of defects.

Finally, let us note that the sensitivity of metal oxides associated with conductor polymers is used in the electronic 'noses' to characterize the flavours (see, for instance: Gardner *et al.* 1990; Moy 1994).

In these analyses, monodetector or revolutionary multisensor array technology applying the technique of pattern recognition associated with artificial neural networks are used (see, for examples, Ichinose 1985; Ketron 1989; Göpel 1990; Gardner 1991; Moseley *et al.* 1991; Gardner *et al.* 1992; Shuk *et al.* 1993; Chadwick *et al.* 1994; and literature therein). It is obvious that all problems related to environmental conditions will give rise to large development of these techniques of detection and monitoring.

7 Local and collective information on surface layers of metal oxides

It would be incorrect to conclude this subjective and non-exhaustive survey without some words about the extraordinary development, during the past 10 years, of the experimental techniques for the analysis of the surface properties. This remarkable progress is due not only to rapid expansion of original techniques and to improvement of both resolution and sensitivity of former techniques but also to use of powerful computers for complex calculations, enabling us to extract complete and accurate information from experimental data.

Specific problems arise with surface analysis of metal oxides, including the following.

- Clean surfaces are difficult to prepare, except by cleavage in ultrahigh vacuum. Hydroxylated or carbonaceous impurities easily adsorb and, as seen above, are difficult to remove. Specific cleaning procedures must be found by associating slight ion sputtering or laser interaction and controlled thermal oxygen treatments. Moreover, analysis can induce defects at, and damage the oxide surface much more than for monoatomic metal surfaces.
- Because of the often insulating property of metal oxides, surface charging effects can hinder or even prevent the analysis by electron or ion microscopies or spectroscopies. These difficulties can be solved by using ultrathin oxide films grown on conducting substrates, by operating with auxiliary electron guns in order to increase the electron emission of the solid or by using a very low-pressure neutral gas, which enables the surface to be discharged by a mechanism of Auger emission type (Ohlendorf *et al.* 1991; Hofmann 1992).

Many techniques are well known and described in this book either for high-resolution local analysis or for large area analysis. We want only to extract from recent literature examples illustrating the evolution of some techniques useful for characterizing the metal oxide surfaces.

- The atomic-scale resolution by scanning tunnelling microscopy can be reached (see, for examples, Bertrams *et al.* (1994) for imaging thin oxide films; Lu *et al.* (1993) for imaging the microstructure of a cleaved tungsten bronze; Tarrach *et al.* (1993) for imaging a natural iron oxide; Rohrer *et al.* (1993) for imaging shear plane intersections of Magneli molybdenum oxides; Sander and Engel (1994) for imaging the effect of annealing on rutile). It may be expected that the tip-induced desorption of hydroxyl groups will be soon observed as with silicon (Chen 1993).
- The atomic force microscopy is well adapted to image down to atomic-scale structure and morphology of insulating oxide surfaces. Microstructures of Cr_2O_3, NiO and TiO_2 surfaces can be given as examples (Lad and Antonik 1991; Antonik and Lad 1992; Antonik *et al.* 1992).
- Scanning reflection and transmission electron microscopies remain powerful tools as they associate high lateral resolution to the possibility of giving structural, topographical and chemical information. In favourable cases, nanometric metallic particles of 20–30 atoms on oxides can be chemically characterized by Auger electron spectroscopy (see, for examples, Cowley and Liu 1993; Liu and Cowley 1993; Wang *et al.* 1994).
- Collective information by electron and photoelectron spectroscopies on both chemical composition and chemical state can be quantified. Microanalysis and topographical data with a lateral resolution better than 50 nm and 10 μm are now possible by Auger electron spectroscopy and X-ray photoelectron spectroscopy, respectively.

Three recent examples of the application to oxides of combined techniques of surface analysis are given in the papers of Johansson (1993) for the overlayers of TiO_2, Bergström (1994) for the ceramic powders and Schrott and Frankel (1993) for the oxide films electrochemically formed.

8 Conclusion

As concluding remarks, we focus on the following points in the framework of the purpose of this book.

• Due to the diversity of their properties, metal oxides will play a more and more primordial role in modern technology either pure or with controlled doping, or associated in composite materials.

• As these properties directly depend upon morphological features, the oxide morphology must be controlled and adapted to the required function of the material. In this matter, studying the reactivity must be promoted both for controlled elaboration and in the use of solids.

• In many facets of this science of metal oxides, surface and interface properties must be better studied and understood as they have often a determining role in the overall behaviour and properties of materials.

• Amongst the areas of applications where the surface properties are fundamental, development can be expected, not only in the traditional fields of catalysis, ceramics science and microelectronics, but probably in energetics, tribology and environmental or sensing science.

• In all these fields, nanocrystalline phases and thin films could hold a particular place in achieving the expected functions. Nevertheless, for any form of these metal oxides, investigating the interfacial properties will be necessary in many problems of the life of the material in extreme conditions: preparation, sintering, shaping, chemical reaction in use, ageing, etc. For this purpose, in connection with experimental development, modelling and theoretical aspects will have to be promoted.

9 References

Antonik, M.D. & Lad, R.J. (1992) *J Vac Sci Technol A* **10**, 669–673.

Antonik M.D., Edwards, J.C. & Lad, R.J. (1992) *Mat Res Soc Symp Proc* **237**, 459–464.

Appleby, A.J. (1993) *Fuel Cell Systems* (eds L.J.M.J. Blomen & M.N. Mugerwa), pp. 157–199. Plenum Press, New York.

Arnaud, Y.P. & Bertrais, H. (1994) *Appl Surf Sci* **81**, 69–82.

Barteau, M.A. (1993) *J Vac Sci Technol A* **A11**, 2162–2168.

Bauer, E. (1958) *Z Kristallog* **110**, 372–394.

Bauer, E. (1991) *Ber Bunsenges Phys Chem* **95**, 1315–1325.

Bergström, L (1994) *Surface and Colloid Chemistry in Advanced Processing* (eds R.J. Pugh & L. Bergström), pp. 71–125. Marcel Decker, Inc., New York.

Bertrams, Th., Brodde, A., Hannemann, H., Ventrice, C.A., Wilhemi, G. & Neddermeyer, H. (1994) *Appl Surf Sci* **75**, 125–132.

Birkenheuer, U., Boettger, J.C. & Rösch, N. (1994) *J Chem Phys* **100**, 6826–6836.

Boldyrev, V.V., Bulens, M. & Delmon, B. (1979) *The Control of the Reactivity of Solids*. Elsevier, Amsterdam.

Bonnet, G., Lachkar, M., Larpin, J.P. & Colson, J.C. (1994) *Solid State Ionics* **72**, 344–348.

Busca, G., Lorenzelli, V., Ramis, G. & Willey, R.J. (1993) *Langmuir* **9**, 1492–1499.

Cabané, J. & Cabané, F. (1992) *Vide Couches Minces* **260**, 179–186.

Cantao, M.P., Lourenco, A., Gorenstein, A., Detorresi, S.I.C. & Torresi, R.M. (1994) *Mat Sci Eng B, Solid State Mat Adv Technol* **26**, 157–161.

Cappus, D., Xu, C., Ehrlich, D. *et al.* (1993) *Chem Phys* **177**, 533–546.

Carniti, P., Gervasini, A. & Auroux, A. (1994) *J Catal* **150**, 274–283.

Chadwick, A.V., Russell, N.V., Whitham, A.R. & Wilson, A. (1994), *Sensors Actuators B* **18**, 99–102.

Che, M. & Bennett, C.O. (1989) *Adv Catal* **36**, 55–172.
Chen C.J. (1993) *Scan Microsc* **7**, 793–804.
Colbourn, E.A. (1992) *Surf Sci Reports* **15**, 281–319.
Colbourn, E.A. & Mackrodt, W.C. (1983) *Solid State Ionics* **8**, 221–231.
Collins, G.E., Armstrong, N.R., Pankow, J.W. *et al.* (1993) *J Vac Sci Technol A* **11**, 1383–1391.
Cowley, J.M. & Liu J. (1993) *Surf Sci* **298**, 456–467.
Czerwinski & Smeltzer (1993), *Oxidat metals* **40**, 503–527.
Delannay, F., Broyen, L. & Deruytterre, A. (1987) *J Mat Sci* **22**, 1–16.
Delmon, B. (1986a) *Surf Interf Anal* **9**, 195–206.
Delmon, B. (1986b) *J Chim Phys* **83**, 875–883.
Domenichini, B., Gillot, B., Tailhades, P., Bouet, L., Rousset, A. & Perriat, P. (1992) *Solid State Ionics* **58**, 61–69.
Dowben, P.A. & Miller, A. (eds) (1990) *Surface Segregation Phenomena*. CRC Press, Boca Raton, USA.
Duffy, D.M., Harding, J.H. & Stoneham, A.M. (1994) *J Appl Phys* **76**, 2791–2798.
Dufour, L.C. & Morin, F. (1993) *Oxidat Metals* **39**, 137–154.
Dufour, L.C. & Nowotny, J. (eds) (1988) *External and Internal Surfaces in Metal Oxides, Materials Science Forum*, Vol. 29. Trans Tech Publications, Aedermannsdorf, Switzerland.
Dufour L.C. & Perdereau M. (1989) *Surfaces and Interfaces of Ceramic Materials*, NATO ASI E173 (eds L.C. Dufour, C. Monty & G. Petot-Ervas), pp. 419–448. Kluwer Academic Publishers, Dordrecht, The Netherlands.
Dufour, L.C., Floquet, N. & De Rosa, B. (1985) *Reactivity of Solids, Materials Science Monographs*, (eds P. Barret & L-C. Dufour). **28A** pp. 47–52. Elsevier, Amsterdam.
Dufour, L.C., Monty, C. & Petot-Ervas, G. (eds) (1989) *Surfaces and Interfaces of Ceramic Materials*, NATO ASI E173. Kluwer Academic Publishers, Dordrecht, The Netherlands.
Dufour, L.C., El Anssari, A., Dufour, P. & Vareille, M. (1992) *Surf Sci* **270**, 1173–1179.
Egdell, R.G. & Mackrodt, W.C. (1989) *Surfaces and Interfaces of Ceramic Materials*, NATO ASI E173 (eds L.C. Dufour, C. Monty & G. Petot-Ervas), pp. 185–203. Kluwer Academic Publishers, Dordrecht, The Netherlands.
Egdell, R.G. & Parker, S.C. (1991) *Science of Ceramic Interfaces, Material Science Monographs*, Vol. 75 (ed. J. Nowotny), pp. 41–79. Elsevier, Amsterdam.
Fang, T.T., Hsieh, H.I. & Shiau, F.S. (1993) *J Am Ceram Soc* **78**, 1205–1211.
Fischer, T.E. & Mullins, W.M. (1992) *J Phys Chem* **96**, 5690–5701.
Floquet, N. & Dufour, L.C. (1983) *Surf Sci* **126**, 543–549.
Freitag, J. & Staemmler, V. (1994) *J Electron Spectr Rel Phenom* **69**, 99–109.
Frey, M.H. & Payne, D.A. (1993) *Appl Phys Lett* **63**, 2753–2755.
Freund, H.J. & Umbach, E. (eds) (1993a) *Adsorption on Ordered Surfaces of Ionic Solids and Thin Films*, Springer Series in Surface Science, Vol. 33. Springer, Heidelberg, Germany.
Freund H.J., Dillmann, B., Ehrlich, D. *et al.* (1993b) *J Molec Catal* **82**, 143–169.
Gardner, J.W. (1991) *Sensors Actuators B* **4**, 109–115.
Gardner, J.W., Hines, E.L. & Tang, H.C. (1992) *Sensors Actuators B* **9**, 9–15.
Gardner, J.W., Bartlett, P.N., Dodd, G.H. & Shurmer, H.V (1990) *Chemisensing Information Processing*, Proceedings of NATO ASI H39 (ed. D. Schild), pp. 137–173. Springer-Verlag, Berlin.
Gillan, M.J. (1991) *Computer Simulation in Materials Science* (eds M. Meyer & V. Pontikis), p. 257. Kluwer Academic Publishers, Dordrecht, The Netherlands.
Gillot, B. (1994) *J Solid State Chem* **113**, 163–167.
Gillot, B. & Rousset, A. (1994) *Heterog Chem Rev* **1**, 68–98.
Göpel, W. (1990) *Solid State Ionics* **40/41**, 1009–1016.
Grabke, H.J., Kurbatov, G. & Schmutzler, H.J. (1995) *Oxidat Metals* **43**, 97–114.
Granqvist, C.G. (1994) *Solid State Ionics* **70/71**, 678–685.
Haruta, M. & Delmon, B. (1986) *J Chim Phys* **83**, 859–868.
Henrich, V.E. & Cox, P.A. (1993) *Appl Surf Sci* **72**, 277–284.

Henrich, V.E. & Cox, P.A. (1994) *The Surface Science of Metal Oxides.* Cambridge University Press, Cambridge, UK.
Hofmann, S. (1987) *J Chim Phys* **84**, 141–147.
Hofmann, S. (1990) *Thin Solid Films* **193**, 648–664.
Hofmann, S. (1992) *J Elect Spectrosc Rel Phenom* **59**, 15–32.
Ichinose, N. (1985) *Am Ceram Soc Bull* **64**, 1581–1585.
Johansson, L.S. (1993) *Acta Polytech Scand Appl Phys Ser* **191**, 1–32.
Jug., K., Geudtner, G. & Bredow, T. (1993) *J Mol Catal* **82**, 171–194.
Kartha, S. & Grimes (1994) *Phys Today* **47**, 54–61.
Kern, R. (1978) *Bull Mineral* **101**, 202–233.
Ketron, L. (1989) *Am Ceram Soc Bull* **68**, 860–865.
Kitsunai H., Hokkirigawa, K., Tsumaki, N. & Kato, K. (1991) *Wear* **151**, 279–289.
Köhler, K. & Schläpfer, C.W. (1993) *Chem Unserer Zeit* **5**, 248–255.
Kossel, W. (1927) *Nachr Ges Wiss Göttingen, Jahresber Mathphysik* Klasse, 135–143.
Kuhlenbeck, H., Odörfer, G., Jaeger, R. *et al.* (1991) *Phys Rev B* **43**, 1969–1986.
Kuhlenbeck, H., Xu, C., Dillmann, B. *et al.* (1992) *Ber Bunsenges Phys Chem* **96**, 15–27.
Kung, H.H. (1991) *Transition Metal Oxides: Surface Chemistry and Catalysis, Studies in Surface Science and Catalysis*, Vol. 45. Elsevier, Amsterdam.
Lad, R.J. & Antonik, M.D. (1991) *Ceram Trans* **24**, 359–366.
Lajtar, L., Narkiewicz-Michalek, J., Rudzinski, W. & Partika, S. (1993) *Langmuir* **9**, 3174–3190.
Langell, M.A., Berrie, C.L., Nassir, M.H. & Wulser, K.W. (1994) *Surf Sci* **320**, 25–38.
Lin X.Y. & Arribart, H. (1993) *Vide Couches Minces* **268**, 287–294.
Linford, R.G. (1978) *Chem Rev* **78** 81–95.
Liu, J. & Cowley, J.M. (1993) *Ultramicroscopy* **48**, 381–416.
Lu, W., Nevins, N., Norton, M.L., Rohrer, G.S. (1993) *Surf Sci* **291**, 395–401.
McCune & Wynblatt (1983) *J Am Ceram Soc* **66**, 111–118.
Mackrodt, W.C. (1989) *J Chem Soc Farad Trans II* **85**, 541.
Mackrodt, W.C. (1992) *Phil Trans Roy Soc Lond Series A* **341**, 301–312.
Mackrodt, W.C. & Tasker, P.W. (1989) *J Am Ceram Soc* **72**, 1576–1583.
Mackrodt, W.C., Harrison, N.M., Saunders, V.R. *et al.* (1993) *Phil Mag A* **68**, 653–666.
Majier, J., Holzinger, M. & Sitte, W. (1994) *Proc MRS Meeting.* Boston, December.
Marks, L.D. (1994) *Rep Prog Phys* **57**, 603–649.
Matijevic, E. (1981) *Acc Chem Res* **14**, 22–29.
Matijevic, E. (1988) *Pure Appl Chem* **60**, 1479.
Matijevic, E. (1991) *Chemtech* **21**, 176–181.
Mayer J.T., Lin, R.F. & Garfunkel, E. (1992) *Surf Sci* **265**, 102–110.
Minh N.Q. (1993) *J Am Ceram Soc* **76**, 563–588.
Morin, F., Dufour, L.C. & Trudel, G. (1992) *Oxidat Metals* **37**, 39–63.
Moseley, P.T., Norris, J.O.W. & Williams, D.E. (eds) (1991) *Techniques and Mechanisms in Gas Sensing*, pp. 108–138, 189–197; 281–323; 347–380. Hilger, Bristol, UK.
Moy, L. (1994) *Parfums Cosmét Arômes* **115**, 60–64.
Mukhopadhyay, S.M., Jardine, A.P., Blakely, J.M. & Baik, S. (1988) *J Am Ceram Soc* **71**, 358–362.
Murugesamoorthi, K.A., Srinivasan, S. & Appleby A.J. (1993) In: *Fuel Cell Systems* (eds L.J.M.J Blomen & M.N. Mugerwa), pp. 465–491. Plenum Press, New York.
Néel, L. (1961) *C R Acad Sci* **253**, 203–208, 1286–1291.
Néel, L. (1962) *C R Acad Sci* **254**, 598–602.
Neyman, K.M. & Rösch, N. (1992) *Chem Phys* **168**, 267–280.
Neyman, K.M. & Rösch, N. (1993) *Chem Phys* **177**, 561–570.
Noguera, C. (1995) *Physique et Chimie des Surfaces d'Oxydes.* Eyrolles, Paris.
Nowotny, J. (1989) *Surfaces and Interfaces of Ceramic Materials*, (eds L.C. Dufour, C. Monty & G. Petot-Ervas), pp. 205–239. Kluwer Academic Publishers, Dordrecht, The Netherlands.

Nowotny, J. (ed.) (1991) *Science of Ceramic Interfaces, Materials Science Monographs*, **75**. Elsevier, Amsterdam.
Nowotny, J. (ed.) (1994) *Science of Ceramic Interfaces II, Materials Science Monographs*, **81**. Elsevier, Amsterdam.
Nowotny, J. & Dufour, L.C. (eds) (1988) *Surface and Near-Surface Chemistry of Oxide Materials*. Elsevier, Amsterdam.
Nygren, M.A. & Pettersson, L.G.M. (1994) *J Electron Spectrosc Rel Phenom* **69**, 43–53.
Ohlendorf, G., Koch, W., Kemper, V. & Borchardt, G. (1991) *Surf Interf Anal* **17**, 947–950.
Oliver, P.M., Parker S.C. & Mackrodt, W.C. (1993) *Modell Simul Mat Sci Engin* **1**, 755–760.
Pacchioni, G. (1993) *Surf Sci* **281**, 207–219.
Pacchioni, G. & Minerva, T. (1992) *Surf Sci* **275**, 450–458.
Pacchioni, G., Clotet, A. & Ricart, J.M. (1994c) *Surf Sci* **315**, 337–350.
Pacchioni, G., Neyman, K.M. & Rösch, N. (1994a) *J Electr Spectrosc Rel Phen* **69**, 13–21.
Pacchioni, G., Ricart, J.M. & Illas, F. (1994b) *J Am Chem Soc* **116**, 10 152–10 158.
Passerini, S. & Scrosati, B. (1992) *Solid State Ionics* **53–56**, 520–524.
Pask, J. & Evans, A. (eds) (1981) *Surfaces and Interfaces in Ceramic and Ceramic-Metal Systems*. Plenum Press, New York.
Perriat, P., Niepce, J.C. & Caboche, G. (1994) *J Thermal Anal* **41**, 635–649.
Pettersson, L.G.M. & Pacchioni, G. (1994) *Chem Phys Lett* **219**, 1047–1112.
Plzak, V., Rohland, B. & Wendt, H. (1994) *Mod Asp Electrochem* **26**, 105–163.
Pöhlchen, M. & Staemmler, V. (1992) *J Chem Phys* **97**, 2583–2592.
Poppa, H. (1993) *Catal Rev Sci Eng* **35**, 359–398.
Resch, G. & Gutman, V. (1980) *Z Physik Chem* **121**, 211–235.
Resch, G. & Gutman, V. (1981) *Z Physik Chem* **126**, 223–241.
Rochefort, A. & Le Peltier, F. (1991), *Rev Inst Fr Pétrole* **46**, 221–249.
Rohrer G.S., Lu, W., Smith, R.L. & Hutchinson, A. (1993) *Surf Sci* **292**, 261–266.
Rösch, N., Neyman, K.M. & Birkenheuer, U. (1993) (eds H.J. Freund & E. Umbach). *Adsorption on Ordered Surfaces of Ionic Solids and Thin Films, Springer Series in Surface Science*, Vol. 33, pp. 206–217. Springer, Heidelberg, Germany.
Roy, S.K., Seal, S., Bose, S.K. & Caillet, M. (1993) *J Mat Sci Let* **12**, 249–251.
Sander, M. & Engel, T. (1994) *Surf Sci Lett* **302**, L263–L268.
Sayle, D.C., Sayle, T.X.T., Parker, S.C., Catlow, C.R.A. & Harding J.H. (1994) *Phys Rev B* **50**, 14 498–14 505.
Sayle, T.X.T., Parker, S.C. & Catlow, C.R.A. (1994) *J Phys Chem* **98**, 13 625–13 630.
Schlag, S. & Eike, H.F. (1994) *Solid State Commun* **91**, 883–887.
Schrott, A.G. & Frankel, G.S. (1993) *IBM J Res Dev* **37**, 191–206.
Seike, T. & Nagai, J. (1991) *Solar Energy Mat* **22**, 107–117.
Shluger, A.L., Gale, J.D. & Catlow, C.R.A. (1992) *J Phys Chem* **96**, 10 389–10 397.
Shuk, P., Vecher, A., Kharton, V *et al.* (1993) *Sensors Actuators B* **15–16**, 401–405.
Siebert, E. (1994) *Electrochem Acta* **39**, 1621–1625.
Stefanovic, E.V. & Truong, T.N. (1995) *J Chem Phys* **102**, 5071–5076.
Stoneham, A.M. (1981) *J Am Ceram Soc* **64**, 54–60.
Stranski, I.N. (1928) *Z Physik Chem* **136**, 259–278.
Takeuchi, T., Ado, K., Asai T. *et al.* (1993) *J Am Ceram Soc* **77**, 1665–1668.
Tarrach, G., Bürgler, D., Schaub, T., Wiesendanger, R. & Güntherodt, H.J. (1993) *Surf Sci* **285**, 1–14.
Tasker, P.W. (1979) *J Phys C* **12**, 4977–4984.
Tasker, P.W. & Duffy, D.M. (1984) *Surf Sci* **137**, 91–102.
Titiloye, J.O., Parker, S.C., Osguthorpe, D.J. & Mann, S. (1991), *Chem Commun* **20**, 1494–1496.
Trasatti, S. (1990) *Croata Chem Acta* **63**, 313–329.
Trasatti, S. (1994) *Electrochemistry of Novel Materials, Series: Frontiers in Electrochemistry* (eds J. Lipkowski & P.N. Ross), pp. 207–295. VCH Publishers, New York.

Ventrice, C.A., Bertrams, Th., Hannmann, H., Brodde, A. & Neddermeyer, H. (1994) *Phys Rev B Cond Matter* **49**, 5773–5776.
Vijande-Diaz, R., Belzunce, J., Fernandez, E., Rincon, A. & Pérez, M.C. (1991) *Wear* **148**, 221–233.
Vurens, G.H., Salmeron, M. & Somorjai, G.A. (1989) *Prog Surf Sci* **32**, 333–360.
Wang, L., Liu, J. & Cowley, J.M. (1994) *Surf Sci* **302**, 141–157.
Winkelmann, F., Wohlrab, S., Libuda, J. *et al.* (1994) *Surf Sci* **307–309**, 1148–1160.
Wynblatt, P. & McCune, R.C. (1988) *Surface and Near-Surface Chemistry of Oxide Materials* (eds J. Nowotny & L.C. Dufour), pp. 247–279. Elsevier, Amsterdam.
Zhong, W.I., Wang, Y.G., Zhang, P.L. & Qu, B.D. (1994) *Phys Rev B Cond Matter* **50**, 698–703.

5 Processes at Interfaces During Solid State Reactions

M. MARTIN

Institut für Physikalische Chemie, Elektrochemie, Technische Hochschule Darmstadt, Petersenstrasse 20, 64287 Darmstadt, Germany

1 Introduction

A solid state reaction is defined as a reaction where at least one reactant is solid. This definition includes the classical solid state reactions between two solids, for example the spinel formation $MgO + Al_2O_3 \rightarrow MgAl_2O_4$ (Wagner 1938), the formation of intermetallic compounds (George *et al.* 1994), for example, Ni_3Al, the formation of silicides (Gas 1994), for example, $NiSi_2$, and many others (Schmalzried 1981). However, it also includes reactions between a solid and a fluid. If the fluid is a gas, technically important solid state reactions are the tarnishing reactions (Kofstad 1988), for example, the oxidation or sulphidation of metals and alloys. Solid state reactions between a solid and a liquid are known from crystal growth processes (Wilke 1988) or the dissolution of solids in liquids or melts (Valverde 1976). In all cases, the reactions are heterogeneous, i.e. the different reactants are separated by phase boundaries or interfaces. In order to show that interfaces play a crucial role during solid state reactions we will discuss mainly two questions: (i) in Section 2 the role of interface equilibrium or the interface resistance, compared to the bulk diffusion resistance, will be analysed; and (ii) in Section 3 the morphology of moving interfaces will be discussed. To focus on the essential points, we will limit the following analysis to solid state reactions where the solid is crystalline, i.e. we will not include amorphous and glassy solids, and we will discuss only reactions where transport in the solid is dominated by bulk (lattice) diffusion, i.e. we will not consider grain boundary diffusion and the resulting complications and changes (Kaur and Gust 1989). Also, we will not include the phenomena of multiphase formation and growth (Wagner 1969; Dybkov 1994). There are, of course, many books and review articles discussing the role of interfaces during solid state reactions in general and in detail for specified problems, only a few of which we can give as reference (Adda and Philibert 1966; Schmalzried 1981, 1993, 1995; Dybkov 1992; d'Heurle 1994).

2 Phase growth kinetics

In this section the growth kinetics of a solid product phase during a solid state reaction will be analysed. We will assume that the product forms as a layer, through which one-dimensional diffusion takes place, i.e. we will assume planar interfaces during the overall reaction (interface instabilities will be discussed in Section 3). In Sections 2.1 and 2.2 the results of theoretical considerations leading to linear and parabolic rate laws will be given, and in Section 2.3 we will discuss the conditions for a constant product layer thickness. In Section 2.4 we will describe a few important experimental results, and a general discussion and a critical outlook will be given.

2.1 *Phenomenological growth laws*

Consider a classical solid state reaction A + B→AB, where at least the reactant A and the product AB are solid and where the solubility of B in A is negligible. Let us assume that the product AB grows at the phase boundary AB–B, which means by diffusion of A through AB (Fig. 5.1). To obtain the law for the growth kinetics of AB, i.e. the thickness Δz of AB as a function of time t, we have to specify a flux law for the diffusion of A through AB and boundary conditions at the two phase boundaries A–AB and AB–B (indices I and II). The latter is complicated by the fact that we are dealing with a reacting system, in which we cannot assume local equilibrium at the phase boundaries compared to a system in equilibrium. To continue, we have to distinguish between interfaces at rest (non-moving) and moving interfaces. In our example the interface A–AB is at rest relative to the crystal lattice of AB, i.e. material is only transferred across the interface, whilst at the AB–B interface material is transferred across the interface and the interface is moving.

2.1.1 NON-MOVING INTERFACE

To drive matter across the non-moving interface A–AB, in this case the component A, there has to be a driving force for the following quasi-chemical reaction:

$$\text{A(in A)} \rightleftarrows \text{A(in AB)} \tag{5.1}$$

which is given by the affinity $A_R = \mu^{I}_{A(A)} - \mu^{I}_{A(AB)} = \Delta\mu^{I}_{A}$, where $\mu^{I}_{A(A)}$ and $\mu^{I}_{A(AB)}$ are the chemical potentials of A in A and in AB at the phase boundary I (= A–AB) (see Fig. 5.1). The reaction rate, v_R, for the reaction in Equation (5.1) is given by (Haase 1990):

$$v_R = v_R^{eq}\left(1 - e^{-\frac{A_R}{RT}}\right) \tag{5.2}$$

Here v_R^{eq} is the reaction rate, for example for the forward reaction in Equation (5.1), in dynamical equilibrium ($A_R = 0$), i.e when per time interval equal amounts of A are crossing the interface in both directions. The flux of A across the interface is directly

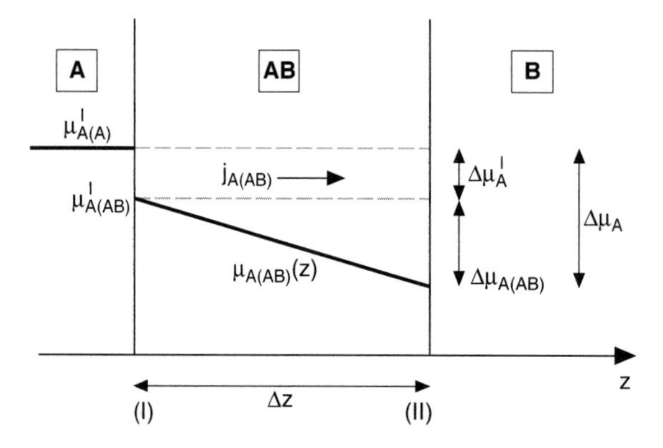

Figure 5.1 Schematic reaction couple A/AB/B and course of the chemical potential of the component A. No equilibrium at the interface A–AB (I), but equilibrium at the interface AB–B (II).

proportional to v_R, resulting in:

$$j_A^I = j_0\left(1 - e^{-\frac{\Delta\mu_A^I}{RT}}\right) \tag{5.3}$$

This relation for the flux of A across the A–AB interface is similar to expressions for the electrical current across an electrode–electrolyte interface in an electrochemical cell (Koryta *et al.* 1993). Thus, $\Delta\mu_A^I$ can be regarded as an overpotential and j_0 as a dynamical exchange current under equilibrium conditions. This means that, in general, we obtain a non-linear relationship between the driving force and the resulting flux across the interface.

According to linear irreversible thermodynamics (de Groot and Mazur 1984), the flux of A across the crystal AB towards the boundary AB–B can be written as a linear function of the driving force:

$$j_{A(AB)} = -L_{AA}\left(\frac{\partial\mu_{A(AB)}}{\partial z}\right) = \bar{L}_{AA}\left(\frac{\Delta\mu_A^I}{\Delta z}\right) \tag{5.4}$$

In the second part of Equation (5.4) the transport coefficient of A in AB, L_{AA}, has been substituted by its average value, $\bar{L}_{AA}$, and the gradient $\partial\mu_{A(AB)}/\partial z$ by $-\Delta\mu_A^I/\Delta z$. Conservation of matter at interface I requires that the fluxes in Equations (5.3) and (5.4) are identical, resulting in a non-linear relationship between the potential drops across the boundary A–AB and the crystal AB:

$$1 - \left(e^{-\frac{\Delta\mu_A^I}{RT}}\right) = \alpha\left(\frac{\Delta\mu_{A(AB)}}{RT}\right)$$

$$\alpha = \frac{RT\bar{L}_{AA}}{j_0}\left(\frac{1}{\Delta z}\right) \tag{5.5}$$

To obtain local equilibrium at the interface, the potential drop across the interface, $\Delta\mu_A^I$, has to be small compared to the potential drop across AB, $\Delta\mu_{A(AB)}$, i.e. $\alpha \ll 1$, which can be obtained by: (i) slow transport in AB (small $\bar{L}_{AA}$); (ii) a large product layer thickness Δz; or (iii) a high exchange current j_0 across the interface. In all other cases the potential drop across the interface cannot be neglected, and the interface possesses an appreciable transport resistance ($\propto 1/j_0$).

The advancement of the AB–B interface, or the growth of the product layer thickness Δz, can be calculated from the mass balance at the interface:

$$v^{int} = \frac{d\Delta z}{dt} = V_{AB}^m j_{A(AB)} = V_{AB}^m \bar{L}_{AA}\left(\frac{\Delta\mu_A^I}{\Delta z}\right) \tag{5.6}$$

and is determined by the potential drop in AB, $\Delta\mu_{A(AB)}$ (V_{AB}^m is the molar volume of AB). For reasons of simplicity we assume now that at the second (moving) interface, AB–B, local equilibrium is established (see Section 2.1.2). Since the total potential drop, $\Delta\mu_A = \Delta\mu_A^I + \Delta\mu_{A(AB)}$ (see Fig. 5.1) is a constant, which is fixed thermodynamically by pure A at one side and the equilibrium AB–B at the other side of the product layer, $\Delta\mu_{A(AB)}$ in Equation (5.5) could now be eliminated, and the interface velocity in Equation (5.6) could be written as a function of Δz only. This is, however, complicated by the fact that Equation (5.5) is a non-linear relationship. For reasons of simplicity we will, therefore, linearize the exponential in Equation (5.5), resulting in $\Delta\mu_A = (1+\alpha)\Delta\mu_{A(AB)}$ and the following rate law for the layer growth:

$$\frac{d\Delta z}{dt} = \frac{\beta}{\gamma + \Delta z}$$

$$\beta = V^m_{AB}\,\bar{L}_{AA}\Delta\mu_A$$

$$\gamma = \frac{RT\bar{L}_{AA}}{j_0} \tag{5.7}$$

Integration yields the integral rate law:

$$\frac{1}{2}(\Delta z)^2 + \gamma(\Delta z) = \beta t \tag{5.8}$$

At small times, $t \ll \gamma^2/\beta$, when the thickness Δz is small compared to γ, the first term on the right-hand side of Equation (5.8) can be neglected, and we obtain a linear rate law:

$$\Delta z = \left(\frac{\beta}{\gamma}\right)t = V^m_{AB}\Delta\mu_A j_0 t \tag{5.9}$$

The linear rate constant β/γ is determined by the molar volume of the product, V^m_{AB}, the driving force for the reaction, $\Delta\mu_A$, and the exchange current, j_0, across the interface, i.e. the interface resistance ($\propto 1/j_0$) dominates, compared to the bulk transport resistance ($\propto 1/\bar{L}_{AA}$). At larger times, $t \gg \gamma^2/\beta$, the second term on the right-hand side of Equation (5.8) can be neglected, and we obtain a parabolic rate law:

$$\Delta z = \sqrt{2\beta t} \tag{5.10}$$

where the parabolic rate constant β is determined only by the bulk transport coefficient $\bar{L}_{AA}$, i.e. now the bulk transport resistance dominates.

2.1.2 MOVING INTERFACE

In this section we analyse the influence of a phase boundary resistance at the moving interface AB–B (Fig. 5.2). Here the flux of A is no longer continuous (in contrast to the

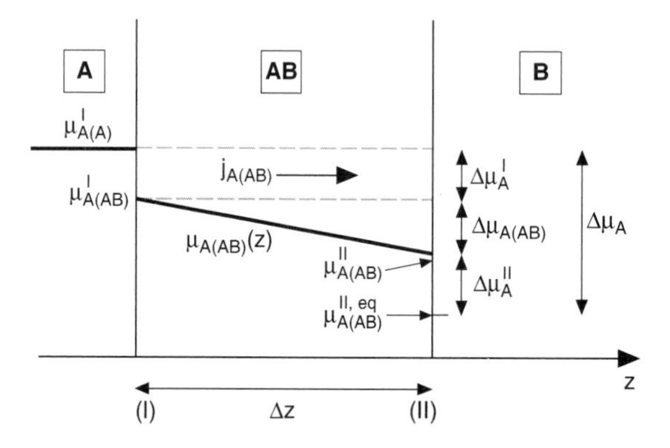

Figure 5.2 Schematic reaction couple A/AB/B and course of the chemical potential of the component A. No equilibrium at both interfaces A–AB (I) and AB–B (II).

non-moving interface A–AB), since matter is used up now to shift the interface with the velocity, v^{int}, which we obtain from the local mass balance in Equation (5.6). The interface velocity, however, is also determined by the deviation of the chemical potential of A at the interface from its equilibrium value, $\Delta\mu_A^{II} = \mu_{A(AB)}^{II} - \mu_{A(AB)}^{II,eq}$. In the linear regime (see Section 2.1.1) we may write:

$$v^{\mathrm{int}} = \kappa\Delta\mu_A^{II} \tag{5.11}$$

where the interface mobility κ is determined by the exchange current j_0 of A across the non-moving interface II in equilibrium (Schmalzried 1978). Equations (5.6) and (5.11) again yield a linear relationship between the two potential drops in the bulk and at the interface, and with $\Delta\mu_A = \Delta\mu_A^I + \Delta\mu_{A(AB)} + \Delta\mu_A^{II}$ (Fig. 5.2) we obtain a growth law as in Equation (5.7) where γ is now given by $\gamma = RT\bar{L}_{AA}/j_0 + V_{AB}^m(\bar{L}_{AA}/\kappa)$. As before, the growth law is linear in time if one or both interface resistances ($\propto 1/j_0$ and $\propto 1/\kappa$) are larger than the bulk transport resistance ($\propto 1/\bar{L}_{AA}$), whilst it is parabolic in time when the bulk transport resistance dominates.

2.2 *Defect relaxation*

In the preceding sections we have distinguished between interface control, leading to linear growth, and diffusion control resulting in parabolic growth. In both cases we have used phenomenological quantities to describe transport in the crystal or across the interface. There is, however, an additional case that is dependent on the behaviour of the crystal defects. To illustrate this point we take our example A + B→AB and assume A to be a metal and B a non-metal, for example oxygen. Equation (5.1) for the transfer of metal across the metal–oxide interface can now be written in terms of the defects in both crystals (Martin 1991):

$$A_{A(A)}^x + V_{A(AB)}'' + 2h^\bullet \rightarrow A_{A(AB)}^x + V_{A(A)}^x \tag{5.12}$$

Here $A_{A(A)}^x$ and $A_{A(AB)}^x$ are metal ions on A-sites in the metal A and in the compound AB, respectively (Kröger–Vink notation), $V_{A(A)}^x$ and $V_{A(AB)}''$ are cation vacancies in A and AB, respectively, and $h^\bullet$ is an electron hole in the semiconducting compound AB. This type of transfer reaction for the component A across the A–AB interface describes the jump of a cation A in the metal lattice into a cation vacancy in the compound metal sublattice (Fig. 5.3), resulting in a cation vacancy in the metal and a cation in the compound. For charge neutrality electron holes also have to be consumed. Thus, cation vacancies are injected into the metal, thereby producing a vacancy supersaturation and a corresponding deviation of the chemical potential of the component A in the metal from its equilibrium value, $\mu_{A(A)}^{eq}$, which for a pure metal is only a function of temperature T and pressure p. This deviation can be calculated as follows. In the metal the vacancy supersaturation can be decreased by lattice diffusion of vacancies towards the interior of the metal and by annihilation of vacancies at appropriate sinks, for example dislocations (for reasons of simplicity in this section the symbol V will be used for the metal vacancies $V_{A(A)}^x$):

$$\frac{\partial[V]}{\partial t} = D\left(\frac{\partial^2[V]}{\partial z^2}\right) - k_V([V] - [V]^{eq})[\mathrm{sink}] \tag{5.13}$$

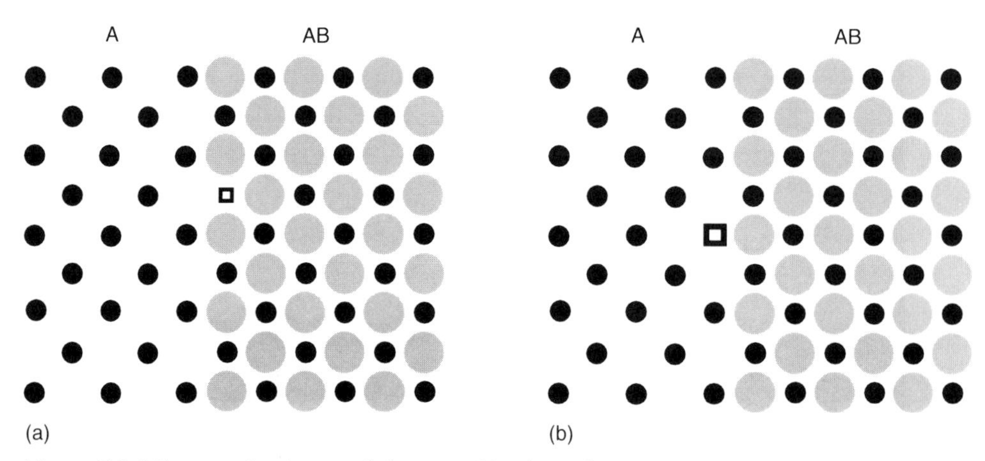

Figure 5.3 Microscopic picture of the crystal lattices of A and AB at the coherent interface A–AB (I). (a) Before jump of $A_{A(A)}$ into vacancy $V_{A(AB)}$. (b) After jump.

The first term describes lattice diffusion of vacancies (diffusion coefficient D) and the second term their annihilation at vacancy sinks, according to $V + \text{sink} = 0$ ($[V]^{eq}$ is the equilibrium vacancy fraction in the metal and k_v is the reaction rate constant). If we assume a homogeneous sink density in the metal A, the quasi-steady state solution of Equation (5.13) is:

$$[V]_z = [V]^{eq} + \Delta V e^{\lambda z}$$

$$\lambda = \sqrt{\frac{k_v\,[\text{sink}]}{D}} \tag{5.14}$$

where ΔV is the vacancy supersaturation in the metal at interface I, $\Delta V = [V]_{z=0} - [V]_{eq}$. The corresponding deviation of the chemical potential of the component A in the metal at interface I from its equilibrium value, $\Delta\mu^{I}_{A(A)} = \mu^{I,eq}_{A(A)} - \mu^{I}_{A(A)}$, can be calculated by using the relation $A = A^{x}_{A(A)} - V^{x}_{A(A)}$ (see Equation (5.12)) and by assuming ideal behaviour for the vacancies, $\mu_v = \mu^0_v + \text{RT}\ln[V]$, resulting in $\Delta\mu^{I}_{A(A)} = \text{RT}(\Delta V/[V]^{eq})$ (Fig. 5.4). From the continuity of the A-flux at the non-moving interface I, $j_{A(A)}(z=0) = -j_{V(A)}(z=0) = j^{I}_{A}$, we obtain the following relation between the vacancy supersaturation, ΔV, and the potential drop across the A–AB interface, $\Delta\mu^{I}_{A}$:

$$\frac{D\Delta V\lambda}{V^{m}_{A}} = j_0\left(\frac{\Delta\mu^{I}_{A}}{\text{RT}}\right) \tag{5.15}$$

Now the total potential drop for A across the system, $\Delta\mu_A$, consists of four parts:

$$\Delta\mu_A = \Delta\mu^{I}_{A(A)} + \Delta\mu^{I}_{A} + \Delta\mu_{A(AB)} + \Delta\mu^{II}_{A} \tag{5.16}$$

If we assume, for reasons of simplicity, that at the moving interface AB–B, equilibrium is established, the potential drop $\Delta\mu_{A(AB)}$, which determines, via Equation (5.6), the growth velocity, v^{int}, can be written as:

$$\Delta\mu_{A(AB)} = \frac{\Delta\mu_A}{1 + \alpha\left(1 + \dfrac{j_0 V^{m}_{A}}{[V]^{eq}\sqrt{k_V[\text{sink}]D}}\right)} \tag{5.17}$$

Here, two limiting cases can be distinguished.

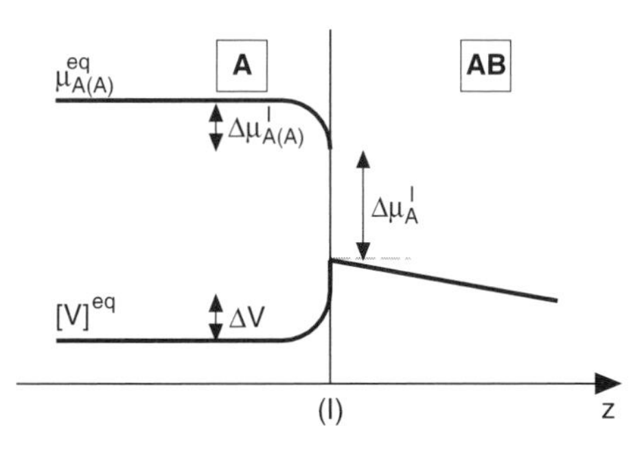

Figure 5.4 Deviation ΔV of the vacancy fraction $[V]$ in the metal from the equilibrium vacancy fraction $[V]^{eq}$ and course of the chemical potential of the component A near the A–AB interface (I).

1 If the transport of the (supersaturated) vacancies away from the boundary and/or the annihilation of the vacancies at sinks is fast compared to the injection rate j_0, $\rho = j_0 V^m_A/[V]^{eq}\,(k_v[\text{sink}]D)^{1/2} \ll 1$, equilibrium is established inside the metal. Then, we obtain from Equation (5.17), $\Delta\mu_{A(AB)} = \Delta\mu_A/(1+\alpha)$, as in Section 2.1.1, and the growth kinetics is linear for small times (dominance of the interface resistance) and parabolic for larger times (dominance of the bulk transport resistance).

2 If the transport of the supersaturated vacancies away from the interface is slow and/or the sink concentration for vacancy annihilation is too low, $\rho \gg 1$, we obtain $\Delta\mu_{A(AB)} = \Delta\mu_A/(1+\alpha\rho)$, which is independent of the exchange current j_0. This means that we have, as before, a linear law for small times, but which is caused now by the finite relaxation distance $1/\lambda$ of the vacancies in the metal and not by the interface resistance. For larger times a parabolic law is obtained again, caused by the transport resistance in the product layer.

This result, obtained here for the special case of supersaturated vacancies, can be generalized. If a chemical component is transferred across an interface between two crystalline solids, regular and irregular structural elements are produced and annihilated in both lattices (as shown, for example, in Equation (5.12)) resulting in non-equilibrium defect concentrations or supersaturation of defects. In addition, structure elements could be injected into wrong sublattices, i.e. supersaturation may build up in different sublattices, for example in the regular and the interstitial sublattices. Defect relaxation may then take place by exchange reactions between different sublattices, for example Frenkel reactions, and by annihilation or production of defects at defect sinks and sources (Schmalzried 1993). In addition, there is the possibility of defect relaxation at the interface itself, by which pores may be formed (see Section 2.4). Finally, it has to be mentioned that up to now we have always assumed the interface under consideration to be coherent, i.e. that there is no lattice misfit between the two crystalline solids at the interface. Implications of deviations from this assumption (semi-coherent or non-coherent interfaces) will be discussed in Section 2.4.

2.3 *Constant product layer thickness*

In addition to the growth kinetics we have already discussed (linear and parabolic in time), there is also the possibility of a constant product layer thickness, with chemical reactions at both sides of the product layer. To exemplify this case we consider an already grown layer of a semiconducting compound AB which is now exposed to different chemical potentials of B (or A) on both sides of the layer. These chemical potentials might be established by different vapour pressures of the component B, or in the case of gaseous B (for example, B = oxygen) by different partial pressures of B in reactive gas mixtures (for example, CO/CO_2). To be specific, let us consider a non-stoichiometric binary transition metal oxide, $A_{1-\delta}O$ (A = nickel, cobalt, iron, manganese), where oxygen is practically immobile whilst cations A are mobile via cation vacancies $V''_{A(AB)}$ (with cation fraction $\delta \ll 1$). If this oxide is chemically reduced by lowering the chemical potential of oxygen in the surrounding atmosphere, cation vacancies $V''_{A(AB)}$ and electron holes $h^{\bullet}$ diffuse to the crystal surface where reduction takes place:

$$AO + V''_{A(AB)} + 2h^{\bullet} \rightarrow A^{x}_{A(AB)} + \frac{1}{2} O_2(g) \tag{5.18}$$

The quasi-chemical reaction in Equation (5.18) shows that the reduction process corresponds to the arrival of a vacancy and two electron holes at the surface and the dissociation of oxygen from the crystal, i.e. the removal of a structural unit, composed of a cation vacancy and an anion, from the crystal whilst the number of cations is conserved. The crystal surface acts as vacancy sink until the new equilibrium state is reached. Thus, we have complicated growth kinetics for the shrinking oxide crystal. In contrast to this non-stationary situation, a stationary state with a constant oxide layer thickness can be established by exposing two parallel crystal surfaces of the oxide to a gradient of the chemical potential of oxygen, which results in reduction at the low-oxygen potential side and oxidation (the reversal of the reaction in Equation (5.18)) at the high-oxygen potential side (Fig. 5.5). After a short time, a stationary flux of vacancies and a corresponding flux of A-ions in the opposite direction occur, which are fed by the chemical reaction in Equation (5.18), and the reverse of it at the opposite interface. As a result of this 'vacancy wind', both crystal surfaces move (relative to the immobile oxygen sublattice) towards the side of higher oxygen activity. This means that now the product layer thickness Δz remains constant during the reaction. The corresponding diffusion problem can be solved easily, provided the following assumptions are made:

1 the crystal surfaces are assumed to be planar;

2 the chemical diffusion coefficient D describing the diffusion processes in the binary oxide is (approximately) constant;

3 local equilibrium is established at the boundaries, i.e. phase boundary reaction kinetics are very fast compared to bulk diffusion (see Section 2.1).

Then, we can calculate a stationary solution by transforming the diffusion equation for the vacancies (or the cations A) and the mass balances at the oxide–gas boundaries to a moving reference frame, $0 \leq z \leq \Delta z^{stat}$ (Δz^{stat} = stationary sample thickness), in which both interfaces are at rest (Martin 1991). We obtain the following stationary solution for the vacancy fraction, $[V]$ (= $V''_{A(AB)}$]), and the velocity of the surface motion, v^{stat}:

$$[V]^{stat}_{z} = 1 - (1 - [V]_{I})(e^{-v^{stat}(z/D)})$$

$$v^{stat} = \frac{D}{\Delta z^{stat}} \ln\left(\frac{1 - [V]_{I}}{1 - [V]_{II}}\right) \tag{5.19}$$

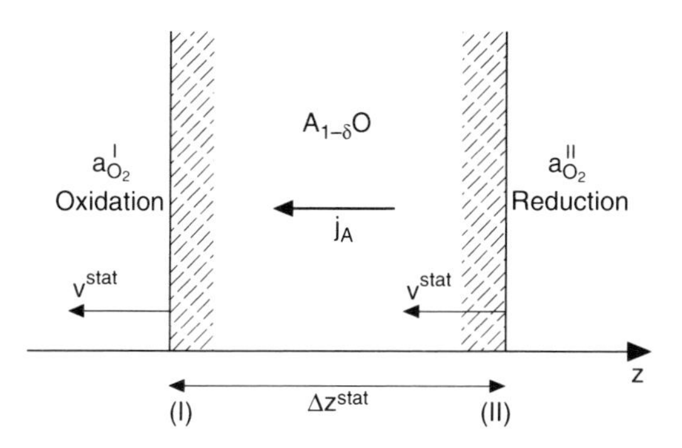

Figure 5.5 Schematic representation of an oxide in an oxygen potential gradient, with oxidation at the high-oxygen potential side (I) and reduction at the low-oxygen potential side (II).

Here, $[V]_{I}$ and $[V]_{II}$ are the vacancy fractions at boundaries I ($z = 0$) and II ($z = \Delta z^{stat}$), fixed by the oxygen partial pressures in the ambient gas phases. A typical value of the drift velocity of the surfaces is $v^{stat} = 3$ μm/hour (for experimental conditions see Section 3.1.2).

The situation becomes more complex if we also consider deviations from local equilibrium at the phase boundaries. In the case of oxides these deviations may be caused by the use of reactive gas mixtures which are used to establish the oxygen activities on both sides of the crystal. The exchange of oxygen between a transition metal oxide $A_{1-\delta}O$ and CO/CO_2 mixtures is often governed by a slow phase boundary reaction as shown experimentally by ^{18}O-exchange experiments (Grabke 1965). Compared to this the exchange of oxygen between oxides (for example, magnetite (Fe_3O_4)) and nitrogen/oxygen mixtures is much more rapid. Independent of the exact reaction mechanism, the rate for the oxygen exchange between an oxide and CO/CO_2 mixtures can be written as (Dieckmann *et al.* 1981):

$$\frac{dn_O}{dt} = FP_{CO_2}k\left(1 - \left(\frac{a^*_{O_2}}{a_{O_2}}\right)^{1/2}\right) \tag{5.20}$$

where n_O is the number of moles of oxygen, F is the surface area, P_{CO_2} is the partial pressure of CO_2 in the gas mixture, k is the reaction rate constant, a_{O_2} is the oxygen activity in the gas phase and $a^*_{O_2}$ is the oxygen activity on the oxide surface. As found experimentally, the rate constant, k, is again a function of the oxygen activity $a^*_{O_2}$ and can be described by a temperature-dependent function, $k_O(T)$, and by an exponent, m:

$$k = k_O(T)(a^*_{O_2})^m \tag{5.21}$$

In the stationary state the exchange rates of oxygen on both sides of the crystal must be identical:

$$\left(\frac{dn_O}{dt}\right)_I = \left(\frac{dn_O}{dt}\right)_{II} \tag{5.22}$$

which connects the two unknown surface oxygen activities $(a^*_{O_2})_I$ and $(a^*_{O_2})_{II}$. Using these relationships, the transport problem can be solved easily, showing that the type of growth kinetics is not changed by the slow surface reaction. As before, the crystal thickness remains constant and the crystal surfaces shift with a constant velocity to the side of higher oxygen potential. However, due to the slow surface reactions the crystal shift velocity is smaller than under equilibrium conditions at the boundaries, since the effective driving force for the crystal drift has decreased (Fig. 5.6).

2.4 *Some experimental results and discussion*

In this section a few important experimental results will be discussed: first, to illustrate the applicability of the theoretical results in the previous sections, and second, to point to a number of open theoretical and experimental questions and problems.

The earlier mentioned formation of spinels AB_2O_4 is a well-studied problem (Wagner 1936; Schmalzried 1981; Pfeiffer and Schmalzried 1989), but there are only few experimental studies showing a transition from linear to parabolic behaviour. For the formation of $ZnAl_2O_4$ spinel according to the reaction $Al_2O_3(s) + Zn(g) + \frac{1}{2}O_2(g) = ZnAl_2O_4$, Duckwitz and Schmalzried (1971) found a linear rate law at small times. This linear law is valid up to thicknesses Δz of the order of 10^{-2} µm from which, via Equation (5.9), the exchange current density j_0 can be estimated, $j_0 = 10^{-6}$ mol/cm^{2}s^{-1}. Becker and co-workers (Becker 1994) studied the formation of $CoAl_2O_4$ spinel from the solid components CoO and Al_2O_3 using *in situ* optical absorption spectroscopy at temperatures above 1000°C. For spinel thicknesses up to 0.5 µm they found linear rate laws. A possible cause for this linear behaviour is the redistribution of cations to the correct sublattice after they have crossed the interface between Al_2O_3 and spinel (see Section 2.2). However, the corresponding relaxation times, which were measured by *in situ* optical spectroscopy (Becker and Rau 1988), are far too small to explain the observed linear rate constants. Varying the orientation of the Al_2O_3 single crystals, Becker and coworkers found a strong anisotropy, i.e. the growth velocity was largest when the interface was parallel to the Al_2O_3 *c*-axis and decreased by a factor of three when the interface was perpendicular to the *c*-axis. This observation points to another cause for the interface resistance during this kind of solid state reaction; this is the necessary transformation of the oxygen sublattice from hcp in Al_2O_3 to fcc in the spinel. The corresponding dislocation network which builds up to decrease the lattice misfit has to move together with the reaction front and may be the microscopic cause for the interface resistance. By transmission electron microscopy (TEM) Carter and Schmalzried (1985) could show for $CoAl_2O_4$ that, as a result of this dislocation network, the crystal lattices of Al_2O_3 and the spinel are tilted by a small angle in order

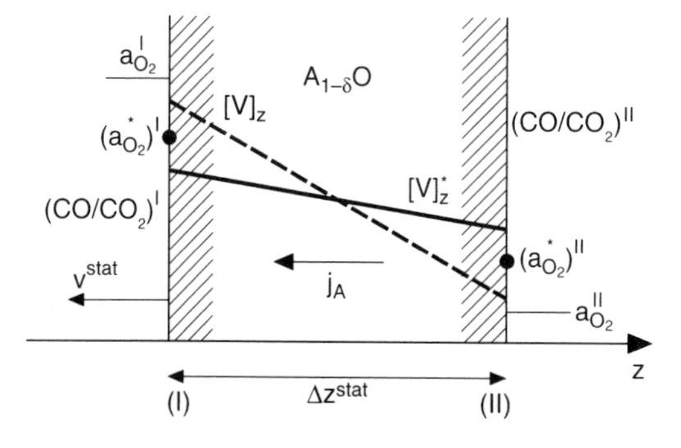

Figure 5.6 Schematic representation of the oxygen activities $a^{*}_{O_2}$ at the oxide surfaces (I) and (II) and changed defect profiles for a slow phase boundary reaction.

to gain a nearly coherent character of the interface. Following these observations, Hesse and co-workers (Hesse *et al.* 1994; Sieber *et al.* 1994) performed detailed studies on a number of spinels ($MgAl_2O_4$, $MgIn_2O_4$, $MgCr_2O_4$, Mg_2TiO_4) with different misfits relative to the MgO substrate, showing that the microscopic interface morphology (atomic steps, microfacets) is of great importance for the reaction kinetics. Similar studies, for example for the formation of $NiSi_2$, which is formed by epitaxial growth on Si (111) (Hesse *et al.* 1993), demonstrate that this mechanism is not restricted to oxide systems and might be a general microscopic cause for an interface resistance.

Linear growth laws are also observed during metal/gas reactions (Kofstad 1988). A well-studied example is the oxidation of silicon (Deal and Grove 1965; Ghez and van der Meulen 1972), where adsorption of oxygen molecules at the surface seems to be the rate-determining step for the linear stage. For the technically important oxidation of iron, different a_{O_2} dependencies have been found for the linear and the parabolic rate constants (Philibert 1994). Processes at the outer surface (adsorption, dissociation) seem to be the rate-limiting step in the linear stage, whilst processes at the internal metal–oxide interface do not seem to be important. In addition, different activation energies are reported for both rate constants, 0.3 eV for the linear rate constant and 1.6 eV for the parabolic rate constant, showing that during iron oxidation the linear stage predominates at lower temperatures.

Whilst the role of interface kinetics during the oxidation of metals can only be seen during the early stages of the reaction, it shows up clearly when an oxide is exposed to an oxygen potential gradient. As discussed in Section 2.3, stationary non-equilibrium values for the surface oxygen activities are established. Experimentally this was shown by Dorris and Martin (1990) for the case of magnetite, Fe_3O_4, by means of a tracer source of radioactive ^{59}Fe which was initially located in the middle of the Fe_3O_4 crystal. The applied oxygen potential gradient causes a broadening and drift of the profile from which diffusion coefficients of the defects (vacancies or interstitials), tracer correlation factors and rate constants for the exchange of oxygen between Fe_3O_4 and CO/CO_2 mixtures can be obtained. If, for example, in the gas atmospheres an oxygen potential gradient corresponding to a ratio $(a_{O_2})_I/(a_{O_2})_{II} = 85$ is applied, the slow phase boundary reactions reduce this ratio to $(a^*_{O_2})_I/(a^*_{O_2})_{II} = 7.4$. Thus, the constant drift velocity of both crystal surfaces was reduced drastically.

There are only few definite experimental indications for defect relaxation processes leading to deviations from local equilibrium at interfaces, as discussed in Section 2.2 for the interface A–AB. One example concerns the interface Ag–AgI between the metal Ag and the Ag ion conductor AgI (Fischbach 1980; Janek and Majoni 1995). By drawing an electrical current through a symmetrical electrochemical cell of the type Ag/AgI/Ag, Ag ions are transported from the metal across the interface into the compound AgI, as described by Equation (5.1). Depending on the current density and the applied mechanical forces, Janek and Majoni (1995) observed periodic oscillations of the voltage drop across the cell. Although a detailed model of this phenomenon does not exist yet, the authors interpret their findings by a combination of defect relaxation processes and geometrical considerations. The supersaturation of vacancies in the metal, which are produced due to the reaction in Equation (5.1), possibly results in pore formation near the interface. Thus, the interface resistance increases (locally), and consequently the voltage that is necessary to drive a constant current increases. Due to relaxation processes, these pores are then annihilated, and the resistance and the voltage drop decrease, which explains qualitatively the observed oscillations.

Another illustrative example of a defect relaxation process with pore formation near interfaces can be found in Pfeiffer and Schmalzried (1989), who studied the formation of $NiCr_2O_4$ spinel. Under certain conditions, strong internal oxygen potential gradients build-up in the course of the reaction. Due to the resulting very high oxygen activities near the $NiO–NiCr_2O_4$ interface, macroscopic pores are formed which extend through the spinel reaction layer and completely destroy the initially one-dimensional reaction geometry.

Linear growth kinetics is also described for the formation of intermetallic compounds, for example Hg_4Pt (Lahiri and Gupta 1980) or Cu_6Sn_5 (Yost *et al.* 1976). Whilst in the former case the reaction at the interface $Hg_4Pt–Pt$ seems to be the rate-limiting step, it is still not clear in the latter case whether it is the proper reaction or the relaxation of defects. More complicated examples of defect relaxation processes (and the resulting enhancements of diffusion processes near interfaces) are discussed by Stolwijk *et al.* (1994).

If we summarize and generalize the preceding analysis and experimental results for the phase growth kinetics during a solid state reaction we can distinguish between five steps which determine the growth kinetics:

1 transport of atoms or ions (through the product phase) to the interface;
2 transport of atoms or ions across the interface;
3 relaxation of defects to decrease supersaturations;
4 relaxation of crystal lattices to minimize lattice misfit;
5 transport of atoms or ions away from the boundary.

Whilst step (1) leads to a parabolic rate law, steps (2–4) result in the simplest cases in linear rate laws. In special cases, however, more complicated rate laws are possible, for example, as a result of defect relaxation processes involving bimolecular or higher reactions. Then, non-monotonous or oscillating behaviour might be possible.

The examples which we have discussed show quite clearly the future necessity for more detailed studies of the processes at interfaces. From a theoretical point of view, there are two problems to be solved: (i) solutions of phenomenological kinetic equations, which are normally non-linear at interfaces; and (ii) microscopic models that describe the atomistic processes near interfaces, i.e. kinetic barriers, defect relaxation, etc. An example is given in a recent study (Deppe *et al.* 1994) where two-dimensional hopping diffusion across interfaces was analysed using Monte Carlo simulation techniques, to obtain the energy barriers at the boundaries. From an experimental point of view, *in situ* studies, particularly, are necessary, since many processes at interfaces are too fast to be frozen in, and to avoid, for example mechanical stresses which build-up during cooling. Another important problem is the measurement of the driving forces near interfaces. This means that the chemical or electrochemical potentials of the atoms or ions have to be measured with high spatial resolution. In contrast to electrochemistry in liquids, these experiments are very difficult when performed in solids, and only few examples exist (see, for examples, Schmalzried and Reye 1979; Gries and Schmalzried 1989; Janek and Majoni 1995). To obtain information about concentrations and dynamics near interfaces, spectroscopic techniques are quite powerful. Again, there is a strong necessity to develop and refine *in situ* techniques, for example Mössbauer or optical spectroscopy (Becker 1994).

3 Morphology of interfaces during solid state reactions

During a solid state reaction where matter is transported across a moving interface, the development of the reaction is strongly determined by the morphology of the interface, since the morphology determines the boundary conditions for the transport problem. One-dimensional diffusion with a planar interface, as discussed in Section 2, is a special case, since morphological instabilities very often result in non-planar interfaces with complicated non-planar structures, which can be classified as cellular, dendritic or fractal. The morphological stability of interfaces is part of the wide class of self-organization or pattern-formation problems in biology, physics, chemistry and geology, in which a non-equilibrium system forms new structures. Examples are complex biological systems, dendrites during the growth of snow flakes, the Bénard instability or spatio-temporal structures during the formation of the so-called Liesegang rings (Langer 1980; Haken 1982; Stanley and Ostrowsky 1986). In all these different systems new structures develop through instabilities from old stationary structures as a result of changed external parameters (control parameters).

It is the aim of this section to demonstrate that similar instabilities determine, to a large extent, the development of diffusion-controlled solid state reactions. A well-known example is the selective oxidation of a binary alloy (Wagner 1956), where the alloy–oxide interface may become morphologically unstable under certain conditions, in contrast to a planar metal–oxide interface during the oxidation of a pure metal. Another widely studied example is the so-called constitutional supercooling during directional solidification of a binary melt, where the stability of the solid–melt interface depends on the applied temperature gradient (Rutter and Chalmers 1953). Morphological instabilities in the form of fractal growth are well known during electrodeposition of a metal from an aqueous solution, which can be described via the so-called diffusion-limited aggregation model (DLA) (Witten and Sander 1981).

In the following sections we will discuss three types of solid state reactions where we have observed morphological instabilities. In all cases we will focus on the role of defects during the reactions. We will start with the solid–gas interfaces between a semiconducting oxide and the ambient atmosphere, which have already been discussed in Section 2.3. Then we will change to a solid–solid interface between two ionic conductors, where the driving force is an applied electric potential gradient. Furthermore, we will analyse the stability of diffusion fronts in an interdiffusion couple between two ionic conductors where we also use an electric field as driving force. Finally, we will discuss some general ideas and principles that might determine the morphological stability of interfaces during reactions.

To determine theoretically whether moving interfaces remain planar during their motion we proceed as follows. It is obvious that a macroscopically planar interface is by no means planar on a microscopic scale. Spatial and temporal fluctuations, for example, in the reaction rates, produce fluctuating values for the positions of the interfaces. If these disturbances do not grow during further transport, the initially planar interface is morphologically stable and retains its macroscopically planar form. If, however, the disturbances grow with time, the planar interface is morphologically unstable, and a new non-planar interface structure may develop. Since the corresponding diffusion problem is described by coupled non-linear differential equations, an exact analytical solution is not known. However, there are two possibilities to circumvent this problem: linear stability analysis (see Section 3.1.1) and Monte Carlo simulations (see Section 3.1.3).

3.1 *Instability of a solid–gas interface*

Our first example concerns the solid–gas interfaces of an oxide crystal, which is exposed to an oxygen potential gradient. In Section 2.3 it was shown that both solid–gas interfaces move with a constant velocity v^{stat}. Now we have to check whether the previous assumption of planar interfaces was correct.

3.1.1 LINEAR STABILITY ANALYSIS

In a linear stability analysis the time evolution of an infinitesimal disturbance of the planar surface is calculated by solving the corresponding non-planar diffusion problem, i.e. the two- or three-dimensional diffusion equation and the mass balances at the non-planar boundaries. If we denote the amplitude of a disturbance by $\Phi(x,t)$ (x is the coordinate perpendicular to z, Fig. 5.7), we obtain the following result for the long-time behaviour of the Fourier components $\Phi(q,t)$ (Martin and Schmalzried 1985):

$$\Phi(q,t) = \Phi(q,t=0)e^{p(q)t} \tag{5.23}$$

This means that a disturbance with wavevector q, or wavelength $\lambda = 2\pi/q$, develops exponentially in time, with a growth rate $p(q)$. If $p(q)$ is positive, the amplitude grows in time, and the surface is morphologically unstable. If, however, $p(q)$ is negative, the amplitude decreases in time and the surface is morphologically stable. This result is

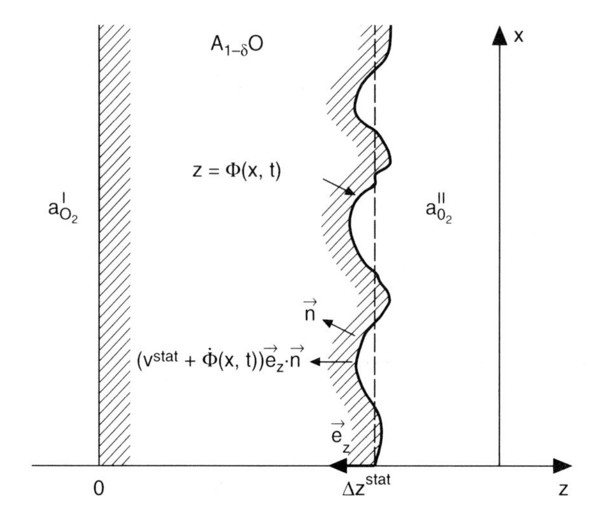

Figure 5.7 Schematic representation of an oxide in an oxygen potential gradient with a disturbed, non-planar surface (II). $\Phi(x,t)$ is the amplitude of the disturbance and $v^{stat} + \dot{\Phi}(x,t)$ is the local surface velocity.

similar to the so-called Mullins–Sekerka instability (Mullins and Sekerka 1964) for the growth of a crystal from a binary melt. But there is an important difference between the normal Mullins–Sekerka instability and the case considered here. In our case we have a finite driven system with two boundaries I and II, moving with the same velocity v^{stat}. In the linear stability analysis (with infinitely small disturbances) we may assume that both surfaces move independently, i.e. we may assume one surface to be planar, whilst we calculate the stability of the other one. If we do so, we obtain the following result: at surface II, where reduction of the oxide takes place, the growth rate $p(q)$ is positive for all q and increases monotonically with q, i.e. this surface is morphologically unstable, whilst surface I, where oxidation takes place, is morphologically stable for all q. Therefore, we find the interesting situation of a diffusion instability with one unstable surface and another stable surface in the same system. In the preceding analysis we have utilized the fact that the curved surfaces of the oxide are iso-concentration lines, i.e. that the boundary concentrations are fixed by the oxygen partial pressure of the ambient gas phase. However, this is not true in reality, due to the surface tension, σ, of the crystal which changes the surface concentrations proportionally to the surface curvature (Gibbs–Thompson effect). Qualitatively, the surface tension tries to minimize the surface area, i.e. it works against the diffusion instability during which disturbances and, therefore, the surface area grow with time. If we include the surface tension in the stability analysis, the growth rate at side I remains negative (as expected), but at side II we obtain a growth rate that is positive for wavevectors $q < q_0$ and negative for $q > q_0$, where q_0 is proportional to the surface tension, σ. This means that now surface II is morphologically unstable only for wavevectors $q < q_0$ or wavelengths $\lambda > \lambda_0$. In addition, $p(q)$ exhibits a maximum in the unstable q-region at $q_{max} = q_0/3^{1/2}$ (Fig. 5.8). Since the Fourier component of a disturbance with $q = q_{max}$ has the largest growth rate, this component will dominate after some time and the initially planar surface is likely to develop a new non-planar structure with a typical dimension λ_{max} which can be written as:

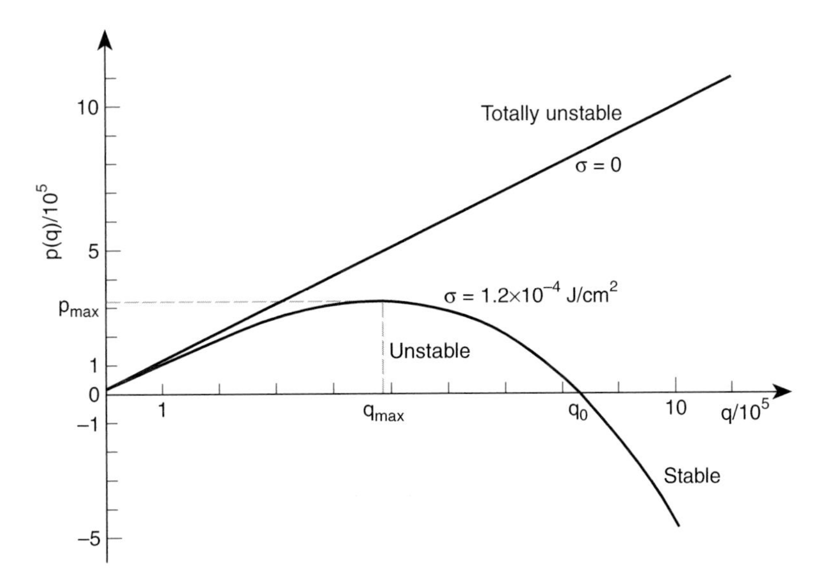

Figure 5.8 Growth rate $p(q)$ of a disturbance at surface II (reduction) as a function of the wave vector q for vanishing surface tension, $\sigma = 0$ and for $\sigma = 1.2 \times 10^{-4}$ Jcm^{-2}.

$$\lambda_{max} = 2\pi\left(\frac{3[V]_{II}\Delta z^{stat}}{[V]_I - [V]_{II}}\left(\frac{\sigma V^m}{RT}\right)\right)^{1/2} \tag{5.24}$$

Here, V^m is the molar volume and RT the thermal energy. For a constant oxygen potential gradient. $([V]_I - [V]_{II}])/\Delta z^{stat}$, the typical wavelength, λ_{max}, is proportional to $([V]_{II})^{1/2}$, i.e. λ_{max} should decrease with decreasing oxygen activity at the reducing side of the crystal. If, in contrast, the oxygen activities are kept constant, λ_{max} is proportional to $(\Delta z^{stat})^{1/2}$. Using typical experimental values (see Section 3.1.2) we obtain predicted values for λ_{max} between 1 and 20 μm.

3.1.2 EXPERIMENTAL RESULTS

Experiments were carried out with the model system CoO using single crystals at a temperature $T = 1200°C$. The oxygen potential gradient was established by flowing different gas mixtures of well-defined oxygen partial pressures through alumina capillaries on both sides of a sample. The morphologies which have developed from the initially planar (100) surfaces are shown in Figs 5.9–5.11 in top view for different experimental conditions (the temperature is always 1200°C and at the oxidizing side air with $a_{O_2} = 0.21$ was always used). In Fig. 5.9 the initial stage of the instability can be seen. Depressions in the form of pyramids have formed within the planar surface, presumably at points where dislocations push through the surface. Starting from these seeds, a nearly regular strongly anisotropic new structure forms after a longer diffusion time. Figure 5.10 shows an overview (a) and an enlarged section (b). The typical dimension of the new structure is about 20 μm, as predicted by linear stability analysis. If the gas mixture at the low oxygen potential side is changed from N_2 to CO_2, both of which contain a comparable amount of residual oxygen, the morphology of surface II

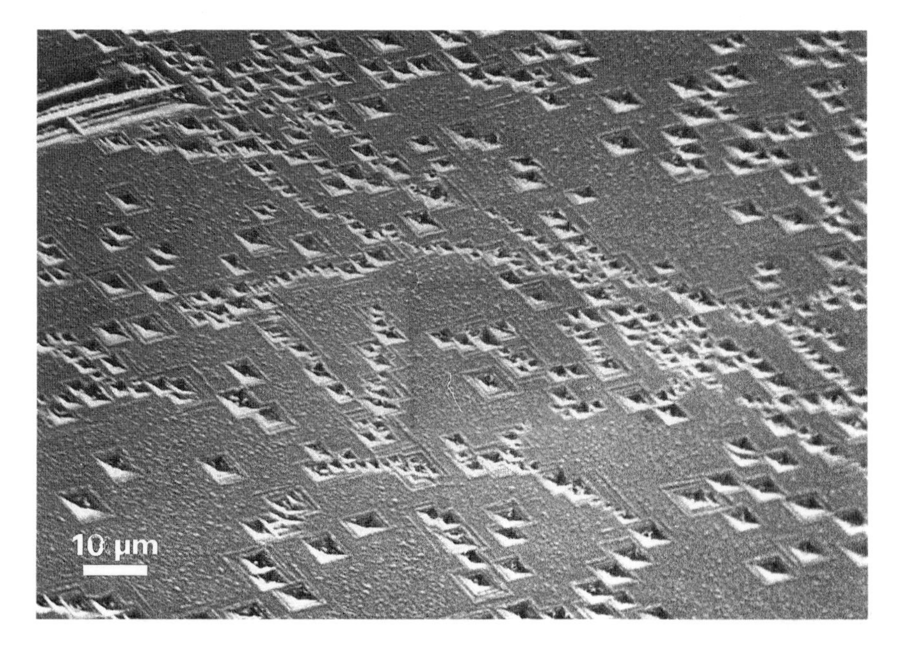

Figure 5.9 Unstable surface (II) of a CoO single crystal ($t = 5$ hours, reducing side: N_2, $a^{II}_{O_2} = 5 \times 10^{-5}$).

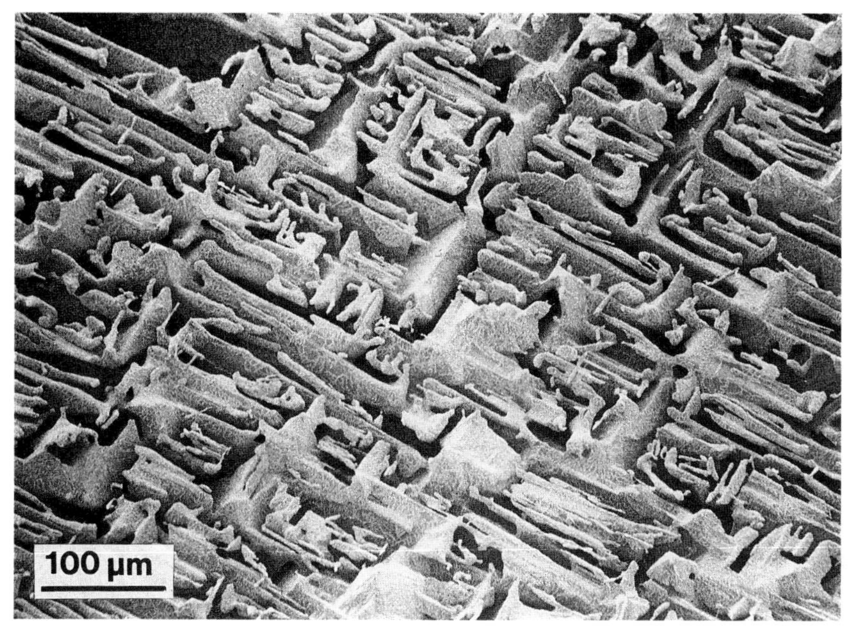

(a)

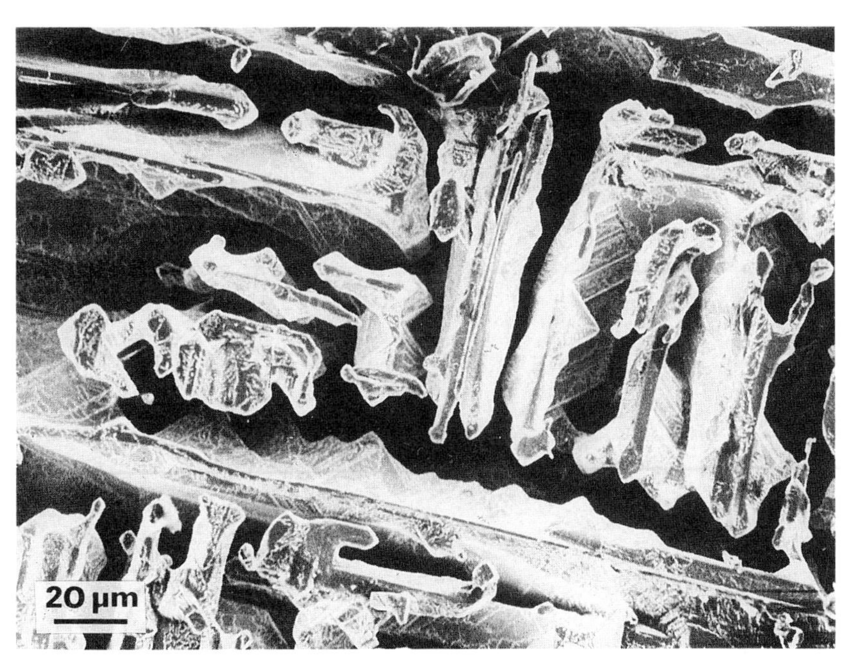

(b)

Figure 5.10 Unstable surface (II) of a CoO single crystal (t = 30 hours, reducing side: N_2, $a^{II}_{O_2} = 5 \times 10^{-5}$). (a) Overview, (b) enlarged section.

changes completely (Fig. 5.11). The typical dimension is still about 20 μm, but now the non-planar structure is completely isotropic. Presumably, the strong anisotropy of the surface tension of CoO is reduced by adsorption, resulting in a nearly isotropic surface tension. However, the changed surface kinetics, as discussed in Section 2.3, might also

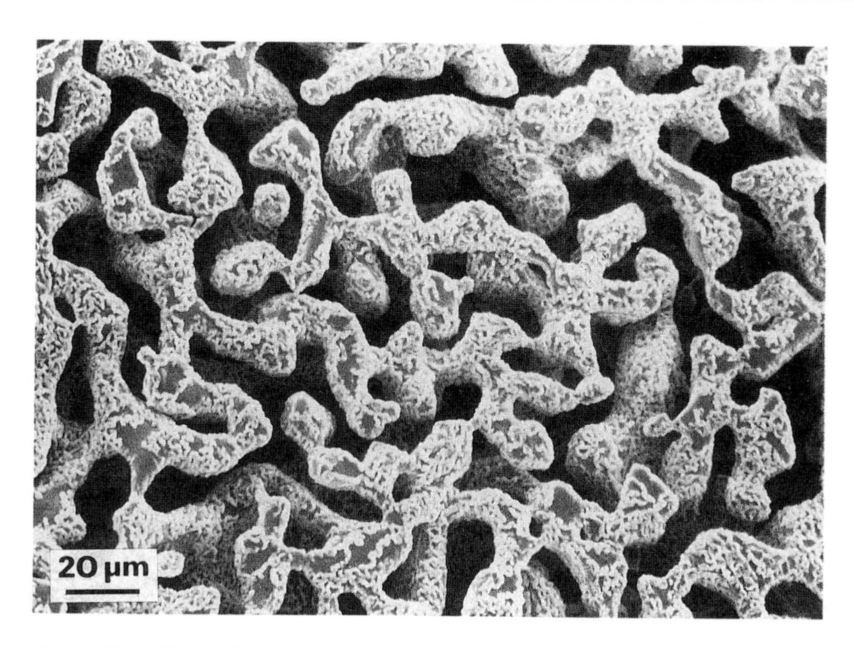

Figure 5.11 Unstable surface (II) of a CoO single crystal (t = 36 hours, reducing side: CO_2, $a^{II}_{O_2} = 5 \times 10^{-5}$).

have an influence. It should be noted that the pattern observed here is very similar to patterns observed in isotropic spinodal decomposition (Cahn 1965), but on a totally different length scale.

3.1.3 MONTE CARLO SIMULATIONS

From the linear stability analysis no predictions of the morphology that develops from the unstable planar surface were possible, nor could the microscopic structure be determined. To investigate the morphology on a microscopic scale, where fluctuations play an important role, Monte Carlo simulations were done of vacancy diffusion in a simple cubic lattice with a free surface. Here, reduction of the oxide was modelled by sticking of vacancies (Martin 1991), and the equivalence to the well-known DLA model (Witten and Sander 1981) was demonstrated. Surface energy was included in a phenomenological way by making the jump probability of a vacancy to a surface position conditional upon the number of nearest neighbours of the new position. In this simple nearest neighbour model, a transition from a fractal surface to a planar surface was found by increasing the value of the phenomenological surface energy.

A more realistic model uses the fact that the oxides, AO, under consideration are ionic crystals bound by Coulombic interaction between cations A^{2+} and anions O^{2-}, and which crystallize in the NaCl structure (Tigelmann and Martin 1992). The local Madelung energy of each ion i (regarded as point charge) is calculated as the Coulombic interaction energy with all other ions j. Since the crystal contains a surface, the binding energy of each ion decreases from the interior of the crystal to the surface. A cation vacancy is introduced at a random position in the crystal interior and then released to perform a random walk within the cation sublattice. Since the typical vacancy fraction is of the order of 10^{-3}, only one vacancy at a time is introduced into the lattice. When this vacancy reaches the crystal surface it may stick there if a neighbouring oxygen ion dissociates from the crystal according to Equation (5.18). A Boltzmann-ansatz is used for the dissociation probability of oxygen:

$$p = e^{-\Delta E_0/kT} = e^{-\beta \Delta \bar{E}_0}$$

$$\beta = e^2/(4\pi\varepsilon r_0 kT) \qquad (5.25)$$

where ΔE_0 is the binding energy of oxygen (identical to the absolute value of its local Madelung energy), k is the Boltzmann constant and T the absolute temperature. Concerning the dimensionless variables, β is the inverse temperature measured in terms of the Coulombic energy of two point charges separated by the nearest neighbour distance r_0, ε is the dielectricity constant of the material and $\Delta \bar{E}_0$ is now the dimensionless binding energy. Which of these surface oxygen ions neighbouring the cation vacancy dissociates (Fig. 5.12), depends on their relative binding energies. If the dissociation is not successful the cation vacancy may jump back into the bulk of the crystal. If, however, the dissociation of oxygen is successful, the vacancy remains at the surface to preserve the crystal structure. In total, a lattice molecule AO, consisting of a cation and an anion, has been removed from the surface resulting in an indentation in the original surface (Fig. 5.12, open circles). Then, the new Madelung energy of the ions in the remaining crystal is calculated, and a new vacancy can be released from a random position in the internal surface. In this way, a non-homogeneous distribution of sticking probabilities for the vacancy along the surface is obtained for $\beta \neq 0$, which stabilizes the surface against the intrinsic diffusion instability. Another important effect is surface diffusion. Since it is several orders of magnitude faster than bulk diffusion, we cannot simulate both effects simultaneously. But the effect of fast surface diffusion can be modelled by permitting the surface ions, anions as well as cations, to relax to positions of greatest local binding energy. The driving force for this process is again the difference in binding energies along the surface. In Fig. 5.12 a situation is shown where after arrival of the vacancy at the surface, oxygen ion 1 dissociates and, subsequently, oxygen ion 2 and cation 3 relax to their new local equilibrium positions.

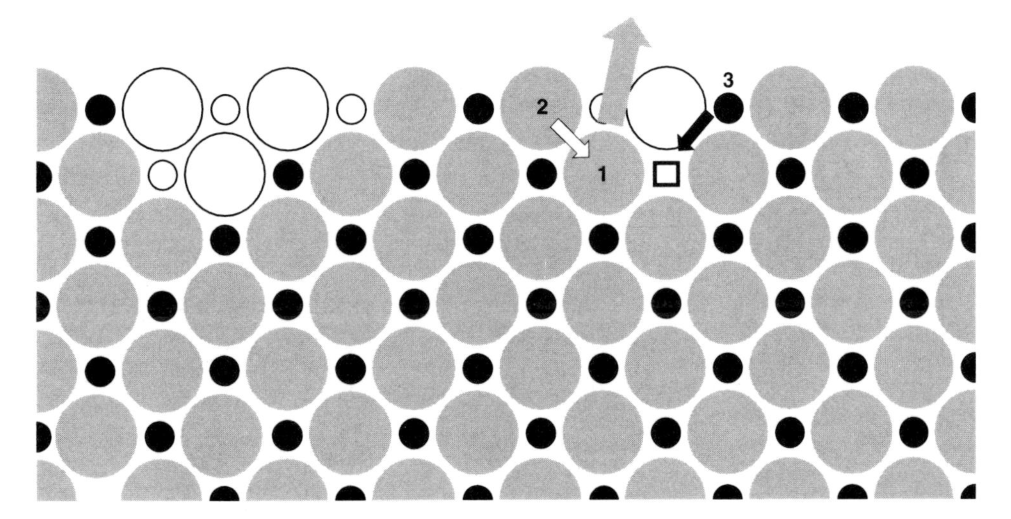

Figure 5.12 Two-dimensional (10) plane of the AO crystal. Large and small circles are oxygen ions and cations, the square represents the vacancy, and the open circles show removed ions.

Figure 5.13 shows structures obtained for different values of the inverse dimensionless temperature, $0 \leq \beta \leq 2.5$. With increasing β, i.e. decreasing temperature, a clear transition from a fractal structure to a still rough, but non-fractal structure can be seen. The root-mean-square thickness, T_{rms}, of the grown object (here, the volume of annihilated lattice molecules) scales with the number N of particles (here, vacancies) as $T_{rms} \propto N^d$ (Meakin 1983). The fractal exponent d changes with increasing β (decreas-

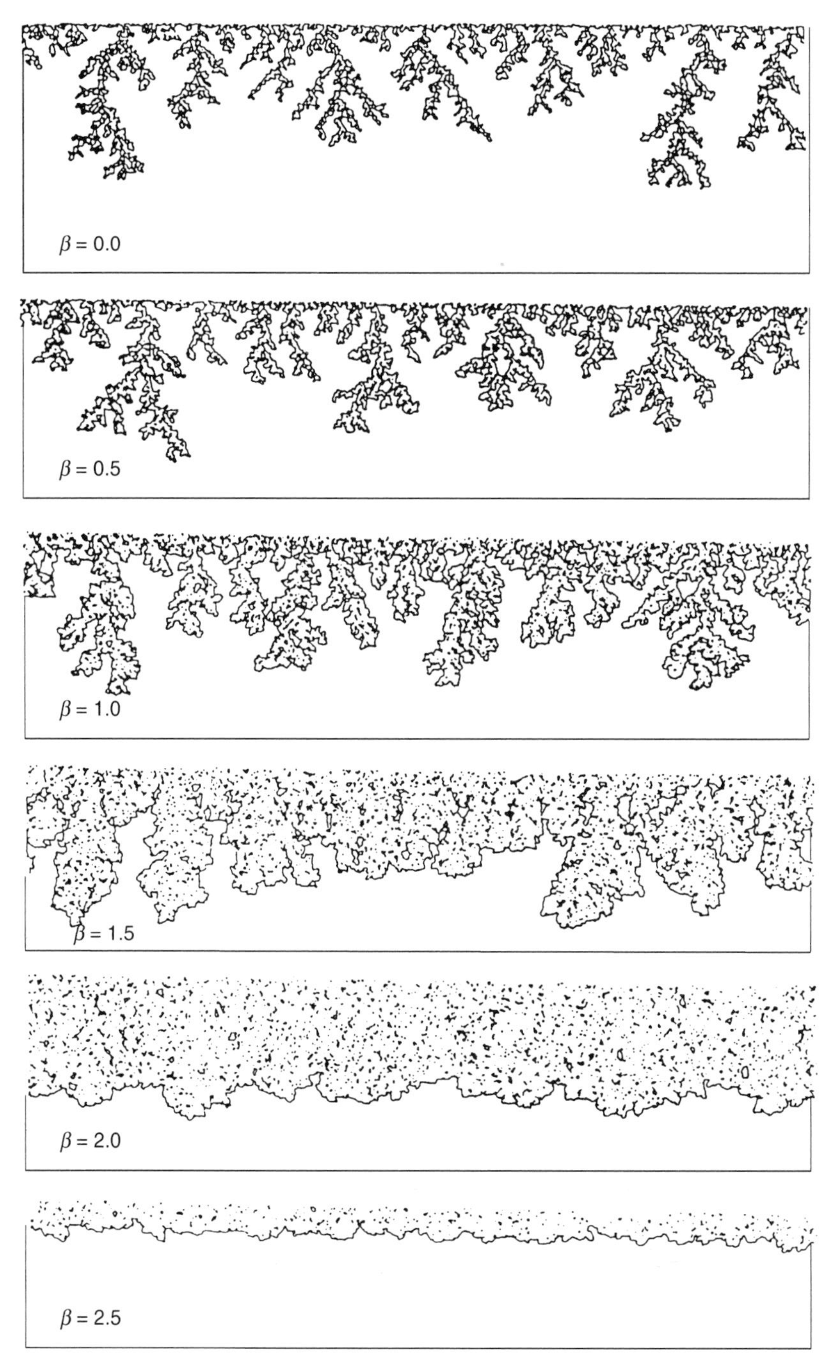

Figure 5.13 Typical structures obtained for different values of the inverse dimensionless temperature, β. Vacancies have moved from the bottom to the top.

ing temperature), clearly showing the transition from a fractal surface with $d \approx 1.42$ at $\beta = 0$ to a planar surface with $d \approx 1$ at $\beta = 2.5$ (Tigelmann and Martin 1992).

Another important feature by which the different surfaces can be characterized is the distribution of facets in the surface. In our two-dimensional model only two types of facets are possible: (10) facets consisting of alternating cations and anions, and (11) facets consisting either of cations or of anions. The facet distribution satisfies a Boltzmann distribution, $f_n^{(ij)} \propto e^{-\alpha(ij)n}$, where n is the length of the facet (ij) and α(ij) is the surface energy per bond measured in units of kT. For $\beta = 1$, we obtain $\alpha(11)/\alpha(10) = 1.57$, showing that the (11) facets have a higher surface energy than (10) facets as a result of the Coulombic repulsion between the identical ions in the (11) facet. The ratio of the surface energies $\alpha(11)$ and $\alpha(10)$ changes from 1.1 at $\beta = 0$ to 2.0 at $\beta = 2.5$, which means that at high temperatures (10) and (11) facets are formed with the same probability, whilst at low temperatures the formation of (10) facets is strongly favoured.

A comparison with experimental studies of the surface instability in CoO (Martin and Schmalzried 1985; Martin 1991) is made difficult because of the very different length scales in both studies. If, however, typical experimental conditions are considered this would correspond to $\beta \approx 1.5$. For this value of β our model predicts a faceted and still rough, but non-fractal surface with an exponent $d \approx 1$ for the root-mean-square thickness. The experimentally observed macroscopic surface structures with a typical dimension of 20 μm are beyond the validity of the microscopic model. However, in a recent TEM study of surfaces of several oxides, including NiO, a fractal analysis was performed at atomic resolution (Bursill *et al.* 1991). The authors report that the exponents in the observed scaling behaviour were only slightly larger than 1 and that the scaling behaviour was in all cases complicated by atomic surface faceting. Although the experimental conditions in the TEM study were different from our simulated experiment, this seems to indicate the importance of atomic surface faceting as suggested by our simple model for the formation of microscopic surface structures during reduction of an oxide.

3.2 *Instability of a solid–solid interface*

To study the morphological development of a phase boundary between two crystalline solids we change from the semiconducting oxides, which we have studied in the previous section, to ionic conductors. As an advantage we can use here externally applied electric potential gradients as driving forces for the motion of the solid–solid phase boundary. To make the situation as simple as possible and to focus on the essential part of the problem we first study a phase boundary between two totally immiscible ionic conductors. For the experiments we choose the quasi-binary system AgCl–KCl as a model system. Below the eutectic temperature, $T_c = 306$°C, there is no mutual solubility, although AgCl as well as KCl crystallize in the rocksalt structure. But the lattice constants of both compounds are very different, resulting in a mismatch of about 12%. Both compounds are nearly pure cation conductors. In AgCl (cation Frenkel disorder) Ag ions are mobile in the interstitial sublattice, whilst in KCl (Schottky disorder) K ions are mobile via cation vacancies. As a result of the different disorder types the ionic conductivity in AgCl is about six orders of magnitude larger than in KCl (Bénière 1972).

Experiments were done with AgCl and KCl single crystals, which were used in an electrochemical cell +/Ag/AgCl/KCl/Ag/– (Schimschal-Thölke *et al.* 1995a). Ag ions are driven by the applied electric potential difference from the Ag anode to the boundary AgCl–KCl. Since Ag is not soluble in KCl, the following exchange reaction takes place:

$$Ag^{+} + KCl \rightleftharpoons AgCl + K^{+} \quad (5.26)$$

This exchange reaction has two consequences: (i) it allows further charge transport via K ions; and (ii) the AgCl phase grows at the expense of the KCl phase. Since transport in AgCl is much faster than in KCl the whole electric potential, ϕ, falls off across the 'slower' KCl crystal, as shown schematically in Fig. 5.14. This means that the rate-determining step for the motion of the phase boundary between the two solids is the flux of K ions in KCl (or the oppositely directed flux of the cation vacancies). If we compare this transport problem to the previously studied problem of a transition metal oxide exposed to an oxygen potential gradient, we can find a complete analogy with the gas–oxide interface II at the reducing side. In both cases, the phase boundary is driven by a vacancy flux towards the boundary, and in both cases the 'faster' phase grows at the expense of the 'slower' phase. (In the previous case of the gas–oxide boundary, the faster phase was the gas phase, in which we had assumed a constant oxygen partial pressure. This means very fast transport of oxygen in the gas phase, compared to slow transport of vacancies in the crystal.) From this analogy we expect the planar solid–solid phase boundary between AgCl and KCl to be morphologically unstable, if we apply the potential difference in such a way that AgCl grows. If we reverse the applied voltage, we expect the boundary to be stable, in analogy to the oxide-gas boundary I. This conception is confirmed by the results of the linear stability analysis (Schimschal-Thölke 1993) for the moving solid–solid interface. A typical experimental result demonstrating clearly that the solid–solid interface is morphologically unstable is shown in Fig. 5.15. We note that AgCl grows into KCl in the form of 'trees', which are typical of fractal growth phenomena (Witten and Sander 1981). Closer inspection shows that these trees are two-dimensional objects with certain orientations in the KCl

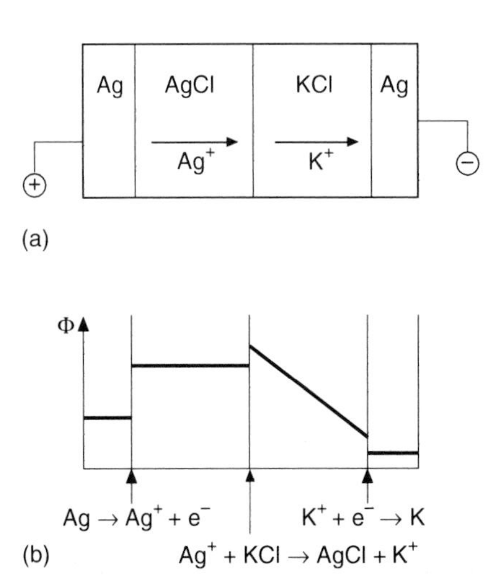

Figure 5.14 (a) Electrochemical cell, (b) electric potential, ϕ, across the cell.

Figure 5.15 Unstable interface AgCl–KCl.

matrix, which are due probably to elastic effects caused by the different lattice constants of both materials. A more detailed discussion can be found in Schimschal-Thölke *et al.* (1995a).

In general, there are mutual solubilities between two ionic conductors and the conductivities might become similar. For this case, Monte Carlo simulations (Schulz 1995) show that: (i) the above simple principle of instability for the growth of the faster phase is only valid above a critical electric field, below which no instabilities occur; and (ii) the morphology of the interface (wavy, dendritic, fractal) depends mainly on the strength of the interaction between the cations.

Finally, it should be noted that the reaction in Equation (5.26) can be regarded as an electrochemical exchange reaction, in analogy to the classical exchange reaction between a metal and a metal oxide, A + BO = AO + B, where non-planar solid–solid interfaces have also been observed (for example, $Fe + Cu_2O = FeO + 2Cu$ (Rapp *et al.* 1973).

3.3 *Instability of diffusion fronts*

In the previous section we demonstrated that the moving interface between two immiscible ionic conductors, which is driven by an external electric field, is morphologically unstable when the 'faster' of both phases grows. In this section we will analyse what happens if we turn to the other extreme, i.e. if the two ionic conductors are totally miscible. A good example is provided by the system AgCl–NaCl. Both compounds form a complete solid solution between the critical temperature, $T_c = 198\,°C$ (below which there is a miscibility gap with mutual solubilities), and the melting temperature of AgCl. As before, Ag ions in AgCl are mobile in the interstitial sublattice, Na ions in NaCl are mobile via cation vacancies, and the ionic conductivity, σ, in AgCl is orders of magnitude larger than in NaCl. In the solid solution, $(Ag_{1-x}Na_x)Cl$, σ decreases approximately exponentially with the composition x (Magistris *et al.* 1974). Experiments were performed in the same way as before, i.e. with an electrochemical cell +/Ag/AgCl/NaCl/Ag/– (Schimschal-Thölke *et al.* 1995b). But, in contrast to the previous experiment, we have now a combination of two effects: interdiffusion of Ag^+ and Na^+, and drift in the externally applied electric potential gradient. Since the diffusion coefficients of Ag^+ and Na^+ depend exponentially on composition x, pure interdiffusion would result in very steep interdiffusion profiles. If we apply, in addition, an electric field in such a way that the faster 'phase' (in this case, the Ag-rich part of the solid solution) grows, we expect morphological instabilities, in analogy to the previously discussed case in Section 3.2 and according to the principle formulated there. Nevertheless, it has to be emphasized that these expected morphological instabilities are instabilities of diffusion fronts in a single-phase system without any interface.

Figure 5.16 shows a typical result of an interdiffusion and drift experiment. The photograph was obtained in an electron microscope using backscattered electrons, which indicate a phase contrast. In the picture, the steep concentration gradient in the interdiffusion zone appears as sharp phase contrast, but microprobe analysis confirms a steep concentration gradient in a single-phase system. Whilst the interdiffusion profile without an external electric field is one-dimensional, i.e. planar as expected, it is by no means one-dimensional with an applied electric field. Instead, one can observe instabilities of the diffusion fronts in the form of 'fingers' (Fig. 5.16).

A theoretical analysis of this instability problem by means of a numerical stability analysis (Martin 1994; Schimschal-Thölke *et al.* 1995b) again shows the existence of a critical electric field, below which no instabilities occur. This can be understood by the fact that for sufficiently strong electric fields the concentration profiles in the interdiffusion zone become very steep and resemble an interface in a two-phase system. Thus, the morphological stability of the diffusion profiles is governed by the same principles as for the case of a real interface.

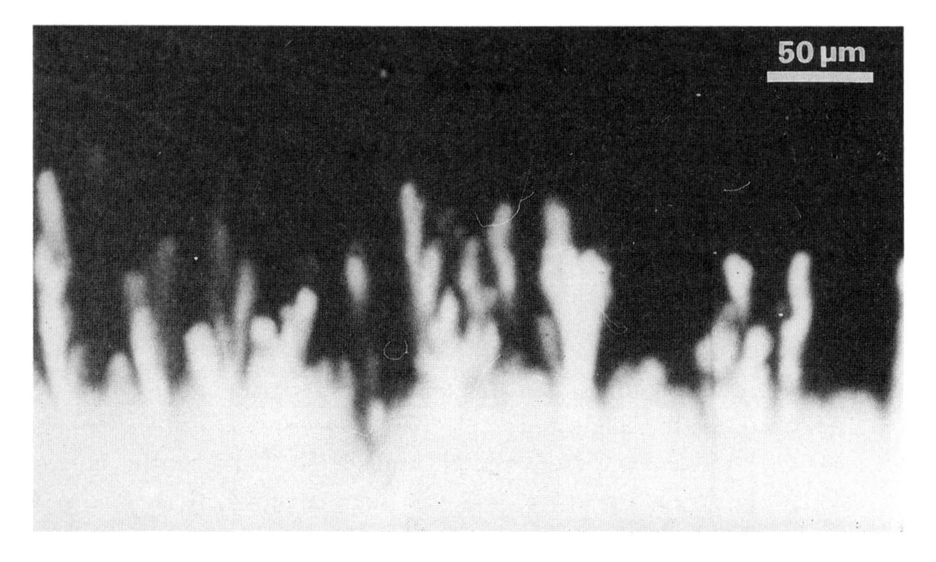

Figure 5.16 Interdiffusion zone AgCl (bright)—NaCl (dark) with applied electric field.

3.4 *General discussion of morphological stability*

In the previous sections we have discussed morphological stability in binary or quasi-binary systems with an external driving force. For the limiting case, where transport in one phase is fast compared to the other phase, we have postulated a principle which says that we obtain morphological instabilities if the faster of the two phases grows. However, this principle might be applicable in general, as can be seen if we turn to systems that are not driven by external forces or to multicomponent systems. The main difference between a multicomponent system and a binary system originates in the fact that a phase boundary is no longer an iso-concentration line. In a binary system the boundary concentrations are fixed thermodynamically, which is no longer the case in ternary or higher systems. This means that in ternary or higher systems the boundary concentrations at a curved boundary are determined by the thermodynamics and kinetics of the system. Since these boundary conditions determine to a large extent the criteria for morphological stability, the situation becomes much more complex than in binary systems.

What happens in a system that is not driven by an external force can be discussed with a two-phase diffusion couple, which is infinitely extend ($-\infty < z < +\infty$). Two phases, α and β, are brought together at time $t = 0$, after which equilibrium between the phases is established by exchange of matter across the boundary and diffusion in both phases. In contrast to the previous driven systems, where the non-equilibrium state was kept by the constant external force, this system is closed and the (internal) driving forces decrease in time. The one-dimensional diffusion problem with planar boundary between phases α and β yields a parabolic rate law for the interface position and error functions for the concentration profiles (for simplicity we use only constant diffusion coefficients). If we plot the concentrations (which depend only on the ratio $z/t^{1/2}$) in the phase diagram we obtain the so-called diffusion paths, along which the position, z, changes from $-\infty$ to $+\infty$ at constant time, t, or the time changes from 0 to $+\infty$ at constant z. If we perform a linear stability analysis for this parabolic diffusion problem, at first for a binary system, we obtain no instabilities if the diffusion paths in both phases are outside the miscibility gap in the phase diagram. In a binary system this is always the case under isothermal conditions (Fig. 5.17a). Only if we quench the sample from a higher temperature into the miscibility gap, the diffusion paths are totally inside the miscibility gap (Fig. 5.17b), and linear stability analysis yields instability of the planar boundary (Martin 1984). A transition between the two cases can be obtained if we expose the interdiffusion couple to a constant temperature gradient, ∇T. Depending on the strength of ∇T, the planar diffusion paths penetrate the miscibility gap or remain totally outside it (Fig. 5.18). Linear stability analysis shows that the transition point corresponds exactly to the point where the planar boundary loses stability (constitutional supercooling (Rutter and Chalmers 1953)). Therefore, we can formulate the following simple principle for a binary closed system:

> In a binary closed system an interface between two phases is morphologically stable if and only if the diffusion paths for the planar solution are in the single-phase regions of the phase diagram. The planar boundary is morphologically unstable if the virtual diffusion paths of the planar solution are (partially) inside the miscibility gap.

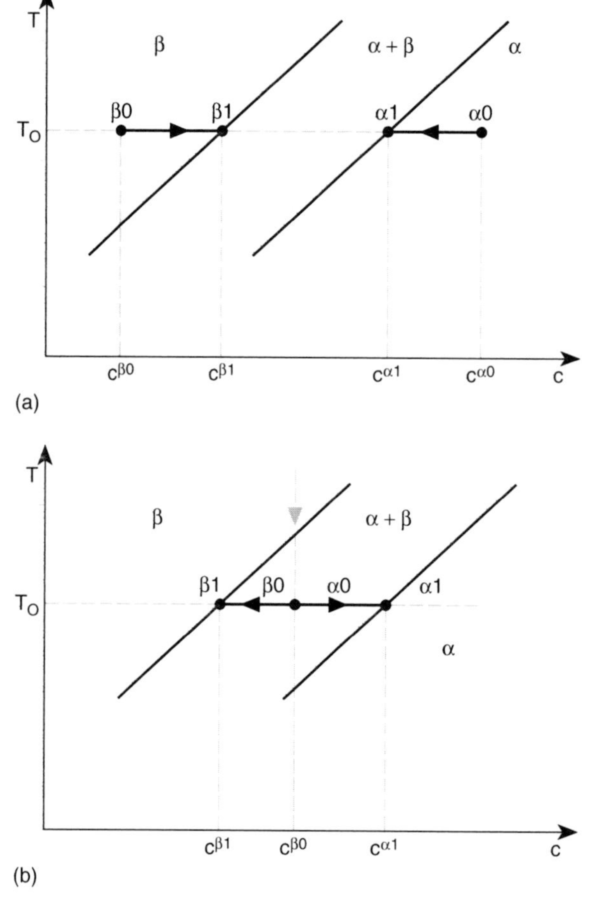

Figure 5.17 Diffusion paths (→—) in a binary two-phase interdiffusion couple (c is the composition). (a) Isothermal interdiffusion, (b) temperature jump into the miscibility gap.

In the second case, the diffusion path has been denoted as a virtual diffusion path, because it corresponds to an unstable solution. If we perform the same analysis, i.e. solution of the one-dimensional parabolic diffusion problem and linear stability analysis, for an interface in a two-phase ternary system, we find that the above principle is only valid in a much weaker form (Martin 1984; Backhaus-Ricoult and Schmalzried 1985):

> In a ternary closed system it is a necessary but not a sufficient condition for morphological instability of a planar phase boundary that the virtual diffusion paths are (partially) inside the miscibility gap.

The formal analysis shows that in the ternary system the magnitude of the diffusion coefficients in both phases and the starting concentrations in the Gibbs triangle determine which of the possible tie lines is chosen and whether the planar interface is stable or not (Fig. 5.19). Experimentally, the possibility of instabilities in a ternary system was first observed during dissolution of a binary alloy (A, B) in liquid metal C (Harrison and Wagner 1959). The interface between (Cu, Ni) and liquid Ag, in which Cu preferentially dissolves, is unstable, whilst the interface between (Cu, Ni) and Bi, in which both Cu and Ni are dissolved similarly, remains planar. In ternary oxide systems morphological instabilities have been observed during preferential oxidation of a

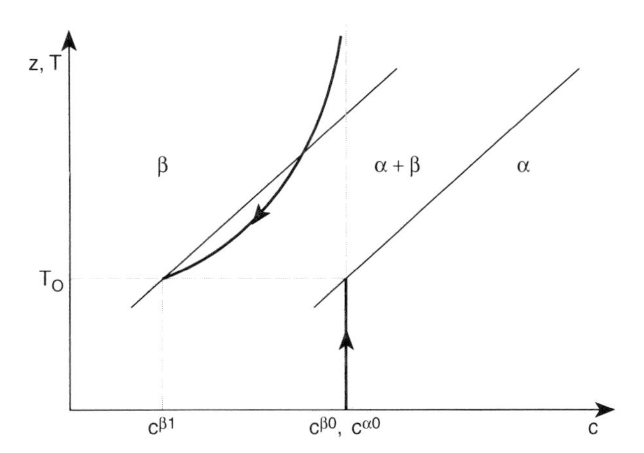

Figure 5.18 Diffusion paths (—▸—) in a binary two-phase interdiffusion couple exposed to a constant temperature gradient.

binary alloy (A,B) (Wagner 1956), and a connection with the diffusion paths has been discussed (Whittle 1973). In agreement with the above principle, the diffusion paths always penetrate the miscibility gaps in the ternary phase diagrams (at least partially) when morphological instabilities are observed. Detailed studies for two-phase diffusion couples in the quasi-ternary oxide system Fe_3O_4–Mn_3O_4–Cr_2O_3 (Backhaus-Ricoult and Schmalzried 1985) or in metal-ceramic systems (Backhaus-Ricoult 1989) show that, depending on the magnitude of the transport coefficients and the starting compositions, all kinds of morphologies, ranging from planar to non-planar and non-planar with a zone of precipitates, can be produced.

The previous discussion shows quite clearly the necessity for more studies concerning the morphology of interfaces during solid state reactions. From an experimental point of view, it seems necessary to find more distinct examples of morphological instabilities in experimentally well-defined model systems. These results may then be helpful in more complicated systems to understand how new morphologies develop via instabilities from old morphologies. From a theoretical point of view modern numerical techniques, bifurcation analysis as well as Monte Carlo simulations have to be used as theoretical tools to study specific problems. More generally, however, there is still the

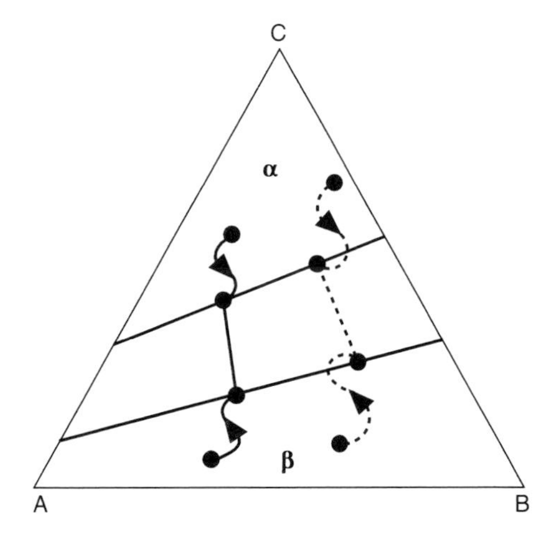

Figure 5.19 Diffusion paths in a ternary two-phase system (—▸—, stable; --▸--, may be unstable).

open question of a general principle governing the morphology and the morphological stability of interfaces during solid state reactions.

4 References

Adda, Y. & Philibert, J. (1966) *La Diffusion Dans les Solides*. Presses Universitaires, Paris.

Backhaus-Ricoult, M. (1989) Metal–Ceramic Interfaces. Acta-Scripta Metallurgica Proceedings Series 4. (eds M. Rühle, A.G. Evans, M.F. Asby & J.P. Hirt), p. 79–92. Pergamon Press, Oxford.

Backhaus-Ricoult, M. & Schmalzried, H. (1985) *Ber Bunsenges Phys Chem* **89**, 1323–30.

Becker, K.D. (1994) *Mat Sci Forum* **155–156**, 71–87.

Becker, K.D. & Rau, F. (1988) *Solid State Ionics* **28–30**, 1290–93.

Bénière, F. (1972) In: *Physics of Electrolytes* (ed. J. Hladik), p. 204–98. Academic Press, London.

Bursill, L.A., XuDong, F. & Julin, P. (1991) *Phil Mag A* **64**, 443–64.

Cahn, J.W. (1965) *J Chem Phys* **42**, 93–9.

Carter, B. & Schmalzried, H. (1985) *Phil Mag A* **52**, 207–29.

Deal, B.E. & Grove, A.S. (1965) *J Appl Phys* **36**, 3770–8.

Deppe, J., Wallis, R.F., Nachev, I. & Balkanski, M. (1994) *J Phys Chem Solids* **55**, 759–66.

Dieckmann, R., Schmalzried, H. & Mason, T.O. (1981) *Arch Eisenhüttenwes* **52**, 211–18.

Dorris, S. & Martin, M. (1990) *Ber Bunsenges Phys Chem* **94**, 721–6.

Duckwitz, C.A. & Schmalzried, H. (1971) *Z Phys Chem NF* **76**, 173–93.

Dybkov, V.I. (1992) *Kinetics of Solid State Chemical Reactions*. Naukova Dumka, Kiev.

Dybkov, V.I. (1994) *Mat Sci Forum* **155–156**, 31–8.

Fischbach, H. (1980) *Z Metallkd* **71**, 115–19.

Gas, P. (1994) *Mat Sci Forum* **155–156**, 39–51.

George, E.P., Yamaguchi, M., Kumar, K.S. & Liu, C.T. (1994) *Ann Rev Mat Sci* **24**, 409–51.

Ghez, R. & Van Meulen, Y.J. (1972) *J Electrochem Soc* **119**, 1100–06.

de Groot, S.R. & Mazur, P. (1984) *Non-equilibrium Thermodynamics*. Dover, New York.

Grabke, H.J. (1965) *Ber Bunsenges Phys Chem* **69**, 48–57.

Gries, B. & Schmalzried, H. (1989) *Solid State Ionics* **31**, 291–99.

Haken, H. (1982) *Synergetik*. Springer-Verlag, Berlin.

Haase, R. (1990) *Thermodynamics of Irreversible Processes*. Dover, New York.

Harrison, J.D. & Wagner, C. (1959) *Acta Met* **7**, 722–35.

Hesse, D., Werner, P., Mattheis, R. & Heydenreich, J. (1993) *Appl Phys A* **57**, 415–25.

Hesse, D., Sieber, H., Werner, P., Hillebrand, R. & Heydenreich, J. (1994) *Z Phys Chem* **187**, 161–78.

d'Heurle, F.M. (1994) *Mat Sci Forum* **155–156**, 1–12.

Janek, J. & Majoni, S. (1995) *Ber Bunsenges Phys Chem* **99**, 14–20.

Kaur, I. & Gust, W. (1989) *Fundamentals of Grain and Interphase Boundary Diffusion*. Ziegler Press, Stuttgart.

Kofstad, P. (1988) *High Temperature Corrosion*. Elsevier, London.

Koryta, J., Dvořák, J. & Kavan, L. (1993) *Principles of Electrochemistry*. Wiley, Chichester.

Lahiri, S.K. & Gupta, D. (1980) *J Appl Phys* **51**, 5555–60.

Langer, J.S. (1980) *Rev Mod Phys* **52**, 1.

Magistris, A., Schiraldi, A. & Schiodelli, G. (1974) *Z Naturforsch*, **29a**, 1330–4.

Martin, M. (1984) *Über die morphologische Stabilität von ebenen Phasengrenzen im binären Kobaltoxid und in Mehrkomponentensystemen*. PhD Thesis. University of Hannover.

Martin, M. (1991) *Mat Sci Rep* **7**, 1–86.

Martin, M. (1994) *Mat Sci Forum* **155–156**, 429–43.

Martin, M. & Schmalzried, H. (1985) *Ber Bunsenges Phys Chem* **89**, 124–30.

Meakin, P. (1983) *Phys Rev A* **27**, 2616–23.

Mullins, W.W. & Sekerka, R.F. (1964) *J Appl Phys* **35**, 444–51.

Pfeiffer, T. & Schmalzried, H. (1989) *Z Phys Chem NF*, **161**, 1–17.
Philibert, J. (1994) *Mat Sci Forum* **155–156**, 15–30.
Rapp, R.A., Ezis, A. & Yurek, G.J. (1973) *Met Trans* **4**, 1283–92.
Rutter, J.W. & Chalmers, B. (1956) *Can J Phys* **31**, 15–39.
Schimschal-Thölke, S. (1993) *Die morphologische Stabilität von Phasengrenzen und Konzentrationsfeldern am Beispiel AgCl/KCl und AgCl/NaCl im elektrischen Feld.* PhD Thesis. University of Hannover.
Schimschal-Thölke, S., Schmalzried, H. & Martin, M. (1995a) *Ber Bunsenges Phys Chem* **99**, 1–6.
Schimschal-Thölke, S., Schmalzried, H. & Martin, M. (1995b) *Ber Bunsenges Phys Chem* **99**, 7–13.
Schmalzried, H. (1978) *Ber Bunsenges Phys Chem* **82**, 273–7.
Schmalzried, H. (1981) *Solid State Reactions.* Verlag Chemie, Weinheim.
Schmalzried, H. (1993) *Polish J Chem* **67**, 167–90.
Schmalzried, H. (1995) *Chemical Kinetics of Solids.* Verlag Chemie, Weinheim.
Schmalzried, H. & Reye H. (1979) *Ber Bunsenges Phys Chem* **83**, 53–9.
Schulz, G. (1995) *Computersimulationen zur Morphologie und Stabilität von bewegten fest/fest Phasengrenzen im ionenleitenden System AX/BX* PhD Thesis. University of Hannover.
Sieber, H., Werner, P., Hillebrand, R., Hesse, D. & Heydenreich, J. (1994) *Solid–Solid Phase Transformations* (eds W.C. Johnson, J.M. Howe, D.E. Laughlin & W.A. Soffa), p. 1201–6. The Minerals, Metals and Materials Society, Warrendale, Philadelphia.
Stanley, H.E. & Ostrowsky, N. (eds) (1986) *On Growth and Form.* Martinus Nijhoff Publishers, Boston.
Stolwijk, N.A., Bracht, H., Hettwer, H.-G. *et al.* (1994) *Mat Sci Forum* **155–156**, 475–92.
Tigelmann, P. & Martin, M. (1992) *Phys A* **191**, 240–7.
Valverde, N. (1976) *Ber Bunsenges Phys Chem* **80**, 333–40.
Wagner, C. (1936) *Z Phys Chem* **B34**, 309–16.
Wagner, C. (1938) *Z Anorgan Allg Chem* **236**, 320–38.
Wagner, C. (1956) *J Electrochem Soc* **103**, 571–80.
Wagner, C. (1969) *Acta Metall* **17**, 99–107.
Whittle, D.P. (1973) In: *High Temperature Corrosion* (ed. R.A. Rapp), p. 171–82. National Association of Corrosion Engineers, Houston, Texas.
Wilke, K.-Th. (1988) *Kristallzüchtung.* Verlag Harri Deutsch, Frankfurt.
Witten, T.A. & Sander, L.M. (1981) *Phys Rev Lett* **47**, 1400–3.
Yost, F.G., Ganyard, F.P. & Karnowsky, M.M. (1976) *Met Trans* **A7**, 1141–8.

6 Modern Tendencies in Heterogeneous Kinetics of Solid State Decomposition Reactions

N. LYAKHOV

Institute of Solid State Chemistry, Siberian Branch of the Russian Academy of Sciences, Kutateladze, 18 Novosibirsk-128, 630128, Russia

1 Introduction

The theory of heterogeneous kinetics of solid state reactions is still incomplete. There are many reasons for this. Nowadays, in contrast to 10 or 20 years ago, most kinetic research work can be reduced to the analysis of $\alpha(t)$ curves. The data obtained from such analysis are often the subject of heated discussions and, usually, cannot be used for the purpose of direct selection of appropriate reaction mechanisms.

The other reason is that heterogeneous kinetics of solid state reactions developed mainly in the direction of formal geometrical description. Since 1969, when B. Delmon's monograph appeared (Delmon 1969) this field of activity may be considered practically covered. The one exception is the introduction of particle size distribution into kinetic models of nucleation–growth in order to modify theoretical expressions or calculations. But, this is a rather technical problem, even though it is very important for the correct interpretation of experimental results.

On the other hand, the problem of molecular mechanisms of interface reactions and corresponding concentration kinetics has been a subject of an unequal number of theoretical and experimental publications. Some of the most recent were stimulated by Prigogine's ideas of self-organization (Nicolis and Prigogine 1977). From the general point of view this is a very promising direction in kinetics and it would be useful to estimate how adequate these ideas are to real solid state chemical processes.

We would define the purpose of this chapter as the examination of possible approaches to the mechanism and kinetics of interface propagation for topochemical reactions. It will not be a review or an overview, but rather a critical analysis of new tendencies in this field of the reactivity of solids.

2 The legend of interface reaction

The basis of heterogeneous kinetics is the assumption of the proportionality of the total reaction rate to the surface area of interface between the solid reagent and the solid product of reaction. This assumption originates from the work of Langmuir (1916) on the thermal decomposition of $CaCO_3$. In his paper he wrote: 'In order that $CaCO_3$ may dissociate and form a phase of CaO (instead of solid solution) it is *necessary that reaction shall occur only at the boundary between two phases.*' This immediately allows that:

$$\frac{d\alpha}{dt} = KS(\alpha) \qquad (6.1)$$

where K is specific rate and $S(\alpha)$ is the surface area of the boundary mentioned by Langmuir.

Later, the combination of this postulate with the concepts of crystallization (or condensation) theory gave rise to a formal kinetic approach, the main purpose of which was to predict the form of function $S(\alpha)$. Equations of this type are well known from the literature (see, for example, Barret 1973).

But, our interest is in what we call the *velocity of interface propagation*, which is also assumed to be constant in all kinetic models where the reaction is controlled by interface processes. In order to understand the connection between this kinetic parameter and the specific reaction rate introduced by Equation (6.1), let us start from the most general admission. This may be written in the form:

$$-\frac{dm}{dt} = km^* \tag{6.2}$$

where m is the mass of reagent in the course of reaction and m^* is the part of reagent that is capable of reacting at the moment of time, t.

Equation (6.2) reflects the fact that for heterogeneous reaction only a small amount of molecules or ions belongs to the reaction zone, which may be surface, or boundary or in some particular cases interface. Further transformations of Equation (6.2) are possible if we postulate that the reaction zone that contains m^* is uniform and has an average effective thickness δ. Then:

$$-\frac{dm}{dt} = k\rho V = k\rho\delta S \tag{6.3}$$

where ρ is the density of reagent (at this stage we do not specify what density), and S is the same as in Equation (6.1). V is evidently the volume of reaction zone ($V = \delta S$).

We can also express the overall reaction rate as a flux of reagent through the surface S using the velocity of propagation ω of this interface:

$$-dm/dt = \rho\omega S \tag{6.4}$$

Comparison of Equations (6.3) and (6.4) gives:

$$\omega = k\delta \tag{6.5}$$

This relationship between ω and k, introduced in Equation (6.2), is associated with the specific rate from Equation (6.1), because:

$$d\alpha/dt = (-1/m_0)dm/dt \tag{6.6}$$

and from Equations (6.4) and (6.1) it follows:

$$K = \rho\omega/m_0 \tag{6.7}$$

Now we are ready to make some conclusions. First of all, if the reaction is fully localized on the boundary between the reagent and the product (according to Langmuir's assumption) one should take:

$$\delta = d \tag{6.8}$$

where d is the interatomic distance in the direction of propagation in the reagent structure. On this condition one obtains from Equation (6.5):

$$\omega = kd \tag{6.9}$$

which is equivalent to the well-known equation of Polanyi and Wigner (1928) if we take into account that:

$$k = k_0 \exp(-E/RT) \tag{6.10}$$

which differs from Polanyi and Wigner's original equation by the factor $2E/RT$.

Furthermore, from Equation (6.8) it follows that the specific reaction rate K may be expressed as:

$$K = \rho\omega/m_0 = k\rho d/m_0 = kC_s \tag{6.11}$$

where $C_s = \rho d$ is the surface concentration of reagent. Equations (6.9) and (6.11) are very important because they show directly that Langmuir's postulate on reaction boundary localization, derived from the phase rule, as well as the Polanyi–Wigner equation and quite simple formulation in Equation (6.11) for surface reaction are *absolutely equivalent* if we only put $\delta = d$. This is an implicit assumption which is contained in any geometrically based kinetic model.

Strictly speaking, the statement of Langmuir cited above restricts the formation of a solid solution of CaO in $CaCO_3$ at equilibrium conditions because the system is monovariant. But, the conclusion on the *proportionality* of reaction rate to the surface area of interface was substituted by the statement that reaction is *localized* on this surface (boundary) which is not, in general, evident. This can be demonstrated by the simple comparison of two situations when $\delta = d$ and $\delta \gg d$, as shown in Fig. 6.1. In both cases the overall reaction rate is proportional to the surface area, whereas concentration profiles are completely different. There is no need to discuss the mechanism of reaction in Fig. 6.1a when $\delta = d$ because all chemical changes are 'compressed' within the layer of thickness d (interatomic). On the other hand, the case $\delta \gg d$ requires concentration

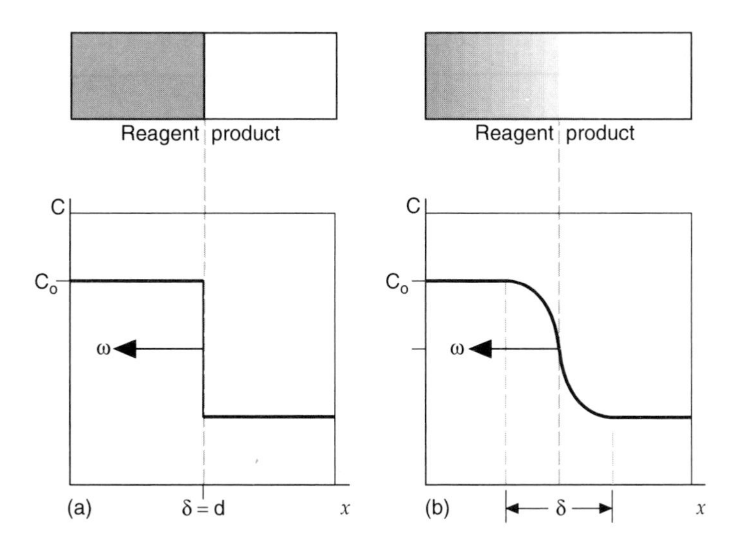

Figure 6.1 Schematic representation of interface zone for the Polanyi–Wigner model when $\delta = d$ (1) and in a more general case when $\delta \gg d$ (b). Graphics show the corresponding $C(x)$ profiles propagating with constant velocity ω in the direction of reagent.

kinetic description, and the profile $C(x)$ determines (or is a result of) the kinetic mechanism of reaction. Nevertheless, both models of Fig. 6.1 are compatible with constant velocity of the concentration front.

From Fig. 6.1a and Equation (6.11) it can be seen easily that the classical approach to the propagation of interface corresponds to evaporation or sublimation of solids. Shultz and Dekker (1955) have analysed all possible formulations of evaporation rate and have finally shown that any of them can be reduced to the equation of Knudsen (Hirth and Pound 1963) for linear velocity of surface displacement in the form:

$$\omega = \frac{\kappa}{\rho}\left(\frac{M}{2\pi RT}\right)^{1/2} P_e \qquad (6.12)$$

where M is molecular mass of evaporating crystal and P_e is equilibrium pressure at the temperature T of experiment. Here ω corresponds to vacuum conditions and κ is a coefficient which should reduce the theoretical velocity to an observable one.

Evidently, when κ is temperature dependent all these expressions are practically the same as Equation (6.9) by Polanyi and Wigner. The only difference may appear in activation energy when κ is constant. For this case Equation (6.12) predicts the activation energy to be equal to the enthalpy of sublimation.

A similar approach to polymorphic phase transitions was used by Bradley (1956), who proposed the formula:

$$\omega = \frac{kT}{h}\delta e^{\Delta S/R} e^{-E/RT}(1 - e^{\Delta G/RT}) \qquad (6.13)$$

which is an accurate expression derived from the theory of absolute rates, and is again exactly the same as Equation (6.8) if $\delta = d$. (Note that usually $\exp(\Delta G/RT) \ll 1$.)

How valuable are these theories? To answer we have to note that all these interface models are rather artificial when applied to solid state reactions, for example to the decomposition of crystals. There is nothing about solid product formation nor about possible ways of gas elimination. The interface between the two solids itself is assumed to be exactly the same as the free surface of evaporating crystal, which is not realistic.

Another argument against this simple approach is a comparison of the activation energies of decomposition with the enthalpies of the same reactions, i.e. a comparison of $\omega(T)$ and $P_e(T)$ dependencies. Of course, such a comparison is also very problematic because of incorrect data processing used for the evaluation of activation energy. But, the values obtained, with all necessary precautions taken (Lyakhov 1990), show that there is no coincidence between enthalpies and activation energies if k is taken to be independent of temperature (Table 6.1).

This brief analysis was necessary in order to show that a positive way to understanding better the mechanism of interface reactions must include the data on the structure of reaction zone.

3 What do we know about the structure of the reaction zone?

It should be noted that the notion of the reaction zone itself is not adopted in solid state kinetics yet. This does not mean that the problem of structural rearrangement has never been tackled. Since 1932 scientists have tried from time to time to propose the sequence of transformations of intermediate solid substance to the final product

Table 6.1 Comparison between activation energies, E, and enthalpies, ΔH, for dehydration of some crystal hydrates

Crystal hydrate	E (kcal mol^{-1})	ΔH (kcal mol^{-1})	References
$CuSO_4.5H_2O$	17.8	12.4	Smith and Topley (1931)
$Li_2SO_4.H_2O$	20.8	14.4	Bach *et al.* (1964)
$K_2C_2O_4.H_2O$	17.8	13.9	Guarini *et al.* (1974)
$K_4Fe(CN)_6.3H_2O$	12.9	11.7	Malcolm *et al.* (1973)
$KNaC_4H_4O_6.4H_2O$	20.9	13.9	Lowry and Morgan (1924)
$CaSO_4.2H_2O$	20.3	14.1	Heide (1969)

Note: references are concerned with the enthalpies ΔH. For the activation energies see Lyakhov (1988) and references therein.

(Topley 1932; Colvin 1938; Macdonald 1938; Searcy and Beruto 1976; Galwey *et al.* 1981). We cannot discuss here all the details of these concepts, but it is important to underline that phase transformations were always considered as secondary with respect to chemical changes. The authors of these models never questioned the basic postulate and assumed the reaction to be strongly localized on the interface boundaries. Nevertheless, they did understand quite well the difficulties in explanation of the mechanism of direct formation of new solid phase from the parent one, i.e. why intermediates such as a gas-like product, an X-ray amorphous product or else a quasi-zeolite product were proposed (and sometimes observed) to explain the absence of correlations between the structures of reagent and product.

Our approach is based partly on our own observations of the intermediate structure in the X-shaped nuclei of dehydration of $CuSO_4.5H_2O$ (Zagray *et al.* 1979), and on the other hand, on the works of Mutin *et al.* (1979) where they followed the continuous phase reorganization in the course of dehydration of $CaH_2(C_2O_4)_2.2H_2O$ (quasi-zeolite reaction). The authors called these intermediate states of the initial substance, with lower than a two water molecules concentration, the *lacunary phase*. In our work (Zagray *et al.* 1979) we have introduced the notion of *vacancy structure* which is, in our opinion, more adequate for the experimentally observed phenomena. In order to better understand this concept, let us consider the simplest reaction of decomposition of the type:

$$(AB)_s \rightarrow A_s + B_g \quad (6.14)$$

where the indices (s) and (g) correspond to the solid and gas phases, respectively. The most important and perhaps the most difficult question concerning the mechanism of this reaction is how to couple (or to join) in time and in some point of crystal the processes of gas product elimination and formation of solid product.

We have proposed a model which would permit us to solve, partially, this problem. This model is based on the very simple assumption that any B molecules escaping from the crystal at the very beginning of reaction form the vacancies in the surface layer of the reactant:

$$(B_s)_{AB} \rightarrow B_s + (V_B)_{AB} \quad (6.15)$$

Here, the index (AB) shows the position of B molecules or V_B vacancies in the corresponding positions of the original structure of reagent $(AB)_s$. Evidently, the further progress of the reaction in Equation (6.15) is possible only if vacancies from the surface layer can diffuse into the bulk of crystal. This is a very important assumption because the diffusion of vacancies may result in the formation of vacancy enriched $(AB)_s$ structure. This situation for the first steps of decomposition reaction is shown schematically in Fig. 6.2. There is no doubt that if mobility of vacancies is low enough reaction will stop after passing few atomic layers, without destroying the $(A_s)_{AB}$ residue of crystal (Fig. 6.2a). This is what we call *vacancy structure* because, for mobile vacancies, when the reaction can proceed, one can not indicate the strong position of interface and for this case definition of phase is not appropriate (Fig. 6.2b).

The reactions of the type in Equation (6.14) seem to be oversimplified, but the formation of the vacancy structures is a more general phenomena. This is an essential element of the reaction mechanism for thermal decomposition of crystal hydrates, hydroxides, oxides, azides and perhaps for nitrates and some other complex compounds. The reduction of metal oxides probably proceeds through vacancy structures which give rise to formation of metal phase. It is of importance to attract attention to the more or less general idea that the solid product cannot appear directly from the reagent.

The limiting concentration of vacancies in $(AB)_s$ will depend on the stability of vacancy structure as defined above. Generally speaking, one can imagine the vacancy structure with 100% of V_B vacancies instead of B_s molecules. In this case, the metastable product $(A_s)_{AB}$ will form a zeolite-like structure, whereas the thermodynamically stable final product will be quite different. But, more often the elimination of B molecules leads to the increase of internal stresses and, finally, to their relaxation by, for example, cracking. This phenomenon is very often observed microscopically. One example is shown in Fig. 6.3. The more or less regular distribution of cracks is evidently the result of minimization of elastic energy.

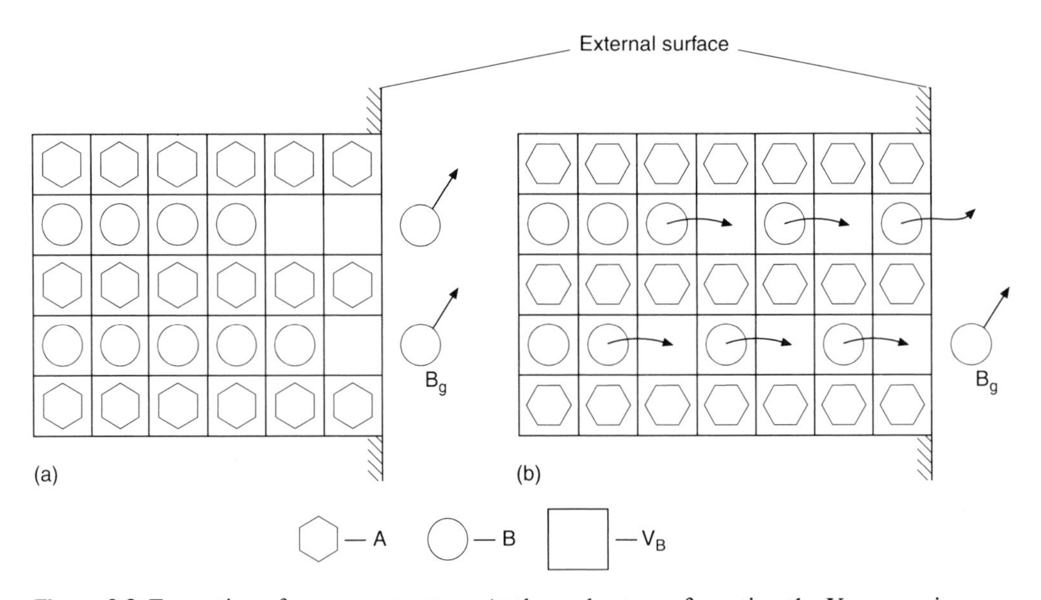

Figure 6.2 Formation of vacancy structure. At the early stage of reaction the V_B vacancies appear (a) in the surface layer of reacting crystal which move into the bulk by a vacancy diffusion mechanism (b).

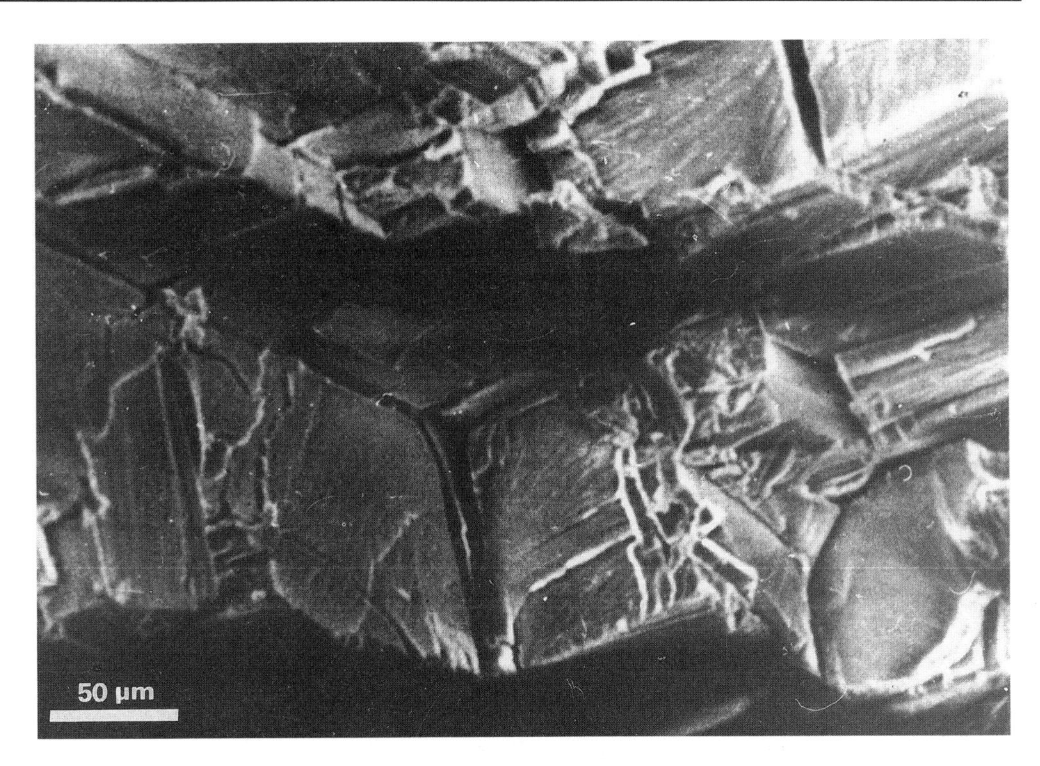

Figure 6.3 The cleaved surface of the reaction product for the dehydration of $CsAl(SO_4).12H_2O$ showing the big domains separated by cracks.

It is important to underline that reactions of the type in Equation (6.14) occur with significant change in molar volume, sometimes, as in the case presented in Fig. 6.3, up to 50% of the volume of the reagent. Strictly speaking, free volume is one of the reaction products. From this point of view, the volume of cracks can be estimated as only a few per cent. This means that the blocks obtained as a result of cracking (primary blocks) are porous, in turn, and this porosity is also the product of chemical and structural transformations.

Figure 6.4 represents schematically the mechanism of steady state reaction front formation. This picture is valid exactly for decomposition of layered compounds. In more complex systems the distribution of internal stresses would provide cracks in all directions, depending on the mechanical properties of a reagent. In Fig. 6.4 the reaction proceeds through diffusion of B_s in the bulk of continuous blocks (for simplicity) of vacancy structure and through desorption and elimination of gas molecules into the free volume of cracks. Since this mechanism includes the formation of a new surface which is the diffusion run-off, the positive feedback loop appears, which maintains the self-regulation of the average flux of the gas product. Note that here stress relaxation is one of the substantial elements of kinetic mechanism, as is now acknowledged by many authors (Dollimore 1992; Galwey 1992).

At first glance the size of primary solid product particles depends exclusively on the mechanical properties of the reagent and, more precisely, on the rate and mechanism of relaxation of internal stresses produced by vacancies within the reaction zone. But, it is not always evident.

Let us examine again the very beginning of the reaction in more detail. Figure 6.5

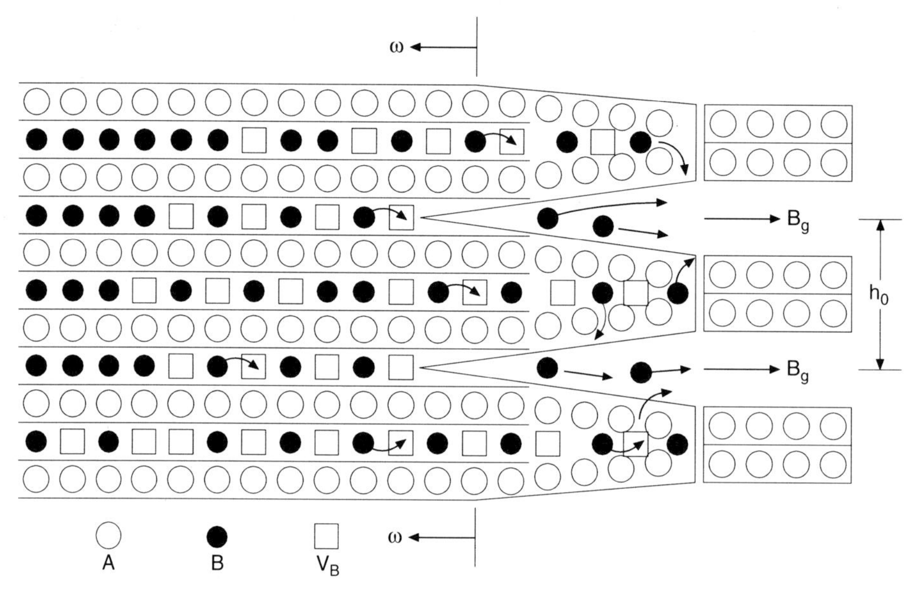

Figure 6.4 The steady state reaction front connected with cracks propagation and diffusion of B atoms within the vacancy structure (h_0 is the average distance between cracks).

represents the formation of concentration profile $C_B(x)$ as a result of diffusion of B from the parent crystal. Three profiles belong to a different reaction time $t_1 < t_2 < t_3$. Suppose that the vacancy enriched near surface layer is stable up to the limiting value of concentration C_B^* (critical value). It is obvious that even for the early moments of reaction one can find a zone where $C_B < C_B^*$, but in a very thin layer. This layer will be stable until the critical thickness is reached, because for a thin layer as for a small particle the necessary level of stresses cannot be attained. This means that to start the destruction of a crystal by cracking, two critical conditions should be satisfied simultaneously: (i) critical concentration C_B^*; and (ii) critical thickness Δx^* of a vacancy

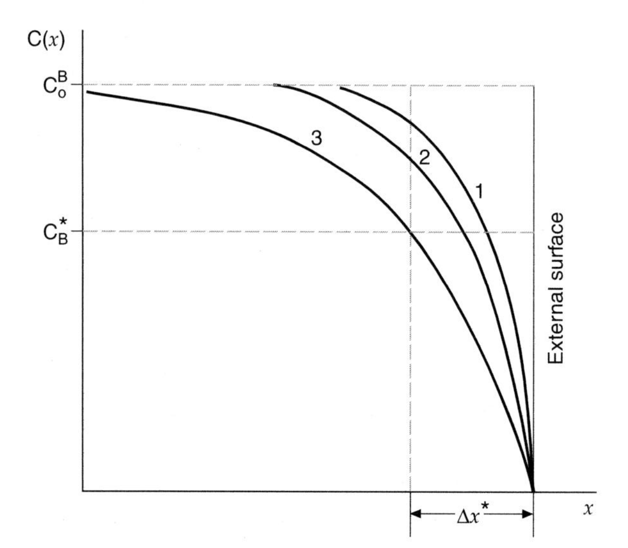

Figure 6.5 Concentration profiles for the diffusion removing of B atoms from $(AB)_S$ for different times $t_1 < t_2 < t_3$ (curves 1, 2, 3, respectively) Here, C_B^* is critical concentration and Δx^* is critical thickness of vacancy structure when it becomes unstable.

structure. This situation is shown in Fig. 6.5 by the concentration profile 3. Up to this moment the concentration of B molecules and that of V_B vacancies are connected by a simple balance equation:

$$C_B = C_B^0 - C_V \tag{6.16}$$

where C_B^0 is the concentration of B in $(AB)_S$ far from the reaction zone.

After destruction of the parent crystal in its near surface layer to many relatively small domains with average size close to Δx^* (see Fig. 6.4), further transformation of these domains containing, as yet, a significant amount of B molecules can proceed in two different ways.

The first one may be cooperative, which is known in literature as topotactic transformation. The second is diffusion reorganization of a vacancy structure to the final product. This separation is rather conventional because the reaction itself can only proceed thanks to vacancy diffusion of B from the bulk of each domain to its surface.

We will consider here both of these mechanisms from the viewpoint of kinetics of interface movement. But, first we reproduce experimental results that give some evidences of the formulated mechanism for steady state propagation of reaction front.

1 The data obtained for thermal dehydration of $CuSO_4.5H_2O$ using X-ray synchrotrone radiation technique with very high space and time resolution (Boldyrev *et al.* 1987) showed the reaction zone was really about 100 μm in depth. This conclusion was made directly from the dependencies of diffraction intensities for reagent on the distance to the microscopically observed interface. The existence of internal stresses within the reaction zone has been proved by the shift of peak positions and their broadening. Both effects are shown in Fig. 6.6.

2 As was mentioned, the model of reaction front represented in Fig. 6.4 is most suitable for layered compounds. This is the case for gypsum, $CaSO_4.2H_2O$. We have studied it by the modified synchrotrone diffraction method when a two-coordinate detector was used to obtain Laue diffraction patterns. The typical picture is shown in Fig. 6.7a. It is seen again that, in the region close to the observable boundary between the reagent and the product, the remarkable distortions of crystalline layers provoke very large elongations of diffraction patterns. The corresponding disordering is presented in Fig. 6.7b (Gaponov *et al.* 1989).

3 Decomposition of $Mg(OH)_2$. This reaction was studied by many authors. Anderson and Horlock (1962) determined the degree of transformation as the volume of the undestroyed part of a single crystal of $Mg(OH)_2$. We have tried to represent in graphic form a comparison of these quantities with those obtained from weight loss measurements. Figure 6.8 shows that the leading front of the reaction connected with cracks propagation is at least 40% ahead of weight loss. The size of primary particles was found within the range of 0.5–1.0 μm, whereas the final product MgO particles were about 0.02 μm in diameter. Moreover, the structures of these particles were different: hexagonal (as for $Mg(OH)_2$) for the primary domains and cubic for final MgO, which is in agreement with our model of decomposition.

Other experimental results, speaking for the vacancy structures role in the mechanism of decomposition of solids, are known, amongst which we may mention the partial or even total reversibility of dehydration followed by electron spin resonance (ESR) spectroscopy on the level of reaction zone (Zyryanov *et al.* 1982) or, repeated

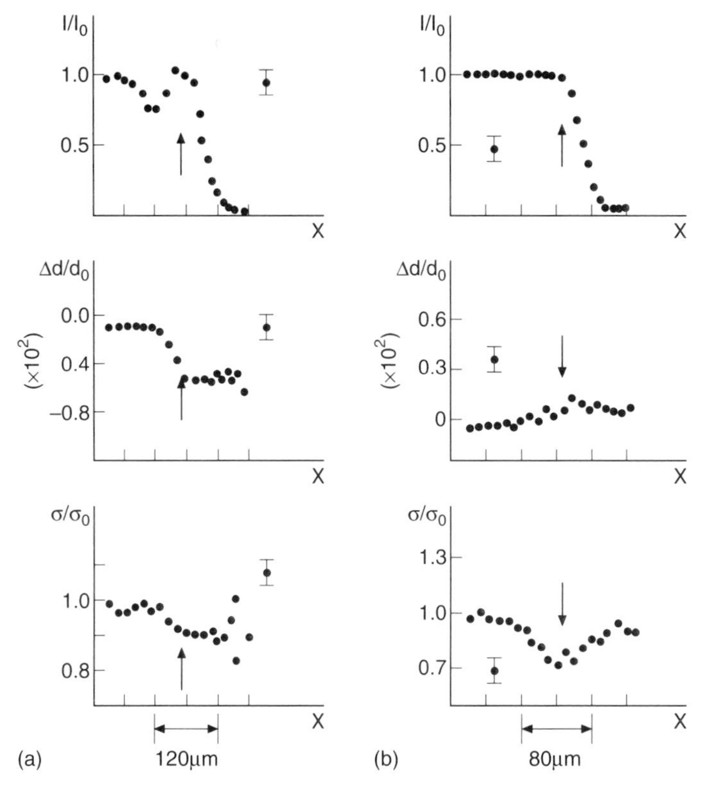

Figure 6.6 Synchrotrone radiation data for the reacting zone scanned along the distance from the observable interface (marked with arrows). The relative intensities I/I_0, change in peak position $\Delta d/d_0$ and broadening σ/σ_0 are shown for the (130) reflection of $CuSo_4.5H_2O$ (a) and for the (002) reflection of $Li_2SO_4.H_2O$ (b). Reagent is always the left side of interface.

many times, reproducible dehydration of single crystals of $Li_2SO_4.H_2O$ (Simakova *et al.* 1995).

Now we would like to discuss the two limiting cases for the structural reorganization mentioned above. For a cooperative mechanism the diffusion elimination of B molecules from the primary domains, formed as a result of the destruction of vacancy structure, may transform *as a whole* to crystalline particles of the final product. Most probably, this transformation will be of topotactic nature because small particles of less than critical size cannot crack further. This idealized mechanism is represented in Fig. 6.9. The average size of the final product particles would be directly connected to the size of the primary blocks through the difference in molar volumes of the reagent and product.

The opposite limiting case corresponds to a relatively high mobility of vacancies. This makes possible the diffusion reorganization of primary domains to final solid product. The diffusion of vacancies results in their coagulation into voids and pores in the bulk of each particle. This process is similar to the decomposition of supersaturated solid solutions with one difference—in our case we observe the segregation of vacancies and elements A_S of a vacancy structure. The result of this process is a solid product that has no crystallographic orientation with respect to the initial phase and which is very porous with an irregular porous structure and is very often X-ray amorphous. The solid products of this type are able to recrystallize spontaneously (Frost *et al.* 1951). The

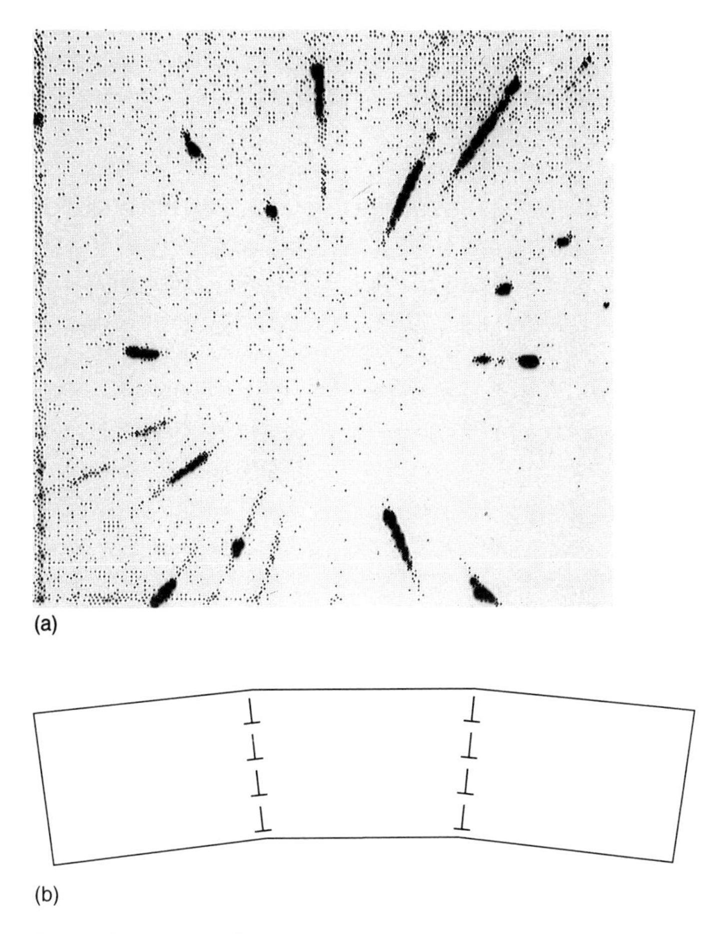

(a)

(b)

Figure 6.7 Laue diffraction picture for $CaSO_4.2H_2O$ in the course of dehydration. The distance from the observable interface is 0.01 mm (a). The corresponding disordering of the crystal layers by polygonization is demonstrated in (b).

examples are presented in Fig. 6.10 for the product of dehydration of $Li_2SO_4.H_2O$ and $Ba(NO_2)_2.2H_2O$. The latter case gives an example of the very strange self-organization of the product.

It should be noted that, probably, the majority of real cases are intermediate between these two limit mechanisms. Formation of the network of cracks may be followed by diffusion reorganization of domains so that even for a regular system of cracks, one observes no crystallographic orientation of product as is the case for $CsAl(SO_4)_2.12H_2O$ (see Fig. 6.2). This means that the formation of blocks itself cannot be taken as the criterion of topotactic reaction. Moreover, since diffusion processes are thermally activated and their rate increases with temperature, the relationship between these two mechanisms may also change with temperature.

The mechanism proposed here permits one to understand the unusual phenomenon when the single crystalline particles of a size less then 30–50 μm react without cracking, whereas the relatively large crystals give the small blocks, usually less than 5 μm (Mutin and Watelle 1979; Shakhtshneider *et al.* 1981). This effect can be explained easily by the relaxation of stresses, not only by cracking but also through diffusion of defects to the external surface of a sample. In such a case the internal stresses are insufficient to destroy the small crystals.

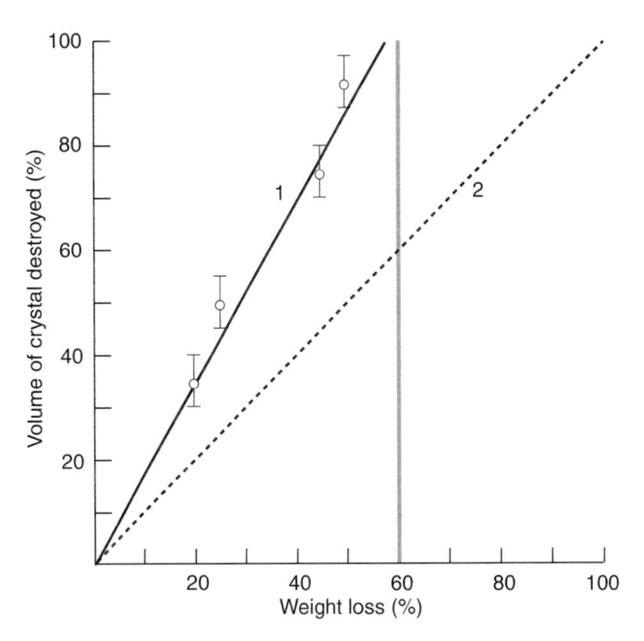

Figure 6.8 The relationship between the volume of destroyed crystal and weight loss for the dehydration of $Mg(OH)_2$: (1), experimental data; (2), theoretical dependence. (From the data of Anderson and Horlock 1962.)

For this reason one should be very careful with comparing the data obtained by direct observations of decomposition by transmission electron microscopy with those obtained for large samples. Because of the extremely small size of samples used for electron microscopy (usual thickness less than 0.1 μm) all transformations would be rather topotactic even though voids formation takes place (see, for example, Fievet and Figlarz 1975; Watari *et al.* 1979).

The model discussed here shows definitely that the kinetic approach based on the Langmuir–Knudsen–Polanyi–Wigner postulates is not satisfactory, because it is not adequate to the real structure of the reaction zone (which is a zone after all, not an interface), or to the probable mechanism of its propagation. This is why new kinetic approaches are necessary, which would take into account morphological features of the reaction zone and the fact that stress relaxation is a new kinetic parameter. Of course, it would not be easy to involve all these interlinked phenomena in one kinetic model, but some positive steps could be taken. This is the subject of the next section.

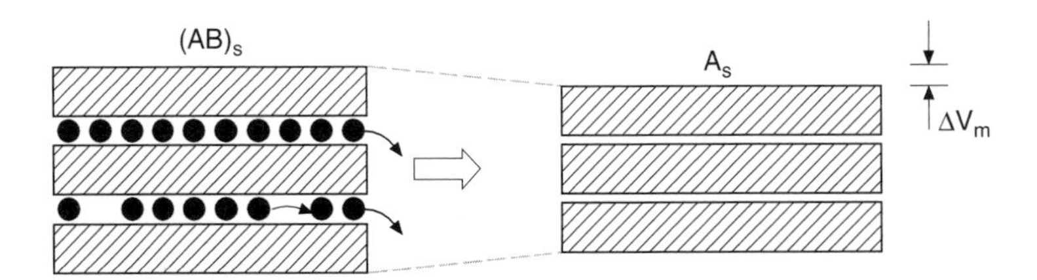

Figure 6.9 Schematic representation for topotactic transformation of layered crystal $(AB)_S$ to A_S through vacancy diffusion of B atoms (●). ΔV is a change of molar volume for this reaction.

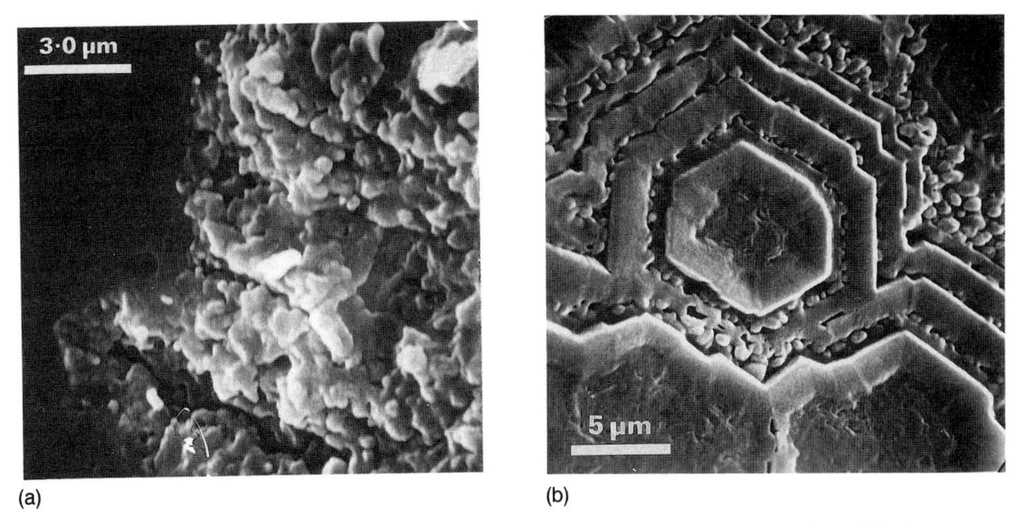

Figure 6.10 Examples of product formation by the mixed mechanism including diffusion reorganization of a vacancy structure; (a) $Li_2SO_4.H_2O$ and (b) $Ba(NO_2)_2.2H_2O$.

4 The kinetic–diffusion models for topochemical reactions

As we can see from Section 3, the process of propagation of the reaction front, at least for decomposition reactions, includes many subprocesses concentrated within the reaction zone: elimination (most probably desorption) of the molecules of the gas product into intergranular space, reciprocal diffusion of B molecules and vacancies in the vacancy structure, formation of a new surface by cracking or through vacancy coagulation, relaxation of internal stresses by different mechanisms. For the steady state process (for the constant velocity of propagation) none of these processes can be taken for the rate-determining step. This means that the problem cannot be reduced to the study of one well-determined reaction step.

It is evident that the situation with propagation mechanism and its qualitative description is, from a general point of view, a macrokinetical one, especially if we remember that all components of this process are temperature dependent and endo- or exothermic.

From the standpoint of macrokinetics, the reaction zone of topochemical reaction is an open system. This follows from Fig. 6.1b, where the reaction zone may be considered as a chemical reactor with constant feed flow on the inlet equal to $\rho\omega$ per unit of surface area, as is written in Equation (6.4). The problem is that we do not know anything distinct about the rates of chemical reactions and other *elementary* processes occurring in this reactor. From the analysis of $\alpha(t)$ curves one may obtain, at best, the same velocity of propagation of the reaction front as whole. This is a unique parameter of comparison with predicted (theoretical) values, if necessary, because even to evaluate the frequency factor from the Arrhenius plots, $\omega(T)$, we need to assume that $\delta = d$, as was shown in Section 1.

Taking into account these preliminary remarks we may select one of two positive ways of developing the kinetics of topochemical reactions. If we are interested in correct $\alpha(t)$ description of a process we have to pay attention to the different kinetic behaviour of the powder systems when the particle size of the reagent is much higher or comparable with the depth of reaction zone. Time–space relations are such that for big particles, having radius $R \gg \delta$, the reaction zone may be taken for the interface separating two regions where $\alpha = 1$ and $\alpha = 0$, respectively. This is the case with geometrical kinetics.

In the opposite case, when $R \cong \delta$, the concentration front will (perhaps) propagate through particle scanning, in time, all phases of chemical and/or structural transformations in succession. Of course, in this case a purely geometrical approach would be inconsistent, because the assumption of ω = constant (steady state condition) is not valid any more.

The other way leads to a macrokinetical concentration description, based on kinetic models of the processes within the reaction zone, and taking into account all but the most specific stages. In this case we can hope to obtain the rate of propagation ω as the solution of a differential equation (or of a system of equations) expressed as a function of some more or less realistic constants. Even though the values of these constants are unknown or cannot be determined from the experiment directly, the analytical results or computer simulation very often give the key to understanding the mechanism of reaction.

To realize this approach we first have to pass from the macroscopic basic Equation (6.1) to the *microscopic* kinetic equation which may be written for one component system in its most general form as:

$$\mathrm{d}C/\mathrm{d}t = F(C) \tag{6.17}$$

where $C = C(\vec{r},t)$ within the reaction zone. For simplificiation we can assume $r = x$ because, for the size of nuclei greater then two or three thicknesses of the reaction zone, the curvature of interface is slow and it can be taken as a plane in a chosen crystallographic direction: $C = C(x,t)$.

As we are sure that diffusion plays a significant role in our model we have to add the corresponding term to the right-hand side of Equation (6.17). One obtains:

$$\frac{\partial C}{\partial x} = F(C) + D\frac{\partial^2 C}{\partial x^2}$$
$$C = C(x,t) \tag{6.18}$$

where D is the coefficient of diffusion. We do not specify at this stage the form of $F(C)$, but if $F(C)$ is a non-linear function, Equation (6.18) falls immediately into the special class of kinetic–diffusion equations well known in chemistry, biology, ecology, etc. We shall see what connection this type of equation may have with the mechanism of interface propagation discussed in the previous section. But, now we would like to recall some important results concerned with self-organization in non-linear chemical systems.

The solution of Equation (6.18) (or of a system of similar equations) with different boundary conditions, including the case when $D = 0$, gives, in general, three types of self-organization (dissipative structures): temporal structures (oscillations), spatial structures (concentration waves and patterns) and isolated concentration fronts (see, for example, Murray 1977; Nicolis and Prigogine 1977). For us the most interesting solutions are those in the form of propagating fronts which may be consistent with that shown in Fig. 6.1b, for some conditions. It was shown by many original works

(Ortoleva and Ross 1974, 1975; Othmer 1975; Hadeler and Rothe 1975) that diffusion plays a decisive role in the stabilization of such fronts. For the model equation of Fisher (Murray 1977), which is based on a simple autocatalytic form of $F(C)$:

$$\frac{\partial C}{\partial t} = kC(1 - C) + D\frac{\partial^2 C}{\partial x^2} \tag{6.19}$$

and the concentration fronts propagate with constant velocity:

$$\omega \geqslant 2\sqrt{kD} \tag{6.20}$$

where the exact equality corresponds to the most realistic initial conditions (Kolmogorov *et al.* 1937). This conclusion is valid practically for any reasonable $F(C)$, the only difference being that ω will be expressed by (Murray 1977):

$$\omega = 2\sqrt{F'(0)D} \tag{6.21}$$

The typical example of a concentration front is presented in Fig. 6.1b. The main characteristic of such a front is that any point on this curve moves with the same velocity ω.

The same general approach, but to the bounded regions, may lead at some conditions to the solutions known as pattern formation. This is also the subject of special interest for topochemistry because the analysis shows the intimate role of diffusion in reacting systems. In an initially uniform medium any small spatial deviations of concentration leads to the formation of patterns that are similar to growing nuclei in topochemical reactions. This phenomenon is known as *diffusion instability*. An excellent example of numerical modelling of this process can be found in the paper of Kernevez *et al.* (1979).

The generality of formulation of the diffusion–kinetic problem compels one to look at the phenomena of reaction fronts from more general positions. If we take into account that non-linearity is practically identical to the existence in the reacting system of a positive feedback, then all qualitative distinctions between different physicochemical systems will be negligible.

For example, propagating waves were observed for the Belousov–Zabotinsky reaction in solution (Smoes 1978). Their behaviour is exactly the same as for reaction fronts in topochemical reactions. They propagate with constant velocity. Estimations based on experimental data (Murray 1977) give the thickness of reaction front $\delta \cong 10^{-3}$ cm for the usual values of $D \cong 10^{-5}\ \text{cm}^2\ \text{s}^{-1}$, which is comparable with our observations (see Fig. 6.6).

Now we return to our problem. The mechanism of positive feedback for solid state reactions may differ from those in autocatalytical gas phase or liquid systems (Boldyreva 1990). As we have seen in Section 3, in our case the feedback loop appears when the critical concentration of vacancies in a vacancy structure is achieved. Vacancies produce internal stresses that relax through crack propagation. As a result a new surface appears which acts as an additional diffusion source (for vacancies) accelerating the overall process of gas elimination. But, this in turn produces more vacancies, more stresses, etc. Of course, for an exact formulation of the system of equations one needs to know, at least, the mechanism of stress relaxation, the dependence of diffusion or kinetic parameters on these stresses, how stresses depend on concentration, etc. But, in order to show the consistency of the model one may take into account only the final result of this feedback mechanism, which is the number of cracks per unit of interface area. The corresponding equation may be written as (Goldberg and Lyakhov 1985):

$$\omega \frac{\partial C}{\partial x} = \frac{\partial}{\partial x} D(C) \frac{\partial C}{\partial x} + \frac{2j}{h_0 C_0^B} \tag{6.22}$$

where the relative concentration of vacancies C is taken in the form:

$$C(x,t) = C(x + \omega t) \tag{6.23}$$

and j is a flux of B molecules from the surface of blocks (Fig. 6.11). Here, h_0 is a morphological characteristic, which is the average distance between cracks, and C_0^B is the initial concentration of B in $(AB)_s$, so that (see Equation 6.16):

$$C = 1 - C_B/C_0^B,$$
$$C = C_v/C_0^B \tag{6.24}$$

In this work j was taken in the form of the first-order reaction:

$$J = kC_B = k(1 - C)C_0^B, \tag{6.25}$$

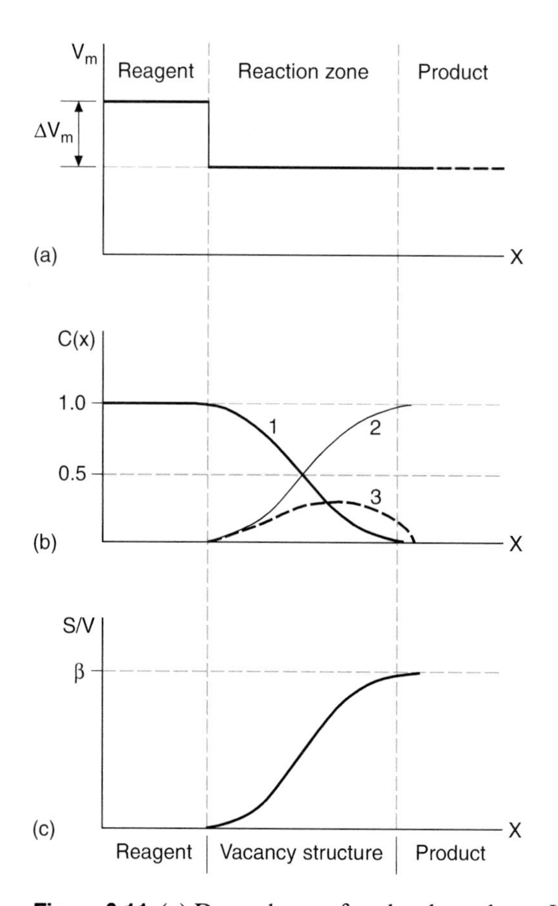

Figure 6.11 (a) Dependence of molecular volume V_m on distance for the model with cracks propagation. (b) The corresponding concentration profiles for B atoms (1) and V_B vacancies (2). Curve (3) represents the real case when vacancy coagulation produces pores and voids. (c) Probable dependence of S/V on distance within the reaction zone (see Equation (6.29)).

but with constant k depending on C due to the above mechanism of feedback: $k = k(C)$. Evidently, under these assumptions Equation (6.22) is slightly different from the Fisher equation and its solution is expressed as (Goldberg and Lyakhov 1985):

$$\omega = \gamma \times 2 \sqrt{h^{-1} \int_0^1 D(C)k(C)C\,dC} \qquad (6.26)$$

where γ depends on the form of $k(C)$. It is seen that Equation (6.26) is equivalent to Equation (6.20) if D is constant.

The same result may be obtained for the diffusion mechanism of reorganization of vacancy structure (Lyakhov 1988). The preceding model is unrealistic in that all changes of specific volume are abrupt and are in line with the necks of cracks (Fig. 6.11). Moreover, the relationship in Equation (6.24) is assumed to be valid until $C = 1$, which means that the product forms as a zeolite-like structure. In a real situation concentration of vacancies falls to zero at some distance from the reagent because the product does not contain the vacancies at all (Fig. 6.11b). The decrease of concentration occurs as a result of coagulation of the vacancies into voids and pores, which are also the additional diffusion source. This process offers another possibility of positive feedback through the formation of an internal surface. For this case we assume:

$$V\frac{\partial C_B}{\partial t} = jS \qquad (6.27)$$

where j is a flux of B from the surface S of all voids and pores in any volume V. If $J = kC_B$ one obtains:

$$dC_B dt = kC_B\,(S/V) \qquad (6.28)$$

where S/V itself is obviously the function of concentration. We accept the linear approximation for this dependence.

$$\frac{S}{V} = \beta\left(1 - \frac{C_B}{C_0^B}\right) \qquad (6.29)$$

which leads to the equation (if we include the diffusion):

$$\frac{\partial C_B}{\partial t} = (k\beta)C_B\left(1 - \frac{C_B}{C_0^B}\right) + D\frac{\partial^2 C_B}{\partial x^2} \qquad (6.30)$$

or, if we introduce dimensionless concentration $C = C_B/C_0^B$:

$$\frac{\partial C}{\partial t} = (k\beta)C(1 - C) + D\frac{\partial^2 C}{\partial x^2} \qquad (6.31)$$

This is again the Fisher equation, which allows us to write the propagation velocity of steady state wave front in the form of Equation (6.20):

$$\omega = 2\sqrt{k\beta D} \qquad (6.32)$$

It is easy to notice that β, which is a limiting value of 'shrinkage' of the reagent in the course of the reaction, here plays the same role as $1/h_0$ in the previous case.

Similar approaches were realized in other models and applications. First, we should mention the work of Goldberg and Kovalenko (1990) where a diffusion–kinetic model was proposed for the catalytic mechanism of propagation of reaction zone in a relatively thin layer of reagent. The system of equations obtained in this work is close to that of Lottka–Volterra with diffusion.

The velocity of the reaction front was found in the form:

$$\omega = 2\sqrt{k\vartheta_0 D} \tag{6.33}$$

where ϑ_0 is the initial concentration of reagent. This relationship is of the same type as in Equation (6.20).

In the work of Chaix and Bertrand (1982), the ideas of non-linearity were used for the explanation of a so-called stratification of the product of oxidation of some alloys. In their equations they introduced non-linear dependence of reaction rate on temperature (which is evident) due to thermal effects produced by reaction. The important point of this work is that experimentally measured rate is oscillating, the period of oscillations being connected with the relaxation of internal stresses. Another important experimental observation that follows directly from theory (Ortoleva and Ross 1975) concerns *multiple fronts* existing in one system. This phenomena was also observed for the Belousov–Zabotinsky reaction (Smoes 1978), for sulphidization of Fe–Mn–Al alloy (Larpin and Bertrand 1985) and for dehydration of $MgSO_4.7H_2O$ (Chupakhin *et al.* 1981). Two, or even more, types of fronts were observed in exactly the same conditions (at the same region of reaction constraints).

One of the bases of our model is the significant role of stress relaxation, which seems to be beyond doubt. This was shown in direct experiments for polymorphic phase transitions in NH_4Cl (Sidel'nikov *et al.* 1985) and for some reactions of isomerization (Boldyreva and Sidel'nikov 1987), and is a subject of intensive research (Boldyreva 1994). The problem is that we cannot propose an explicit dependence of reaction rate or diffusion coefficient on the level of these stresses. In a real system, even for one component and the one-dimensional case, the macrokinetical description will include a set of equations much more complex than was used in our models. First of all we have to take into account the dependence of kinetics on internal stresses σ:

$$\frac{\partial C}{\partial t} = F(C;\sigma) + \frac{\partial}{\partial x} D(C;\sigma) \frac{\partial C}{\partial x} \tag{6.34}$$

Then we need an equation connecting σ with concentration of defects in the general case (in our case with concentration of vacancies):

$$\sigma = \sigma(C;t) \tag{6.35}$$

The relaxation of stresses should also be taken into account.

$$\frac{\partial \sigma}{\partial t} = \phi(\sigma,t\ldots) \tag{6.36}$$

The system in Equations (6.34–6.36) is too complex to be analysed in a general form. That is why it would be better to solve different particular problems in order to understand some specific manifestations of the above mechanism. The recent publications (Knyaseva 1992, 1993, 1995) could serve as an illustration of the fruitfulness of this approach.

As for the near future, the most interesting problem is that of the configuration of the reaction zone in real crystalline structures. It seems that application of the method used for modelling the pattern formation in biological systems would help us to understand the origin of different shapes of nuclei in topochemical reactions. There must be a relatively new type of diffusion–kinetic equations where the diffusion coefficients will depend on direction in a crystalline solid. This will explain, at least qualitatively, why nuclei are sometimes isotropic in a low-symmetry unisotropic crystal structure.

5 Concluding remarks

The heterogeneous kinetics of solid state reactions are not exhausted by kinetics of topochemical reactions. Some quite different problems exist for solid–solid reactions, others for solid–liquid ones, etc. But, we cannot join all the aspects of this science in one chapter.

Nevertheless, we hope this rather brief introduction to the modern kinetics of topochemical decomposition reactions, based on a limited number of experiments, will be useful as a stimulating factor. Because any solid state reaction has some elements discussed above: changes of molar volume, different crystal structures of reagent and product, the necessity of forming a new solid phase from the initial one through a known transport mechanism, most probably, through diffusion, and finally self-initiation and relaxation of stresses. The point is that presently existing kinetic theories do not take into consideration all these phenomena, with the exception of some 'strange' cases such as stratification during oxidation, and the like.

The progress of the last few years in kinetics of solid state reactions is not as good as one would desire, perhaps, because experimental science and the theoretical bases are developing on different planes instead of helping one another.

6 References

Anderson, P.J. & Horlock, R.F. (1962) *Trans Farad Soc* **58**, 1993–2004.

Bach, R.O., Baardman, W.W. & Forsyth, M.W. (1964) *Chimia* **18**, 110–111.

Barret, P. (1973) *Cinetique Heterogene*. Gauthier-Villars, Paris.

Boldyrev, V., Gaponov, Yu., Lyakhov, N. *et al.* (1987) *Nucl Instr Meth* **A261**, 192–199.

Boldyreva, E. (1990) *React Solids* **8**, 269–282.

Boldyreva, E. (1994) *Mol Cryst Liq Cryst Inc Nonlin Optics* **242**, 17–52.

Boldyreva E. & Sidel'nikov, A. (1987) *Izv Sib Otd Akad Nauk USSR Ser Chim* **5**, 139–144.

Bradley, R.S. (1956) *J Phys Chem* **60**, 1347–1354.

Chaix, J.M. & Bertrand, G. (1982) *J Chim Phys* **79**, 791–798.

Chupakhin, A., Sidel'nikov, A., Boldyrev, V., Lyakhov, N. & Chertilina, L. (1981) *Izv Sib Otd Akad Nauk URSS Ser Chim* **5**, 3–12.

Chupakhin, A., Sidel'nikov, A. & Boldyrev, V. (1987) *React Solids* **3**, 1–19.

Colvin, J. (1938) *Trans Farad Soc* **34**, 983.

Delmon, B. (1969) *Introduction a la Cinetique Heterogene*. Edition Technip, Paris.

Dollimore, D. (1992) *J Thermal Anal* **38**, 99–110.

Fievet, F. & Figlarz, M. (1975) *J Catal* **39**, 350–356.

Frost, G., Moon, K. & Tompkins, E. (1951) *Canad J Chem* **29**, 604–632.

Galwey, A.K. (1992) *J Thermal Anal* **38**, 111–130.

Galwey, A.K., Spinicci, R. & Guarini, G. (1981) *Proc Roy Soc Lond* **A378**, 477–505.

Gaponov, Yu., Lyakhov, N., Tolochko, B., Boldyrev, V. & Sheromov, M. (1989) *Nucl Instr Meth* **A282**, 698–700.

Goldberg, E. & Kovalenko, Yu. (1990) *Solid State Ionics* **42**, 153–157.

Goldberg, E. Lyakhov, N. (1985) *Izv Sib Otd Akad Nauk USSR Ser Chim* **1**, 14–20.

Guarini, G., Spinicci, R. & Donatti, D. (1974) *Thermal Analysis, Proceedings of the 4th ICTA, Budapest* Vol. 1, pp. 185–186. Elsevier, Amsterdam.

Hadeler, K.P. & Rothe, F. (1975) *J Math Biol* **2**, 251–263.

Heide, K. (1969) *Siikattechnik* **20**, 232–234.

Hirth, J.P. & Pound, G.M. (1963) *Evaporation and Condensation.* Pergamon Press, Oxford.

Kernevez, J.P., Joly, G., Duban, M.C., Bunov, B. & Thomas, D. (1979) *J Math Biol* **7**, 41–56.

Knyaseva, A. (1992) *Phis Gorenya Vsryva* **28**, 13–18.

Knyaseva, A. (1993) *Phis Gorenya Vsryva* **29**, 3–13.

Knyaseva, A. & Dyukarev, E. (1995) *Phis Gorenya Vsryva* **31**, 27–46.

Kolmogorov, A., Petrovsky, I. & Piskunov, H. (1937) *Bull Moscow State Uni Ser Math Mech* **1**, 1–25.

Langmuir, I. (1916) *J Am Chem Soc* **38**, 2221–2295.

Larpin, J.-P. & Bertrand, G. (1985) *React Solids* **1**, 75–86.

Lowry, H.H. & Morgan, S.O. (1924) *J Am Chem Soc* **46**, 2192–2198.

Lyakhov, N. (1988) *Physical Chemistry. Contemporary Problems* (ed. Ya. M. Kolotyrkin), pp. 221–247. Chimiya, Moscow.

Lyakhov, N. (1990) *React Solids* **8**, 313–322.

Macdonald, J.Y. (1938) *Trans Farad Soc* **34**, 977–978.

Malcolm, I.R., Staveley, A.K. & Worswick, R.D. (1973) *J Chem Soc Farad Trans* **69**, 1532–1540.

Murray, J.D. (1977) *Lectures on Nonlinear-Differential-Equation. Models in Biology.* Clarendon Press, Oxford.

Mutin, J.-C. & Watelle, G. (1979) *J Solid State Chem* **28**, 1–12.

Mutin, J.-C., Watelle, G. & Dusausoy, Y. (1979) *J. Solid State Chem* **27**, 407–421.

Nicolis, G. & Prigogine, L. (1977) *Self-organization in Nonequilibrium Systems.* John Wiley & Sons, New York/London.

Ortoleva, P. & Ross, J. (1974) *J Chem Phys* **60**, 5090–5107.

Ortoleva, P. & Ross, J. (1975) *J Chem Phys* **63**, 3398–3408.

Othmer, H.G. (1975) *J Math Biol* **2**, 133–163.

Polanyi, M. & Wigner, E. (1928) *Z Phys* **A139**, 439–452.

Searcy, A.W. & Beruto D. (1976) *J Phys Chem* **80**, 425–429.

Shultz, R.D. & Dekker, A.O. (1955) *J Chem Phys* **23**, 2133–2138.

Shakhtshneider, T., Ivanov, E., Logvinenko, V. & Boldyrev, V. (1981) *Izv Sib Otd Akad Nauk USSR Ser Chim* **5**, 17–23.

Sidelnikov, A., Chupakhin, A. & Boldyrev, V. (1985) *Izv Sib Otd Akad Nauk USSR Ser Chim* **6**, 56–62.

Simakova, N., Lyakhov, N. & Rudina, N. (1995) *Thermochimica Acta*, **256** 381–389.

Smith, M.L. & Topley, B. (1931) *Proc Roy Soc* **A134**, 224–245.

Smoes, M.L. (1978) *Synergetics. Far From Equilibrium* (eds A. Pacault, C. Vidal), pp. 80–96. Springer-Verlag, Bordeaux, France.

Topley, B. (1932) *Proc Roy Soc* **136**, 413–428.

Watari, F., Van Landuyt, J., Delavignette, P. & Amelinckx, S. (1979) *J Solid State Chem* **29**, 137–150.

Zagray, A., Zyryanov, V., Lyakhov, N. & Boldyrev, V. (1978) *Dokl Akad Nauk USSR* **239**, 872–875.

Zagray, A., Zyryanov, V., Lyakhov, N., Chupakhin, A. & Boldyrev, V. (1979) *Thermochim Acta* **29**, 115–119.

Zyryanov, V., Lyakhov, N. & Boldyrev, V. (1982) *Dokl Akad Nauk USSR* **262**, 1416–1419.

7 Homogeneous Solid State Reactions

E.V. BOLDYREVA

Institute of Solid State Chemistry, Siberian Branch of the Russian Academy of Sciences, Kutateladze, 18 Novosibirsk-128, 630128, Russia

1 Introduction

Homogeneous solid state reactions seem to attract noticeably less attention than the heterogeneous ones. This is not surprising. Liquid or gaseous state reactions are, with rare exceptions, homogeneous. Therefore, the heterogeneity is often considered to be the main peculiarity of solid state reactions. Heterogeneous reactions seem to give a fascinating challenge of controlling not only the rate but also the spatial pattern of solid state processes. This is an attractive and inspiring fundamental problem, and it is no less important for practical applications. Hence, it is, to some extent, natural that so much attention was and is drawn to the problems related to the formation, evolution and propagation of interfaces during solid state reactions. These problems (both experimental and theoretical) are so complicated that, in the shadow of heterogeneous transformations, homogeneous solid state reactions are sometimes considered to be very simple, uninteresting and therefore not worthy of a special study. I shall try to show in this chapter that this is not so.

2 Some terminology

It is more or less generally accepted to label reactions in which there is phase separation as *heterogeneous*, and those where there is no such separation as *homogeneous* (see, for example, Atkins and Beran 1992). However, there is some terminological confusion in the publications on solid state reactions. Sometimes, a solid state reaction is referred to as a homogeneous one if the single crystal of the reactant is transformed to the single crystal of the product with similar dimensions and similar structural perfection (see, for example, Schklover and Timofeeva 1985). However, generally speaking, a homogeneous reaction is not necessarily a single crystal to single crystal reaction, which is not accompanied by the fragmentation of the starting crystal. Conversely, a single crystal to single crystal transformation may be heterogeneous. For example, solid state nitro–nitrito linkage isomerization in Co(III)–nitro–ammine complexes proceeds without phase separation, via the formation of solid solutions of the two isomeric forms (Grenthe and Nordin 1979a,b; Boldyreva *et al.* 1984, 1993). So, this is a homogeneous reaction in the commonly accepted meaning of the term. However, unless special precautions are taken, the crystals are fragmented violently during the isomerization (Boldyreva *et al.* 1984), so that the reaction can hardly be considered as a single crystal to single crystal transformation. An opposite example is the solid state polymerization of $NiBr_2[P(CH_2CH_2CN)_3]_2$, described by Cheng and Foxman (1977). The reaction is clearly heterogeneous, i.e. the reaction interface is formed and its propagation in the crystal is highly anisotropic. However, after the transformation is complete, the product seems to be a single crystal with only slightly deteriorated quality. The same is true for many other *topotactic* reactions (i.e. reactions in which the majority of the atomic positions in the original and transformed material are substantially the same and there is an accord in three dimensions between initial and final lattices (Bernal 1960)): many of them are heterogeneous, even if they proceed in a single crystal to single crystal mode. Martensitic phase transitions are also common examples of heterogeneous single crystal to single crystal solid state transformations.

In this chapter I shall use the term *homogeneous* in the generally accepted meaning, i.e. referring to the reactions proceeding without phase separation via the formation of solid solutions 'reactant–products' independently, whether the reaction is a single crystal to single crystal transformation or not.

3 Why are homogeneous solid state reactions studied?

Homogeneous solid state reactions are worthy of a detailed study due to several reasons.

1 They are obviously important for practical applications. Reversible single crystal to single crystal homogeneous transformations can be used in various devices, for example in photometers (Boldyreva *et al.* 1988), or for information recording, in particular for creating holographic grating (Novak *et al.* 1993). They also allow one to use solid state synthesis to get single crystals of desirable solid products, often in metastable state, which cannot be obtained in any other way (see, for example, Feitknecht 1964; Wegner and Fisher 1970; Kaiser and Wegner 1972; Kaiser *et al.* 1972a,b; Wegner 1977; Nakanishi *et al.* 1977a,b, 1980a–c, 1981; Hasegawa 1983; Bloor 1985; Cohen 1987; Ohashi *et al.* 1993; Ohgo and Arai 1993; Enkelmann *et al.* 1993, 1994; Novak *et al.* 1993, 1994; Gillot and Rousset 1994).

2 Homogeneous solid state reactions are interesting for fundamental solid state science.

First of all, if we want to know why most solid state reactions are heterogeneous, we should study in more detail those exceptional reactions in which *no* interfaces are observed, the product and the reactant forming solid solutions. Thus, a study of homogeneous reactions is helpful for a better understanding of the origin of heterogeneity.

A study of a homogeneous solid state reaction gives some challenges as compared with a study of heterogeneous processes.

It is no longer necessary to know the volume of the interface region for a correct kinetic study. Kinetic data, required to control the reaction and to get a better insight into the reaction mechanism, can be obtained in a much easier way and, more importantly, their interpretation is more unambiguous.

Structural and chemical changes in the crystal during a homogeneous reaction can be followed without special microprobe techniques. An X-ray analysis of the electron distribution in the starting reagent, in the final product and at the intermediate stages can provide information on the reaction's driving force and the atomic movements involved.

As a result, very detailed quantitative studies of homogeneous reactions become possible. One can make a noticeable progress in solving the following problems, central for understanding solid state reactivity:

- how does the crystalline environment *affect* a transformation at a reaction site?
- how does the crystalline environment *respond to* a transformation at a reaction site?

Practical applications of homogeneous solid state reactions were discussed in a number of reviews, which can be recommended to an interested reader (see, for example, Hirshfeld and Schmidt 1964; Kaiser and Wegner 1972; Cohen 1975, 1979, 1987; Baughman and Chance 1978; Baughman and Yee 1978; Bloor 1983, 1985; Hasegawa 1983). Therefore, we shall not consider them any more in this chapter.

The studies of homogeneous reactions in relation to *fundamental* problems of solid state reactivity will be discussed in more detail in Section 6.

4 Examples of homogeneous solid state reactions

If one compares the number of publications on solid state heterogeneous transformations with the number of papers devoted to the solid state reactions proceeding without phase separation, i.e. homogeneously, one can come to the conclusion that homogeneous solid state reactions are rare. To some extent this is true, but one could also ask if more homogeneous reactions were described, if an interest in them would be larger.

There are not very many examples of unambiguously homogeneous solid state reactions described in the literature. Several reactions are described as proceeding in a single crystal to single crystal mode. However, as was mentioned above, this does not necessarily mean that the reactions are homogeneous, and surprisingly many publications, discussing in detail the structures of the reactant and of the product, do not give even a hint whether an interface was formed at intermediate degrees of conversion, or whether the reaction was truly homogeneous. As a result, some misinterpretation is possible when referring to the literature data. The following types of transformation seem to provide examples of homogeneous solid state reactions.

1 'Intramolecular' reactions in molecular solids, such as intramolecular linkage isomerization in Co(III)–ammine complexes (Adell 1952, 1955; Grenthe and Nordin 1979a, b; Boldyreva *et al.* 1984, 1993), racemization and isomerization of a series of cobaloximes and alkyl cobalt complexes (Ohashi *et al.* 1982, 1983, 1993; Schklover and Timofeeva 1985; Uchida *et al.* 1987, 1991; Ohashi 1988; Ohgo and Arai 1993; Takenaka *et al.* 1993).

2 Photodimerizations in organic crystals and their back thermal reactions (Cohen 1975, 1987; Nakanishi *et al.* 1980a–c, 1981; Misra and Prasad 1982; Chang *et al.* 1982; Kearsley and Desiraju 1985; Swiatkiewicz *et al.* 1982; Schklover *et al.* 1986; Ramamurthy and Venkatesan 1987; Wang and Jones 1987; Enkelmann *et al.* 1993, 1994; Novak *et al.* 1993, 1994).

3 Polymerizations in organic crystals (Wegner 1969, 1971, 1977; Wegner and Fischer 1970; Kaiser and Wegner 1972; Kaiser *et al.* 1972a,b; Nakanishi *et al.* 1977a,b; Baughman 1978; Baughman and Chance 1978, 1980; Baughman and Yee 1978; Braunschweig and Bässler 1980; Bloor 1983, 1985; Hasegawa 1983; Braun and Wegner 1983a,b; Basilevsky *et al.* 1985; Bloor *et al.* 1985; Hasegawa *et al.* 1985; Schklover *et al.* 1986; Cohen 1987).

4 Oxidation of some oxides (for example, of Fe_3O_4) (Feitknecht 1964; Gillot *et al.* 1978; Gillot and Rousset 1994).

5 Thermal decomposition, in particular dehydration of crystal hydrates, sometimes only at early stages (Freund 1968; Mutin and Watelle-Marion 1972, 1977, 1979; Niepce and Watelle-Marion 1973; Niepce *et al.* 1973; Giovanoli and Brutsch 1974, 1975; Prodan *et al.* 1976; Mutin *et al.* 1979; Mutin and Dusausoy 1981; Ben Amor and Mutin 1982; Guarini and Dei 1983; Ben Amor *et al.* 1983a,b; Guarini and Magnani 1988; Galwey *et al.* 1994; Sidel'nikov *et al.* 1994).

6 Early stages of reduction of some oxides, such as, for example, MoO_3 and $V2O_5$ (Anderson 1972).

7 Some other reactions, for example the solid state rearrangement of orthocyclohexadienone into paracyclohexadienone (Lamartine *et al.* 1986).

5 How are homogeneous solid state reactions studied? (techniques)

A variety of techniques can be applied for studying solid state homogeneous reactions.

5.1 *Optical microscopy*

Optical microscopy is one of the techniques that can show if an interface is formed during a solid state reaction. Observations of the fragmentation and/or deformation of crystals during the reaction can give evidence that there *is* a response of the crystalline environment to the chemical transformation at a reaction site. (As examples, see Wegner 1977, Nakanishi *et al.* 1977b, Boldyreva *et al.* 1984; Kearsley and Desiraju 1985; Berrehar *et al.* 1989).

5.2 *Spectroscopic techniques*

Spectroscopic techniques (including infrared (IR), Raman, optical and, more rarely, X-ray spectroscopy) are most commonly used to identify the products of a homogeneous reaction. They are also the main tools for following the kinetics of the transformation. It is worth noting once again that the kinetic study of a homogeneous reaction has essential advantages compared with heterogeneous reaction studies. It is not necessary to follow the evolution of the interface region in order to get correct kinetic parameters. It is not necessary to use special microspectroscopy techniques allowing one to follow chemical and structural changes locally at various sites of the crystal, in particular at the reaction interface. The following examples of the application of 'routine' spectroscopy for kinetic studies of homogeneous solid state isomerizations can be cited: Beattie and Satchell (1952); Adell (1955); Wang and Jones (1987); Uchida *et al.* (1987, 1991); Phillips *et al.* (1990); Dulepov (1992); Dulepov and Boldyreva (1992, 1994).

Spectroscopy can be used to study the effect of the crystalline surrounding on the reaction, not only being applied for kinetic measurements, but also as a tool to trace the effect of environment on the molecule in the crystal prior to the reaction. Changes in the environment of a molecule in the crystal manifest themselves in IR, Raman, optical or X-ray spectra. For example, IR-vibration frequencies of the nitro-group in $[Co(NH_3)_5ONO]XY$ crystals were shown to be very sensitive to the changes of the outer-sphere anions, X and Y (Siebert 1958, Dulepov 1992; Dulepov and Boldyreva 1992, 1994; Boldyreva *et al.* 1992). They were shown also to be noticeably different for different polymorphs of the same complex (Dulepov 1992). X-ray spectroscopy has revealed measurable differences in the electronic structure of the $[Co(NH_3)_5NO_2]XY$ and $[Co(NH_3)_5ONO]XY$ complexes with different anions (Kravtsova *et al.* 1996).

On the other hand, spectroscopic techniques can also be helpful for a study of the response of the crystalline environment to the intramolecular reaction, since the response can manifest itself in changes in the positions of band maxima or/and in band shapes. Presently, however, only a few examples of the application of spectroscopic measurements for this purpose can be cited and, with rare exceptions, not for homogeneous reactions (McBride 1983; Berlyand *et al.* 1989; Chukanov *et al.* 1986; McBride *et al.* 1986; Hollingsworth and McBride 1990; Dulepov 1992; Dulepov and Boldyreva 1992, 1994).

Raman spectroscopy is worth a special mentioning when spectroscopic techniques are discussed. This is a technique that allows one to distinguish between the homogeneous and heterogeneous character of the reaction (Misra and Prasad 1982; Swiatkiewicz *et al.* 1982). The method also contributes to the elucidation of the reaction mechanism. In particular, Raman spectra provide information on electron–phonon interactions in crystals, on the phonon-assisted reactivity, on the nature of excited states and on local lattice configurations which are precursors of the parent lattice.

5.3 *X-ray diffraction*

X-ray diffraction (both powder and single crystal variants) is one of the techniques most commonly used for studies of homogeneous solid state reactions. First of all, it gives evidence that the process is homogeneous and that no phase separation takes place. It can also be applied for the identification of the products. In particular, when a homogeneous reaction takes place in a molecular crystal, the technique is often used to follow the changes in the *molecular structure* as a result of the reaction. X-ray diffraction gives direct data, in contrast to a spectroscopic technique. Solving the *crystal structure* of the product makes it possible to analyse if there are correlations between the crystal structure of the starting reactant and the molecular (for molecular crystals) and crystal structure of the product (Wegner and Fischer 1970; Kaiser and Wegner 1972; Kaiser *et al.* 1972a,b; Giovanoli and Brutsch 1974, 1975; Mutin and Watelle-Marion 1977; Nakanishi *et al.* 1977a,b, 1980a–c, 1981; Grenthe and Nordin, 1979a,b; Mutin *et al.* 1979; Mutin and Dusausoy 1981; Ben Amor and Mutin 1982; Chang *et al.* 1982; Hasegawa 1983; Braun and Wegner 1983b; Ben Amor *et al.* 1983a,b; Cohen 1987; Enkelmann *et al.* 1993, 1994; Novak *et al.* 1993, 1994; Ohashi *et al.* 1993; Masciocchi *et al.* 1994). For example, the molecular structure of the products of solid state dimerizations and polymerizations, as well as the very possibility of the reaction, were explained in many publications by the mutual juxtaposition of molecules in the crystal prior to reaction (see, as examples, Schmidt 1967; Cohen 1987).

Many such topotactic homogeneous reactions, as oxidation or thermal decomposition, were observed to proceed homogeneously only when the size of the reacting crystals was small enough. Therefore, electron and not X-ray diffraction, was applied for their study in some cases. High-resolution electron microscopy also was used to follow the early stages of the reduction of some oxides, and to reveal their homogeneous character (Anderson 1972).

In quite a number of publications, X-ray diffraction was used (in various ways) to get a better insight into the reaction mechanism. One of the variants of such applications is to solve and refine the structure of the same crystal repeatedly, at different time intervals as the reaction proceeds, so as to follow the changes in the lattice parameters and in the fractional coordinates of atoms in the elementary cell. It is worth reminding once again that it is the homogeneous character of the transformation that makes it possible not to use local microdiffraction with a high spatial resolution. It was possible to apply single crystal techniques when the reaction did not result in the fragmentation of the crystal (a single crystal to single crystal type) or in considerable deterioration of its quality (Grenthe and Nordin 1979a,b; Nakanishi *et al.* 1980a,b,c, 1981; Wang and Jones 1987; Enkelmann *et al.* 1993, 1994; Novak *et al.* 1993, 1994; Ohashi *et al.* 1993). For relatively slow transformations, 'routine' four-circle diffractometers were used, whilst special equipment (a specially designed IPD-WAS system operating with an 'ordinary' X-ray source (Kamiya *et al.* 1993)) was required for more rapid reactions. High-resolution powder diffraction and profile analysis with Rietveld refinement proved to be a good alternative to single crystal studies when it was not possible to prevent the fragmentation of the crystals in the reaction course (Masciocchi *et al.* 1994).

In most publications the main attention was drawn to the changes in the relative positions of the molecular fragments in those molecules which themselves took part in the reaction. For some intramolecular isomerizations, the positions of the neighbouring atoms (prior to reaction) were also considered, in order to discuss the effect of the environment of the molecule in the crystal on the reaction (Ohashi *et al.* 1982, 1983, 1993; Ohashi 1988; Podberezskaya *et al.* 1990, 1991; Kubota and Ohba 1992; Takenaka *et al.* 1993). At the same time, a diffraction study can also give information on the response of the crystalline environment to the reaction: lattice distortion, the changes in the juxtapositions of the adjacent atoms and/or molecules during the reaction.

Lattice distortion can be followed by measuring the changes in lattice parameters in the course of the reaction. These data become available after any successful diffraction study. The data can be used, at least, to calculate the integral changes in the molar volume of the crystal. Moreover, a detailed analysis of the anisotropy of lattice distortion induced by the reaction is possible. The corresponding strain tensor can be calculated, similar to how it is done when lattice distortion induced by temperature, pressure or chemical substitution is studied (Nye 1957; Hazen and Finger 1982; Zotov 1990; Jessen and Küppers 1991; Zotov and Petrov 1991; Boldyreva *et al.* 1994, 1996). Although this analysis does not require any additional experimental data, it is very rarely found in the publications presently available (but, see Boldyreva 1994; Boldyreva *et al.* 1994, 1995; Masciocchi *et al.* 1994). Most often, the changes in the lattice parameters are reported without proper (or even without any) discussion. Sometimes, the measurements of the changes in the lattice parameters are used for calculating the kinetic constant of the transformation (see, for example, Ohashi *et al.* 1993; Takenaka *et al.* 1993). It is necessary to note, however, that in general this is not correct. The changes in lattice parameters characterize nothing else but the crystal structure distortion induced by the reaction, and they are not necessarily directly proportional to the degree of transformation. For kinetic studies, because of the pronounced anisotropy of lattice distortion, it is especially incorrect to use the measurements of the changes in only one of the lattice parameters chosen without proper justification.

If the experiment gives the changes in lattice parameters and also the changes in the fractional coordinates, it becomes possible to complete the analysis of the anisotropy of lattice distortion by a detailed analysis of the relative shifts of all the atoms in the course of the reaction. It is this analysis that is required to understand how the chemical transformation at a reaction site interrelates with the response of the crystal structure. Almost any publication comparing the structures of the parent and product crystals reports experimental data necessary for such an analysis. However, there are only very few examples of studies in which this analysis was, to some extent, done (Mutin and Watelle-Marion 1972; Mutin *et al.* 1979; Nakanishi *et al.* 1980a–c, 1981; Mutin and Dusausoy 1981; Ben Amor *et al.* 1983a,b; Enkelmann *et al.* 1993, 1994; Novak *et al.* 1994; Masciocchi *et al.* 1994).

6 Studies of homogeneous reactions and fundamental problems of solid state reactivity

A discussion of solid state homogeneous reactions usually relates to one of the following problems.

1 What makes a particular solid state reaction homogeneous or heterogeneous? How can one make the reaction proceed homogeneously without fragmentation of single crystals?

2 Does a feedback arise in the course of a homogeneous solid state reaction?

3 What is the interplay between intra- and intermolecular interactions in the course of a homogeneous solid state reaction?

Actually, all three problems are closely interrelated.

6.1 *When is a solid state reaction homogeneous and when heterogeneous?*

Will a particular solid state reaction proceed homogeneously or heterogeneously under the chosen experimental conditions? This is not an easy question to answer. It may seem to be obvious that a reaction in which a solid is brought into contact either with another solid or with a liquid (or a gas) must be heterogeneous, since the interface region is the only part of the system where the reactants can meet each other. However, the diffusion of reactants and/or products sometimes, under special experimental conditions, allows these reactions to proceed homogeneously. It is even less obvious that types of solid state reactions such as ‘intracrystalline rearrangements’ (isomerizations, dimerizations and polymerizations) should necessarily be heterogeneous.

One can claim that a solid state reaction should proceed homogeneously if the solid solution ‘product–reactant’ is thermodynamically preferable as compared with a two-phase system. This may be the case, for example, when the affinity of the reactant and the product is high enough, and their structures allow a solid solution to be formed. In the case of molecular crystals, formation of a solid solution should be consistent with the molecular structures of the components. In order to enable formation of the solid solution, one of the components may even adopt an unusual conformation which it never has when it crystallizes as an individual phase. This phenomenon was termed by

Cohen a 'structural' or 'conformational mimicry' (Cohen 1975, 1987). Thus, for example, polydiacetylene molecules formed during solid state photopolymerization remain in the planar extended chain conformation only as long as the system is homogeneous. Once the solid solution 'reactant–product' decomposes and the phases are separated, it is not posible to re-establish this conformation (Cohen 1987).

If the solid solution is stable for all the concentration range, the reaction remains homogeneous until completion. However, it is also possible (and, at least presently, seems to be more common) that the solid solution is stable only for low concentrations of product. In this case the reaction may start as a homogeneous one, and then as the solid solution decomposes and the product(s) and reactant form separate phases, it becomes heterogeneous. The hypothesis that many heterogeneous solid state transformations start as homogeneous ones was discussed in a number of publications, although it was not always unambiguously proved experimentally (see, for example, Garner 1955; Hillert 1961; Cahn 1962, 1968; Schmidt 1967; Freund 1968; Anderson 1972; Mutin and Watelle-Marion 1972, 1977, 1979; Niepce and Watelle-Marion 1973; Niepce *et al.* 1973; Giovanoli and Brutsch 1974, 1975; Cohen 1975, 1987; Prodan *et al.* 1976; Mutin *et al.* 1979; Scheffer 1980; Mutin and Dusausoy 1981; Ben Amor and Mutin 1982; Misra and Prasad 1982; Guarini and Dei 1983; Ben Amor *et al.* 1983a,b; Guarini and Magnani 1988; Galwey *et al.* 1994; Sidel'nikov *et al.* 1994). One of the existing points of view on the origin of heterogeneity of solid state transformations is that the observed phase separation and spatially modulated composition may result not necessarily from the formation and growth of product nuclei, but from a spontaneous spinodal decomposition of originally formed metastable solid solutions (Hillert 1961; Cahn 1962, 1968; Anderson 1972). Thus, if the researchers were really interested in homogeneous solid state reactions, they could, probably, find more experimental examples studying the initial stages of the reactions, which are traditionally considered to be heterogeneous.

As early as 1916, Langmuir had noted that:

> when, according to the phase rule, separate phases of constant composition are present, the reaction must take place exclusively at the boundaries of the phases. This kinetic interpretation of the phase rule indicates clearly the distinction between reactions in which solid solutions are formed and those in which separate phases appear (Langmuir 1916).

The heterogeneity of solid state reactions can be considered as a manifestation of the self-organization phenomenon (see, for example, Anderson 1972; Barret 1975; Bertrand *et al.* 1986; Schmalzried 1990; Martin 1991). Positive feedback is responsible for the so-called 'autolocalization' of solid state reactions, when the process continues preferably near the site where it has started (Boldyrev 1973). Positive feedback during solid state reactions may arise due to various reasons, which were summarized in a number of reviews (Boldyrev 1973, 1977; Boldyrev *et al.* 1979; Boldyreva 1988, 1990, 1992a). Amongst the most important of these reasons one can mention: (i) the generation of mechanical stresses and their relaxation; (ii) the changes in the local concentrations of chemical species taking part in the reaction; (iii) the local temperature changes; and (iv) the local shifts in the electron-hole equilibria because of the electric potential jump over the interface.

Additional factors that make a solid state reaction more likely to be heterogeneous are the inhomogeneity of the starting crystal (for example, an inhomogeneous distribution of defects), as well as the initiation of the reaction non-uniformly throughout the bulk of the crystal (local heating, mechanical action, irradiation strongly absorbed by the surface layer). Actually, highly local initiation of the process can make even gas phase reactions heterogeneous if positive feedback is strong enough (one of the most common examples is the propagation of the combustion front).

This brief review of the factors responsible for the heterogeneous character of a solid state reaction also helps to predict under which conditions a solid state reaction can proceed homogeneously.

1 Reaction is more likely to be homogeneous if it is initiated uniformally throughout the bulk of the crystal.

2 It is desirable that either the reaction is not sensitive to the presence of crystal defects or, at least, the defects are distributed homogeneously in the bulk of the crystal.

3 It seems natural, and was also confirmed by the results of a computer simulation (Boldyreva and Salikhov 1985; Boldyreva 1987a,b, 1988), that a solid state reaction can be homogeneous if feedback either does not exist at all, or if it is negative. Positive feedback usually is consistent with the homogeneity of a reaction, being either very weak (Boldyreva 1987b), or very long range (Baughman 1978; Baughman and Chance 1980). If the positive feedback is strong and relatively short range, the reaction inevitably becomes heterogeneous (Boldyreva 1987b, 1988). As far as negative feedback is concerned, the stronger the feedback the more likely it is that a solid solution 'product–reactant' is formed (Boldyreva 1987a,b, 1988).

4 A solid state reaction is likely to proceed homogeneously if the formation of solid solution product–reactant results in minimizing the free energy of the system. The minimizing of mechanical stresses induced in the reacting crystal is of particular importance. It was repeatedly noted in the publications of different authors that the homogeneous character of a solid state transformation is interrelated with the ability of the lattice to accumulate mechanical strain or to relax without phase separation (see, for example, Baughman 1978; Baughman and Chance 1978; Braun and Wegner 1983a,b; Ben Amor *et al.* 1983a,b; Basilevsky *et al.* 1985; Cohen 1987; Novak *et al.* 1993, 1994; Enkelmann *et al.* 1994; Sidel'nikov *et al.* 1994). For example, solid state polymerization was noticed to procced homogeneously if 'bulky substituents formed a fixed matrix in the crystal and this matrix was able to accommodate considerable atomic displacements of the reacting fragments without being destroyed' (Novak *et al.* 1993). According to Cohen, 'a solid-state polymerization tends to be homogeneous if the monomer molecules are flexible or their crystals contain solvent of crystallization, thus providing ready modes for strain relaxation' (Cohen 1987, and his reference to Baughman and Chance 1978 therein). The necessity to minimize mechanical stresses and crystal structure distortion in some cases, made it preferable that the solid state reaction proceeded homogeneously, even if this resulted in the formation of a thermodynamically unstable polymorph. Basilevsky *et al.* (1985) have proposed a model for the mechanism of diacetylene polymerization in the solid state, according to which the whole system can be considered as a strained solid solution of polymer chains in monomer crystal. The metastable strained solid solution is stabilized by the interaction of polymer chains with the monomer lattice. Relaxation processes, according to the model of Basilevsky *et al.* (1985), lead to a destruction of solid solution. Basilevsky *et al.* (1985) claim the interconversion of chemical and mechanical energy to be a general rule inherent to solid state polymerizations.

In general, in all the solid state reactions which were ever observed to proceed homogeneously, the crystalline structure was able to relax as the reaction proceeded and mechanical stresses and strain were unavoidably induced. Ability of the structure to resist mechanical stresses without phase separation does not mean, however, that the structure is not distorted. On the contrary, lattice strain is one of the ways of relaxation of mechanical stresses, and a continuous lattice distortion was observed, for example, for homogeneous isomerizations (Grenthe and Nordin 1979a,b; Boldyreva *et al.* 1993; Ohashi *et al.* 1993, references therein; Takenaka *et al.* 1993; Masciocchi *et al.* 1994), or for dimerizations and polymerizations (Wegner and Fischer 1970; Nakanishi *et al.* 1977a,b, 1980a–c, 1981; Braun and Wegner 1983a,b; Bloor *et al.* 1985; Wang and Jones 1987; Enkelmann *et al.* 1993, 1994; Novak *et al.* 1993, 1994). If the reaction was reversible, so was the lattice distortion. If mechanical stresses induced the crystal deformation (for example, bending, as in the case of the crystals of $[Co(NH_3)_5NO_2]Cl(NO_3)$ during photoisomerization (Boldyreva *et al.* 1984; Boldyreva and Sidel'nikov 1987)), the deformation was also reversible as the back reaction proceeded. In some cases the structural distortion did not lead to the fragmentation of the crystals, and the homogeneous reaction proceeded in a single crystal to single crystal mode (Nakanishi *et al.* 1977a,b, 1980a–c, 1981; Grenthe and Nordin 1979a; Braun and Wegner 1983b; Enkelmann *et al.* 1993, 1994; Novak *et al.* 1993, 1994; Ohashi *et al.* 1993). In other cases, however, the crystals were fragmented (Nakanishi *et al.* 1977b; Grenthe and Nordin 1979b; Mutin and Watelle-Marion 1979; Braun and Wegner 1983a,b; Boldyreva *et al.* 1984; Kearsley and Desiraju 1985; Berrehar *et al.* 1989; Kubota and Ohba 1992).

The generation and relaxation of the mechanical stresses during a solid state reaction are influenced by the conditions under which the reaction proceeds (temperature, intensity and wavelength of irradiating light for photochemical transformations, pressure of water vapour for dehydration reactions, etc.). Therefore, by varying the conditions one can switch the reaction from a heterogeneous to homogeneous mode. Since the relaxation of mechanical stresses is highly dependent upon the size of the crystals, so is the probability that a solid state reaction will proceed homogeneously.

Most of the presently known homogeneous solid state reactions can proceed either homogeneously or heterogeneously, depending on the experimental conditions and on the size of the crystals.

For example, the oxidation of Fe_3O_4 by oxygen at 120°C proceeds homogeneously and gives a metastable polymorph, γ-Fe_2O_3 only if the reactant particles are smaller than 3000 Å. If the particles of the starting Fe_3O_4 are larger than this, or the temperature is higher, the reaction is heterogeneous, and another polymorph of Fe_2O_3, i.e. the thermodynamically stable α-Fe_2O_3, is produced (Feitknecht 1964). Thermal decompositions of crystalline solids (and thermal dehydration, in particular) also proceed homogeneously only under very special experimental conditions and when reacting particles are small enough.

Solid state photodimerizations and polymerizations usually proceed heterogeneously if irradiation is carried out with broad-band irradiation in the maximum of the absorption (Cohen 1987; Novak *et al.* 1993). However, the same reactions become homogeneous when induced by irradiation in the tail of absorption, i.e. when light is absorbed less by surface layers of the crystal and, therefore, the reaction is initiated more uniformly throughout the bulk of the crystal. This was proved in a series of experiments by Wegner and coworkers (Braun and Wegner 1983b; Enkelmann *et al.* 1993, 1994; Novak *et al.* 1993, 1994).

6.2 *Does a feedback arise in the course of a homogeneous solid state reaction?*

As was discussed in the previous section, it is, to a large extent, the type of feedback that determines whether a solid state reaction proceeds homogeneously or heterogeneously. Conclusions concerning the feedback in a solid state reaction are usually based on the kinetic analysis. Self-acceleration is considered to be a manifestation of the positive feedback, whilst self-retardation is assumed to indicate that the feedback is negative.

Only one example of a strongly autocatalytic homogeneous solid state reaction has been described up to now, i.e. the polymerization of diacetylenes (Baughman 1978; Baughman and Chance 1980). If this reaction is really homogeneous, as it is claimed to be in the literature (and not a single crystal to single crystal heterogeneous transformation), then this should mean that the strong positive feedback is very long range, so that no autolocalization of the process takes place. This can be possible if both the parent and the product crystals are able to relax very easily, and a perturbation (mechanical stress) is transferred over large distances in the crystal. Basilevsky *et al.* (1985) suggested that the positive feedback may arise during the polymerization of crystalline diacetylenes due to the interconversion of chemical and mechanical energy. According to the model of Basilevsky *et al.* (1985):

> The formation of a diacetylene dimer at the initial step of the chain generation produces a considerable mechanical strain in the crystal lattice which pushes the dimer towards its partner in the next addition reaction, the neighboring monomer molecule. This is expected to lower the barrier of the chain propagation reaction in the crystal as compared to that in the gas or liquid phase. The multi-step process of polymer chain formation is facilitated by the directed transformation of the chemical energy, emerging at generation and successive propagation steps, into mechanical strain energy. The latter results in a lowering of the potential barrier at the next step of the chain propagation.

More commonly, kinetics of homogeneous solid state reactions is reported to follow the first-order law, or to be characterized by self-retardation. Quite often the reaction does not go to completion, but stops at some intermediate product/reactant ratio. In these cases, kinetic analysis alone does not allow a distinction to be made between a real equilibrium and a so-called 'kinetic stop' of the reaction due to the negative feedback or, alternatively, due to the so-called 'kinetic non-equivalence' of different sites in the starting crystal, some of them being more reactive than others (Boldyreva 1988; for more on the kinetic non-equivalence and polychromatic kinetics see the following publications: Roginskii 1948; Lebedev 1978; Vorobiev and Gurman 1982; Burstein *et al.* 1984; Doba *et al.* 1984; Kutyrkin *et al.* 1984; Sieberand and Widman 1986; Tolkachev 1996). In general, kinetic analysis is not a very sensitive method to study feedback; only extremes, i.e. very strong positive or negative feedback, are usually detected (Boldyreva 1988), which seems to be not very typical for homogeneous reactions. An experimentally observed first-order kinetic law does not prove that a feedback exists, but neither does it prove that there is no weak negative feedback in the system. Therefore, in order to make unambiguous conclusions concerning the feedback in a solid state homogeneous reaction, in addition to a kinetic analysis, one also needs special experiments in which the changes induced in the system by the reaction and the effect of these changes on the further reaction course are followed directly (Boldyreva 1988, 1990, 1992a,b). For example, the most common 'change' in the crystal induced by a homogeneous reaction is mechanical strain. One can also strain the crystals artificially, for example by external loading, and then follow the reaction in the strained crystals. Such experiments were carried out for solid state nitro–nitrito isomerization in the elastically bent single crystals of $[Co(NH_3)_5NO_2]Cl(NO_3)$ (Boldyreva and Sidel'nikov 1987; see also an English-language description of these experiments in the review of Boldyrev *et al.* 1990). Quantum yield of nitro–nitrito isomerization was shown to be less in compressed parts of the crystals. On the other hand, the molar volume decreases as a result of the reaction, and the crystals are compressed. Therefore, one can suppose that the reaction can be characterized by a weak (negative) feedback (Boldyreva 1988).

6.3 *The interplay between intra- and intermolecular interactions in the course of homogeneous solid state reactions*

Most of the homogeneous solid state reactions (at least, amongst those described up to now; see Section 4) are reactions in molecular crystals. It is the interplay between the intra- and intermolecular interactions in the crystals that determines if a particular reaction proceeds homogeneously, how large the structural strain is and if the crystals are fragmented in the reaction course. Hence, it is not surprising that the role of intra- and intermolecular interactions in homogeneous reactions in molecular crystals was (and is) the focus of so many studies.

The problem can be formulated in the form of two basic questions.

1 Does the environment influence the reaction at a reaction site?

2 Does the reaction at a reaction site influence the surroundings in the crystal?

The questions should be more precisely formulated in the following way: is the effect of the environment on the reaction and the effect of the reaction on the environment large enough to be measurable? Any species will have an effect on another, but the magnitude of the interactions and their consequences may be imperceptibly small. Experimental data give evidence that both effects are measurable, and can determine the reaction rate, the quality of the product crystals or, sometimes, the very possibility of the reaction, and not only the crystal structure, but also the molecular structure of the product(s).

6.3.1 THE EFFECT OF THE ENVIRONMENT ON THE REACTION

The effect of the environment on the reaction course can be considered as a sort of extreme 'solvent effect' (Gavezzotti and Simonetta 1982; Hollingsworth and McBride, 1990). The arrangement of atoms surrounding a reaction site in a crystal is not fluctuating with time, and, before the reaction has started, it is usually more or less uniform throughout the bulk of the crystal (the only obvious exceptions are the defect areas). The environment can be characterized in detail using direct data of X-ray diffraction experiments. This makes solid state systems advantageous in many respects for a consideration of solvent effects as compared with the liquid state, where the environment is mobile and can be characterized only with some degree of certainty on the basis of indirect spectroscopic or kinetic data (Zamaraev 1994). In the case of crystals with a well-defined reaction environment, one can speculate more deeply about the constraints imposed by the 'reaction cavity', or about cooperative motion of adjacent molecules (see below). If a solid state reaction proceeds homogeneously, one can also get direct experimental data on the changes of both the structure of the reacting unit and its environment in the course of the reaction.

Systems. To understand the effect of the environment on the reaction course in molecular crystals, it often proved to be advantageous to study a series of related compounds. By way of example, one can cite the studies of photoracemization and photoisomerization in the series of cobaloxime complexes (Ohashi *et al.* 1993), linkage nitro–nitrito isomerization in Co(III)–ammine complexes with different ligands in the complex cation (Grenthe and Nordin 1979a,b) or the photodimerization in the series of related organic compounds (Bloor 1983; Hasegawa 1983; Ohgo and Arai 1993). The compounds in the series had identical fragments, which took part in the reaction directly, but differed in other fragments of the molecule. In the studies of linkage isomerization in a series of $[Co(NH_3)_5NO_2]XY$ and $[Co(NH_3)_5ONO]XY$ complexes, the cation was kept the same, but the anion not participating directly in the reaction was varied (Adell 1952; Dulepov 1992; Dulepov and Boldyreva 1992, 1994; Boldyreva 1994). Photodimerization in a series of acridizinium salts with different anions was studied by Wang and Jones (1987).

Facts. The above-mentioned studies have shown the kinetics of the reactions to be quite sensitive to the presence of the 'non-participating' atoms and to the environment of the reacting site in the crystal (Adell 1952; Wang and Jones 1987; Dulepov 1992; Dulepov and Boldyreva 1992; Ohashi *et al.* 1993; Boldyreva 1994). Sometimes, the reactivity of a molecule was affected by the environment to such an extent that a molecule reacted in one environment and remained inert in another. One of the most convincing examples is provided by studies of photoisomerization in some molecular crystals that contained crystallographically non-equivalent molecules (Ohashi *et al.* 1982; Uchida *et al.* 1984). The effect of the environment was so strong that when the crystals were irradiated, only one of the two non-equivalent molecules transformed, the second remaining inert.

The environment was also shown to influence the strain induced in the crystals in the reaction course and, hence, one could (to some extent) control the deterioration of quality and fragmentation of the crystals (compare, for example, Grenthe and Nordin 1979a,b).

There is some evidence that in the course of an intramolecular reaction, intramolecular motions and the intramolecular geometry of the product may be affected by the environment in the crystal structure. For example, Grenthe and Nordin (1979a,b) analysed the steric hindrances for the particular intramolecular motions in the crystals of some nitrito-isomers, and came to the conclusion (based on the data of single crystal diffraction studies) that the nitrito-ligand goes out of its original plane in the course of linkage nitrito–nitro isomerization in $[Co(NH_3)_5(ONO)]Cl_2$, but remains in this plane if the isomerization proceeds in the crystals of *trans*-$[Co(en)_2(NCS)(ONO)]X$ ($X = ClO_4^-$

or I^-). Later on, a similar analysis was undertaken for nitro–nitrito isomerization in $[Co(NH_3)_5NO_2]Cl(NO_3)$ (Podberezskaya *et al.* 1990, 1991) and, more recently, for nitro–nitrito isomerization in $[Co(NH_3)_5NO_2]Cl_2$ (Kubota and Ohba 1992).

Another example is provided by the study of Masciocchi *et al.* (1994). The complex cation in $[Co(NH_3)_5ONO]Br_2$ isomer, obtained by irradiation of solid $[Co(NH_3)_5NO_2]Br_2$ seems to have a somewhat unusual geometry, as compared with the geometry reported previously by Grenthe and Nordin (1979b) for the same cation in the crystals of chloride grown from aqueous solution. The same result was obtained for the photochemically synthesized chloride. The crystal structures of $[Co(NH_3)_5ONO]Br_2$ (or Cl_2): (i) grown from aqueous solution; and (ii) obtained by a solid state photochemical reaction are different. The different geometry of the complex cation can be explained well by different interactions of the cation with the environment in two crystal structures (Masciocchi *et al.* 1994). Formation of N–H--O hydrogen bonds between the nitrito-ligand of one cation and an ammine-ligand of a neighbouring cation can be supposed to be of particular importance.

When non-participating atoms are varied, both the chemical composition and the packing arrangements in the crystal are changed simultaneously. Therefore, sometimes it is advantageous to study not a series of related substances, but different polymorphs of the same compound. Such examples can also be found in the literature (Uchida *et al.* 1987, 1991; Dulepov 1992; Dulepov and Boldyreva 1992; Ohgo and Arai 1993; Boldyreva 1994). The environment of the reacting site is often more different in two polymorphs of the same compound than in the structures of different substances. As a result, the reactivity of different polymorphs may differ more than the reactivity of related compounds of the same series.

One more alternative is to keep both the chemical composition and the polymorph the same, and to distort the crystalline environment of a reacting fragment in a continuous way, applying hydrostatic pressure (Boldyreva *et al.* 1992, 1994; Boldyreva 1994). These studies, however, are not very common yet.

Interpretations. The effect of the surroundings on the reacting species in solution is usually interpreted in terms of one of the two essentially different models.

In the first model, the environment is simulated as an isotropic continuum which can be characterized by some particular viscosity and dielectric properties. The species under study are considered to be embedded in a 'reaction cage', or in a 'reaction cavity' of a continuous dielectric, of a particular size and shape and with a particular surface charge. This cage can affect the reaction either restricting the escape of the products and, thus, facilitating their secondary reactions (the 'cage effect'; see Rabinowitch and Wood 1936, or Bamford and Tipper 1969 as a more recent reference), or due to the phenomenon of dielectric polarization (Bishop 1994, and references therein).

An alternative model does not consider the surroundings as a continuum, but rather interprets the effect of the environment in terms of formation of outer-sphere complexes, which are treated as supramolecules (Zamaraev 1994).

The models used when discussing the effect of the surroundings on the reaction species in the solid state are, in many respects, similar to those applied to the processes in liquids.

One of the most common models used when interpreting the effect of crystalline environment on the reactivity in crystals is also termed the model of 'reaction cavity'. the term seems to have been first used when discussing solid state reactivity by Schmidt (1967), and was then modified by Cohen (1975). A 'reaction cavity' was originally defined by these authors as 'the space of certain size and shape in the starting crystal, occupied by the molecules which are going to participate directly in a solid-state reaction' (Cohen 1975). The 'walls of the cavity' were considered to be formed by the atoms belonging to neighbouring molecules. 'The reaction cavity is surrounded by the contact surface of the molecules inside it with the surrounding molecules' (Cohen 1975). The main idea of this model was to consider the surroundings as a rigid template which can restrict the intramolecular motions within the reacting molecule and, in this way, can affect the reaction, determining the products of solid state photodimerizations in organic solids (as well as the very possibility of the reaction).

Approximation of a 'rigid reaction cavity' was used in many publications devoted to homogeneous solid state reactions in organic solids. For example, Ohashi and co-workers have studied systematically photoracemization and photoisomerization in a series of cobaloxime complexes, trying to find experimentally correlations between the size of the reaction cavity and the reaction rate (Ohashi *et al.* 1982, 1983, 1993; Ohashi 1988).

A reaction cavity is not an inert template accommodating in a passive way the shape changes that occur as the reactants are transformed to products. The 'walls' of the cavity are formed by atoms, and these atoms interact with the atoms of the incorporated molecule. Interactions may vary from weak van der Waals forces to strong hydrogen bonds or electrostatic forces between charged centres. Sometimes, the directionality of interactions can be quite significant, for example if hydrogen bonds are formed. Ramamurthy and co-workers (Ramamurthy *et al.* 1993; Weiss *et al.* 1993) have defined a reaction cavity with the directional interactions between the walls and the incorporated molecule as 'active'. The model of the 'active reaction cavity' resembles, in some respects, the model of an outer-sphere complex used in solution chemistry. When considering a solid state process, one should pay special attention to the fact that interactions in crystals, including those between a molecule in the cavity and the walls of the cavity, are essentially anisotropic.

To understand the effect of the reaction cavity on the reaction, it is necessary to take into consideration all possible interactions between the incorporated molecule and the walls of the cavity. Not very many examples of such studies can be cited. For some of the simplest cases (without directional or strong Coulombic interactions) computer simulations were carried out, and the changes in the interaction energy between the atoms of the reacting molecule and the atoms forming the 'walls of the reaction cavity' were calculated (Gavezzotti and Bianchi 1986; Gavezzotti 1987a,b, Ariel *et al.* 1987; Uchida and Dunitz 1990; Braga *et al.* 1990). The results of the calculations were used to suggest a preferable 'molecular mechanism' of the reaction.

6.3.2 THE EFFECT OF THE REACTION ON THE ENVIRONMENT

The studies in which chemical reactivity of molecular crystals is correlated with the *initial* (prior to the reaction) crystal structure and, in particular, with the initial spatial distribution of 'free volume', are very common. It is nowadays more or less generally accepted that the free volume must be present at a site in sufficient quantity if reactant molecules are to be able to undergo the shape changes required for their transformation to products. The initial structure is certainly very important for the reaction course in the crystal, and this is the reason why the concept of rigid reaction cavity seems to work in quite a number of cases. However, there is experimental evidence that not only the crystal structure can affect a homogeneous reaction in a molecular crystal, but also the reaction can induce noticeable distortions of the crystal structure. The effect of the reaction on the environment can manifest itself in spectra or in diffraction patterns. Experimental evidence in favour of its existence can be found in almost any publication on homogeneous solid state reactions in molecular crystals, even if the authors of the publication themselves stick to the model of rigid reactive cavity. Thus, for example in the above-mentioned studies of Ohashi and his co-workers, the anisotropic lattice distortion induced by the intramolecular reaction was measured experimentally (Ohashi *et al.* 1993, and references therein; Takenaka *et al.* 1993). For most of the reactions studied by Ohashi and co-workers, the kinetic rate constants actually calculated were from the changes in the lattice parameters.* Any change within a molecule in the molecular crystal (not even a reaction, but simply light absorption or a high-spin–low-spin transition) also affects the positions of the adjacent molecules. The process can be termed as a 'distortion of the reaction cavity', or 'relaxation of the reaction cavity' (see, for example, Ramamurthy and Venkatesan 1987; Murthy *et al.* 1987; Ramamurthy *et al.* 1993; Weiss *et al.* 1993). Craig and co-workers (Collins and Craig 1981; Craig and Mallett 1982; Norris *et al.* 1983; Craig *et al.* 1984) have suggested a model according to which a localized electronic excitation of a molecular crystal produces a particular type of instability of the lattice configuration, leading to large molecular displacements:

> Localized excitation means an existence of an excited molecule which on account of its altered properties is seen by its neighbors as an impurity. The creation of this 'impurity' molecule introduces a local instability in the lattice configuration and leads to relaxation. This relaxation process can involve large displacements from the original structure and in that sense far from the equilibrium configuration of the unexcited crystal.

In order to characterize the relaxation of the environment, Ramamurthy and co-workers (Ramamurthy *et al.* 1993; Weiss *et al.* 1993) have suggested distinguishing between 'rigid' and 'flexible' reaction cavities.

The 'flexibility of the reaction cavity' can be interpreted also in terms of the cooperative motion of the species adjacent to the reacting one. The role of this cooperation in solid state reactivity was discussed in a number of publications (see, as examples, Boldyreva 1982; Basilevsky *et al.* 1985; Hollingsworth and McBride 1990). Some computer simulations, for example the one carried out by Gavezzotti (1988), give an illustration of the importance of cooperative motions of molecules in a molecular

*This is, in general, not correct (see Section 5.3). As a matter of fact, it is not the reaction rate and the size of the reaction cavity that were correlated in these studies, but the size of the reaction cavity in the starting crystal *prior to* reaction and the rate of structural distortion *during* the reaction. It would be most interesting to use already available data in order to follow the changes in the size of the reaction cavity in the reaction course.

crystal even for such relatively simple (as compared with a reaction) processes as molecular rotations. Gavezzotti (1988) has calculated the potential energy barriers for the rotation of naphthalene in its molecular plane around the axis of maximum inertia in two models: (i) zero cooperation, whereby the molecule moves in the static field of the surrounding stationary molecules; and (ii) full cooperation, when the first shell of molecules surrounding the rotational site cooperate by tilting motions around their own inertial axes to make way for the rotating molecule. The full cooperation model proved to give a better agreement with the rotation barrier found from experimental data.

One can also get a better insight into the problem of the role of lattice relaxation in photochemical homogeneous reactions by the experimental studies of the effect of temperature on the transformation. A study of Hasegawa *et al.* (1985) may be cited as an example.

Summing up, one can say that the most general model describing the effect of the crystalline environment on a reaction site and the effect of the reaction on the environment could be termed as the model of 'flexible and active reaction cavity'. This model could adequately describe the effect of the environment on the reaction and the relaxation of the environment in the course of the reaction. It would allow one to find a proper balance when considering the 'steric' and the 'electronic' aspects of the interaction between the molecule and the environment. A detailed study of intermolecular interactions in the crystals, and of the cooperative motions which affect not only the intramolecular transformation but also the structural relaxation, should be central for the proper understanding of solid state reactivity.

7 An example: linkage isomerization $[Co(NH_3)_5NO_2]XY \rightleftharpoons [Co(NH_3)_5ONO]XY$ ($XY = Cl_2$ Br_2, I_2, $Cl(NO_3)$, $(NO_3)_2$)

In order to illustrate some of the most important statements made in previous sections, I would like to consider the results of a study of one particular homogeneous solid state reaction, i.e. the nitro–nitrito linkage isomerization in the crystals of Co(III)–penta-ammine complexes. I have chosen this reaction because we have been studying it in our group in various aspects over several years. Discussing our own results, I avoid the risk of misinterpreting someone else's data. Moreover, the reaction seems to be a representative example of reversible homogeneous solid state reactions (Boldyreva 1982, 1994).

Nitro- and nitrito-isomers of Co(III)–penta-ammine complexes are considered to be the first known examples of linkage isomers (Jorgensen 1893; Hitchman and Rowbottom 1982). They differ in the type of coordination of the nitro-ligand to the central atom, Co. In nitro-isomers the nitro-group is linked to Co via nitrogen, and in nitrito-isomers via oxygen (Hitchman and Rowbottom 1982). The change in the coordination of the ligand to the central atom results in a change of colour and can be followed visually or by optical spectroscopy (Wendlandt and Woodlock 1965; Hitchman and Rowbottom 1982). Another common technique allowing the linkage isomers to be distinguished is IR spectroscopy (Beattie and Satchell 1952; Penland *et al.* 1956; Hitchman and Rowbottom 1982). The two isomeric forms also have different X-ray absorption and X-ray emission spectra (Timonova *et al.* 1987; Kravtsova *et al.* 1996).

Nitrito-isomers are less favourable thermodynamically (Doron 1968), and when stored, convert into nitro-isomers (see, for example, Adell 1952, 1955; Grenthe and

Nordin 1979b; Boldyreva *et al.* 1993). Irradiation of the nitro-isomers with visible (blue) or ultraviolet (UV) light induces a reverse process: nitro–nitrito isomerization (Adell 1955; Wendlandt and Woodlock 1965; Scandola *et al.* 1974; Rose and McClure 1981). The isomerization was observed both in solution and in the solid state, and was claimed to be an intramolecular process, i.e. there is supposed to be no exchange of the ligands between different complex cations during the isomerization (Murmann and Taube 1956; Grenthe and Nordin 1979b).

Optical microscopy studies did not reveal any interface formation during the solid state nitro–nitrito photoisomerization (Boldyreva *et al.* 1984). X-ray powder diffraction proved the thermal nitrito–nitro reaction to be homogeneous; no phase separation was observed (Grenthe and Nordin 1979b; Boldyreva *et al.* 1993; Masciocchi *et al.* 1994). To illustrate this, Fig. 7.1 shows schematically, by way of example, a series of X-ray powder diffraction diagrams (Boldyreva *et al.* 1993). Figure 7.1a corresponds to the sample of $[Co(NH_3)_5NO_2]Br_2$; Fig. 7.1b to the same sample after irradiation, resulting in the formation of $[Co(NH_3)_5ONO]Br_2$; Figs 7.1c–e correspond to the same sample (a solid solution of $[Co(NH_3)_5ONO]Br_2$ and $[Co(NH_3)_5NO_2]Br_2$) slowly transforming back to nitro-form during storage at room temperature in the dark (at different time intervals). Similar results were obtained for the isomerization in the whole series $[Co(NH_3)_5NO_2]XY$ ($XY = Cl_2$, Br_2, I_2, $Cl(NO_3)$, $(NO_3)_2$) (Boldyreva *et al.* 1993).

The reaction seems to be very promising for the studies of the effect of the crystalline environment on an intramolecular transformation in the solid state.

The isomerization in solution is known to be sensitive to the presence of Lewis acids and bases, pH and the dielectric properties of the solvent (Jackson *et al.* 1982). The isomerization in the solid state was shown to be one to two orders of magnitude slower than the same reaction in solution; this gives evidence that the crystalline environment of a complex cation is also important for the intramolecular reaction.

The environment of complex cations in the crystal can be varied by different means: (i) substituting outer-sphere anions; (ii) preparing different polymorphs of the same complex; or (iii) compressing the crystals by an external load (Boldyreva 1994). Adell (1952) was the first to show the effect of the outer-sphere anion on the kinetics of nitrito–nitro solid state thermal isomerization. Rose and McClure (1981) have also reported the photochemical nitro–nitrito isomerization in the solid state to be sensitive to the substitution of the anion. In our studies (Boldyreva and Sidel'nikov 1987; Dulepov 1992; Dulepov and Boldyreva 1992, 1994; Boldyreva 1994) we have used several methods of modifying the crystalline environment of complex cations (substitution of anions, preparation of different polymorphs, application of hydrostatic pressure to the sample or elastic bending of the crystal) in order to affect the kinetics of solid state isomerization. An extended (as compared with the experiments of Adell (1952)) series of $[Co(NH_3)_5ONO]XY$ complexes was studied ($XY = Cl_2$, Br_2, I_2, $(NO_3)_2$, $Cl(NO_3)$, Cr_2O_7, C_2O_4). The noticeable effect of the anion on the rate of the solid state intramolecular thermal isomerization was confirmed. We have also compared the isomerization rate in different polymorphs with the same chemical formula, and the effect was even more pronounced (Dulepov 1992). Our preliminary experiments have shown that a continuous distortion of the surroundings of a complex cation in the structure of the same polymorph by applying high pressure to the sample, also resulted in a change in the equilibrium ratio of nitro/nitrito-coordinated cations in the

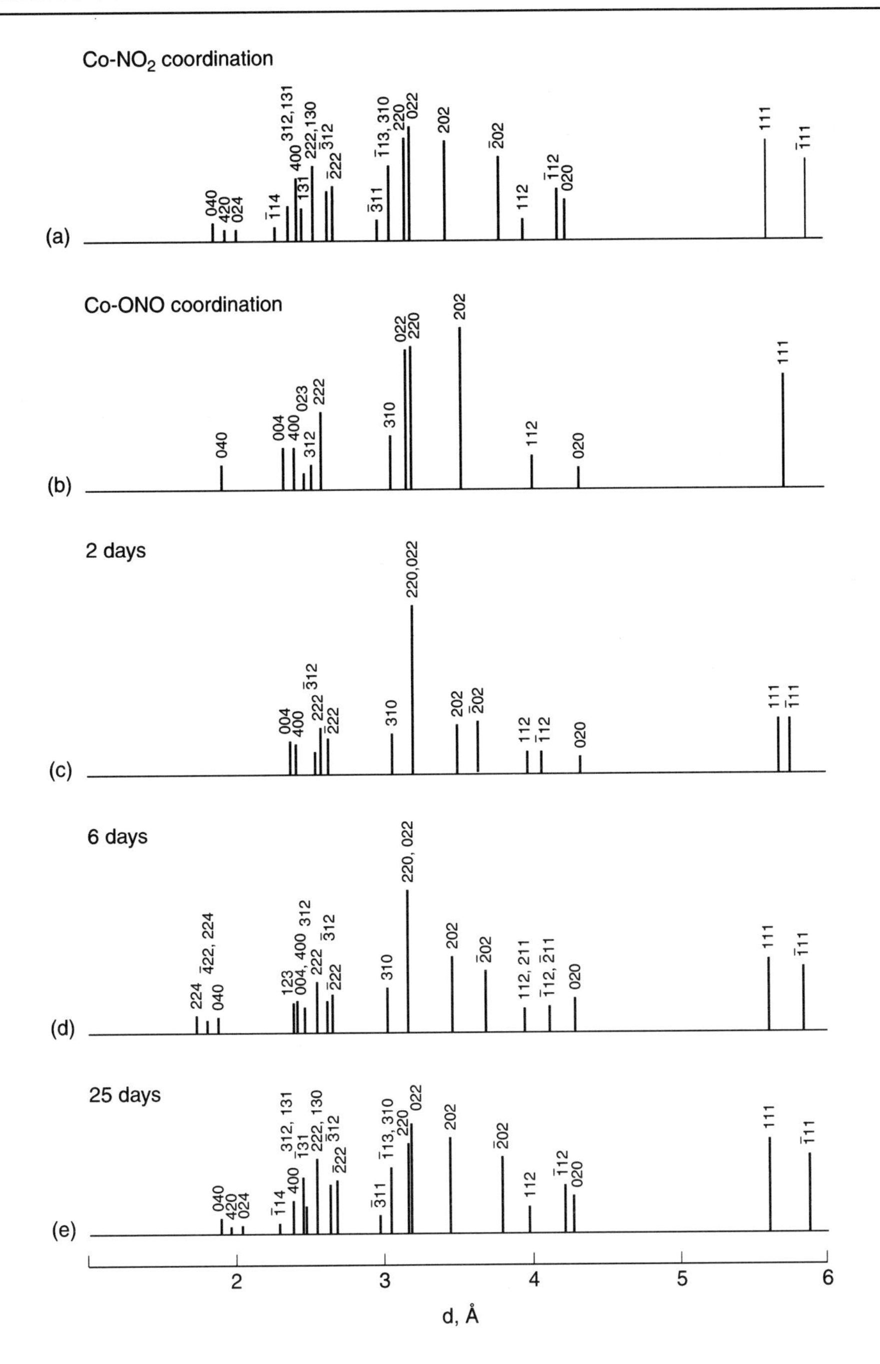

Figure 7.1 Powder X-ray diffraction diagrams for nitro–nitrito–nitro isomerization in $[Co(NH_3)_5NO_2]Br_2$. (a) Starting nitro-isomer, (b) nitrito-isomer, (c)–(e) solid solutions of $[Co(NH_3)_5NO_2]Br_2$ and $[Co(NH_3)_5ONO]Br_2$ after (correspondingly) 2, 6 and 25 days of storage of nitrito-isomer in the dark at room temperature. (From Boldyreva *et al.* 1993.)

crystals, achieved by the irradiation of solid nitro-isomer, as well an increase in the nitrito–nitro isomerization rate (E.V. Boldyreva and H. Ahsbahs unpublished results). Similar results were obtained when studying nitro–nitrito photoisomerization in single crystals of $[Co(NH_3)_5NO_2]Cl(NO_3)$. The effective quantum yield of the photoisomerization was shown to decrease in elastically compressed crystals (Boldyreva and

Sidel'nikov 1987; see also an English-language description of these experiments in the review of Boldyrev *et al.* 1990).

The effect of the environment of the isomerization in solution was interpreted in terms of the interactions between the complex cation and the outer-sphere species in an 'outer-sphere complex' (Jackson *et al.* 1982). The nitrito-ligand may interact with the outer-sphere species directly. An alternative is that the outer-sphere species interact with the ammine-ligands, in particular with *trans*-ammine and this also results in a redistribution of the electron density in the reacting nitro-group (the '*trans*-effect' well known in coordination chemistry, see Chernyaev 1926 as the original reference, or Basolo and Pearson 1958; Wilkinson *et al.* 1987 as more recent references).

When discussing the mechanism of the effect of the environment on the linkage isomerization in the solid state, one can also start with an analysis of the 'geometry of the outer-sphere complex', i.e. of the contact distances between the atoms of a complex cation and their nearest neighbours in the crystal structure, which form the 'walls of the reaction cavity'.

As an illustration, I can briefly comment on the results of a comparison of the crystalline environment of complex cations in nitro-isomers with different anions, $[Co(NH_3)_5NO_2]XY$ ($XY = Cl_2, Br_2, I_2, Cl(NO_3)$). Chloride and bromide are monoclinic (C2/c; see Börtin 1968; Cotton and Edwards 1968; and more precise recent data by Boldyreva *et al.* 1996). Iodide and chloride-nitrate are orthorhombic (Pnma; see Podberezskaya *et al.* 1991; Virovets *et al.* 1992; Boldyreva and Naumov 1996; Boldyreva *et al.* 1996). Despite different space symmetry of the crystal structures, in all four complexes the nearest environment of a complex cation turned out to be qualitatively similar. Negatively charged halide anions are located straight above and below the nitrogen atom of a nitro-group, whilst the oxygen atoms of nitro-ligands tend to form weak hydrogen bonds with *cis*-ammine-ligands of the neighbouring complex cations. *trans*-Ammine-ligands do not interact with the nitro-groups, but participate in the formation of hydrogen bonds with anions. As an example see in Fig. 7.2 the nearest environment of a complex cation in the crystals of $[Co(NH_3)_5NO_2]Cl_2$ (Fig. 7.2a) and $[Co(NH_3)_5NO_2]Cl(NO_3)$ (Fig. 7.2b). It is worth noting that the environment of the nitro-group in nitropenta-ammine complexes turned out to be quite similar to the environment of a nitro-group in chemically very different organic crystals (see composite crystal field environment for this group in the paper by Taylor *et al.* 1990, or the analysis of hydrogen-bond patterns formed between nitro- and ammine-groups in organic crystals by Panunto *et al.* 1987).

By changing the outer-sphere anions one can affect the distribution of electron density in the complex cation (and, in this way, the linkage isomerization) in two ways.

1 The anions can interact with the complex cation directly. An anion, hydrogen bonded to *trans*-ammine-ligand, inevitably also affects the electron-density distribution on the Co–Nitro bond and within the nitro-ligand (the *trans*-effect). Electrostatic interactions between the nitrogen of a nitro-ligand and a halide anion is another type of 'direct' cation–anion interaction.

2 A change in the outer-sphere anion affects the packing density of the crystal structure, and in such a way influences the interaction contacts, in particular the N–H--O interactions between the nitro-ligand of one cation and the *cis*-ammine-ligand of another (Table 7.1).

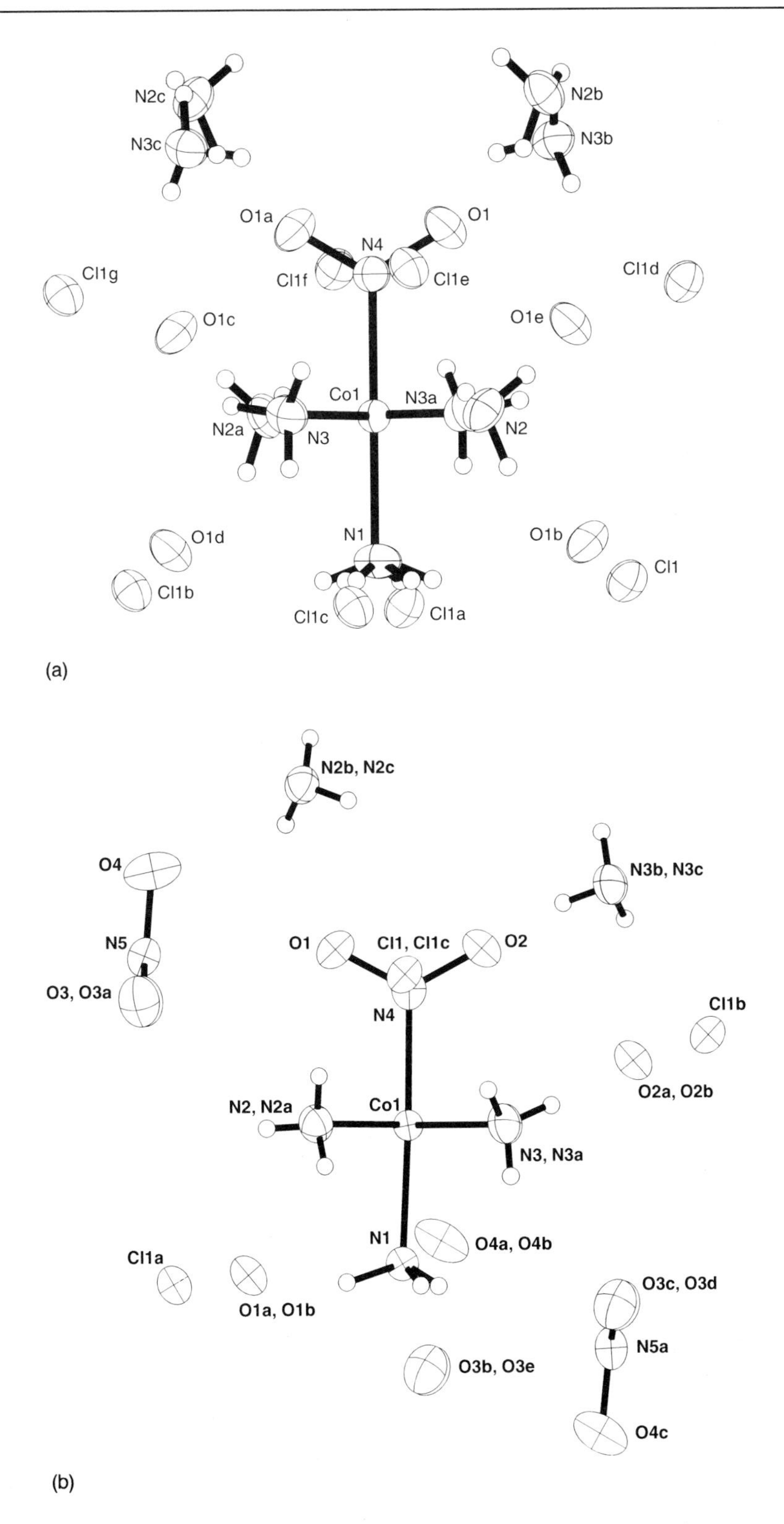

Figure 7.2 Environment of a complex cation in the crystal structures of (a) $[Co(NH_3)_5NO_2]Cl_2$ and (b) $[Co(NH_3)_5NO_2]Cl(NO_3)$, projection on the plane of the nitro-group. More explanations concerning generation of symmetry equivalent atoms and the lengths of the contacts can be found in the Appendix.

Table 7.1 Selected contacts of a complex cation with the nearest crystalline environment in the structures of $[Co(NH_3)_5NO_2]XY$ ($XY = Cl_2, Br_2, I_2, Cl(NO_3)$)

	$[Co(NH_3)_5NO_2]XY$			
	Cl_2	Br_2	I_2 (in Pnma)	$Cl(NO_3)$
Contacts of a nitro-ligand				
WITH HALIDE ANIONS				
N–Hal (Å)	0.3776	0.3917	0.4459	0.3661
O–Halammine	0.3816	0.3949	0.3510	0.3814
	0.4098	0.4234	0.5414	0.3838
WITH *cis*-LIGANDS				
O–N (Å)	0.2973	0.3078	0.3014	0.3096
	0.3089	0.3242	0.3191	0.3284
H–O (Å)	0.2526	0.2668	0.2568	0.2729
	0.2520	0.2668	0.2629	0.2607
N–H–O (°)	115.4	123.2	119.4	110.1
	128.2	110.3	132.6	144.4
Contacts of a trans-ammine-ligand				
N–Hal (Å)	0.3459	0.3542	0.3738	0.3391
	0.3485	0.3604	0.3750	
N–O (Å) (from NO_3 anion)	–	–	–	0.3016
				0.3348
				0.3713
*Packing coefficient**				
All the atoms in the structure	0.70	0.72	0.67	0.68
Only cations	0.51	0.49	0.39	0.46

*Packing coefficient was calculated as described in Virovets, 1992. It characterizes the ratio of ‘occupied space’ to ‘free space’ in the crystal structure. Calculating the packing coefficient for ‘cationic substructure’ only (the last line in the table) provides information on the role of ‘cation–anion interactions’ in the formation of the crystal structure.

One can see the similarity in such an interpretation of the effect of the environment on the isomerization in the solid state with the model proposed by Jackson *et al.* (1982) for an explanation of the effect of pH and metal ions on the same reaction in solution.

When considering a crystal (as compared with the solution), we have an advantage of knowing the precise geometry of an outer-sphere complex. These data can be used, for example, in the quantum chemistry calculations aimed to simulate the effect of the environment on the intramolecular geometry of the complex cation, on its vibration and electronic spectra and, finally, on its reactivity. These studies are at the very beginning, but there are already some preliminary experimental data that allow one to hope that such calculations might be fruitful. For example, there is experimental evidence that, small as they are, changes in the geometry of complex cation in different environments *are* observed. By way of example, Table 7.2 compares the bond lengths and bond angles in $[Co(NH_3)_5NO_2]$-cation in some nitro-complexes with different anions. One can see that the intramolecular bond lengths in $[Co(NH_3)_5NO_2]Cl(NO_3)$ are slightly, but reliably different from those in bromide and chloride. The effect is even

Table 7.2 Selected bond lengths and bond angles in complex cations in three complexes at 290 K*

	$[Co(NH_3)_5NO_2]XY$			
	Cl_2	Br_2	I_2 (in Pnma)	$Cl(NO_3)$
Bond Lengths (Å)				
Co(1)–N(1) (*trans*)	0.1995(3)	0.1992(4)	0.1976(6)	0.1988(2)
Co(1)–N(2) (*cis*)	0.1962(2)	0.1962(3)	0.1961(4)	0.1971(2)
Co(1)–N(3) (*cis*)	0.1956(2)	0.1965(3)	0.1969(7)	0.1959(2)
Co(1)–N(4) (nitro)	0.1921(2)	0.1931(4)	0.1927(1)	0.1961(2)
N(4)–O(1)	0.1237(2)	0.1235(3)	0.126(1)	0.1221(3)
N(4)–O(2)			0.117(1)	0.1214(3)
Bond angles				
Co(1)–N(4)–O(1)	120.5(1)	119.8(2)	122.5(6)	118.0(2)
Co(1)–N(4)–O(2)			118.7(7)	119.1(2)
O(1)–N(4)–O(1A) (O(1)–N(4)–O(2))	119.1(3)	120.3(4)	123.1(2)	122.9(2)

*The values from a recent redetermination of the crystal structures by Boldyreva *et al.* 1996.

more pronounced for $[Co(NH_3)_5NO_2]I_2$. For nitrito-isomers the few data already available, although not very precise and reliable because of the disordered orientation of the nitrito-groups in the crystals, seem to indicate that the internal geomertry of the $[Co(NH_3)_5ONO]$ cation is also sensitive to the crystalline environment. In particular, the value of the Co–O–N angle in a polymorph synthesized by irradiation of the corresponding nitro-isomers in the solid state, differs essentially from that in another polymorph of the same nitrito-isomer but crystallized from aqueous solution (Masciocchi *et al.* 1994). IR spectra and X-ray absorption spectra were also shown to be sensitive to the crystalline environment, whatever the 'tool' of modifying the environment was: changing the anion, crystallizing another polymorph or applying hydrostatic pressure (Siebert 1958; Dulepov 1992; Boldyreva *et al.* 1992; Kravtsova *et al.* 1995).

Until now we were discussing the possible mechanism of the effect of the crystalline environment on the intramolecular linkage isomerization as if the reaction cavity were rigid, and did not relax in the course of the reaction. However, this approximation turns out to be oversimplified.

Kinetics of the nitrito–nitro isomerization in the solid state deviates from the first-order kinetic law and does not proceed until full completion, i.e. a small amount of nitrito-isomer is always left in the sample (Beattie and Satchell 1952; Dulepov 1992; Phillips *et al.* 1990; Dulepov and Boldyreva 1994). As was noted in Section 6.2, this self-retardation of the reaction and a stop at some intermediate product/reactant ratio may be caused by several reasons: the existence of a thermal equilibrium, the 'kinetic non-equivalence' of different sites in the starting crystal or the negative feedback due to a change in the state of the reacting crystal in the course of the reaction. Kinetic analysis alone does not allow one to distinguish between these three different possibilities, and additional direct experimental data are required.

There is evidence in favour of the existence of a thermal nitrito–nitro equilibrium (Doron 1968); heating of specially prepared pure nitro-isomers in the dark was in fact shown to produce a small amount of nitrito-isomer in the sample (Dulepov 1992).

There is also the evidence that the crystals do 'respond' to the linkage isomerization in the complex cations and, hence, the state of the crystal and the effective rate constant may change in the course of the reaction. Optical microscopy, IR spectroscopy and X-ray diffraction show clearly that the crystal structure distorts continuously as the intramolecular linkage isomerization proceeds. The change in the coordination of the nitro-ligand to Co is a perturbation large enough to induce considerable strain in the crystal. Let us discuss the experiments giving this evidence in more detail.

Visual observations even without a microscope, but with the naked eye, show that the crystals are fragmented violently during the transformation (Fig. 7.3a). Under special experimental conditions, elastic bending of the crystals was observed (Fig. 7.3b) (Boldyreva *et al.* 1984). This indicates clearly that mechanical stresses arise in the crystal as a result of the reaction, and that the crystalline environment does respond to the intramolecular transformation in complex cations. Fragmentation, of course, is not reversible i.e. a crystal cannot be restored from the fragments by a reverse thermal nitrito–nitro isomerization. Deformation, however, *is* reversible, and a crystal bent in the course of nitro–nitrito photoisomerization restores its shape after storage at room temperature or after heating, as a reverse nitrito–nitro isomerization proceeds (Fig. 7.3c). Moreover, the properties of a crystal do not change even after a large number of cycles of irradiation–heating (Boldyreva and Sidel'nikov 1987; Boldyreva *et al.* 1988).

IR-frequency shifts of the symmetrical stretching vibrations of the nitrito-group were measured as the crystals of nitrito-isomers were undergoing the nitrito–nitro isomerization, see Fig. 7.4a as an example of such a shift in $[Co(NH_3)_5ONO]Br_2$ (Dulepov 1992; Dulepov and Boldyreva 1992, 1994; Boldyreva 1994). It is important to note that these were not the frequency shifts related to the change in the coordination type from nitrito to nitro, but the shifts of the vibration frequencies of the oxygen-coordinated ligands, observed as the ratio of nitrito/nitro-isomers in the crystal decreased. Similar frequency shifts were observed earlier when studying various solid state decomposition reactions, and were interpreted in terms of local stresses induced in the crystals (McBride 1983;

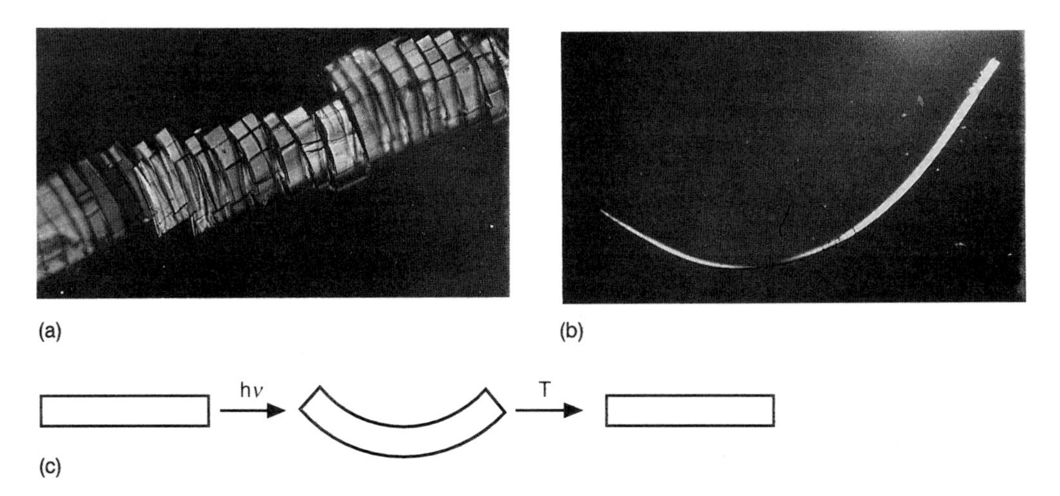

Figure 7.3 Fragmentation (a) and bending (b) of single crystals of $[Co(NH_3)_5NO_2]Cl\ (NO_3)$ in the course of photoisomerization. Reversibility of the deformation (c). (From Boldyreva *et al.* 1984, Boldyreva 1992b, 1994.)

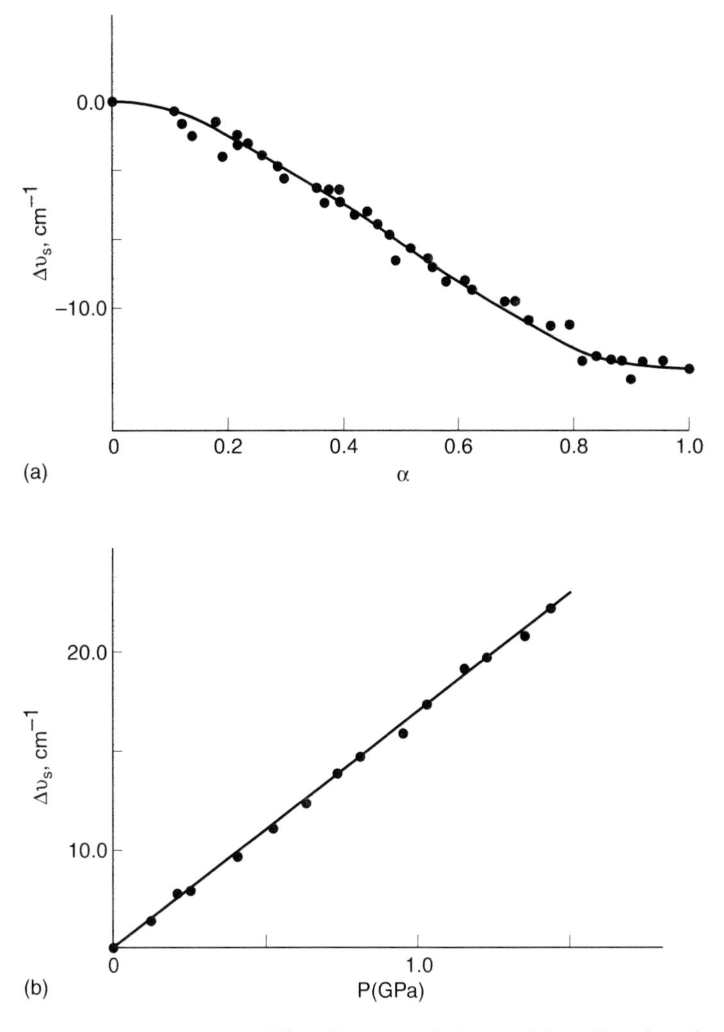

Figure 7.4 Frequency shifts of symmetrical stretching vibrations in the IR spectra of nitrito (oxygen-coordinated) ligand during (a) nitrito–nitro isomerization and (b) with pressure.

Chukanov *et al.* 1986; McBride *et al.* 1986; Berlyand *et al.* 1989; Hollingsworth and McBride 1990). Even some quantitative estimates of stress value (in terms of effective pressure) were made, based on a comparison of the frequency shifts observed in the course of reactions, with the corresponding pressure-induced frequency shifts. A comparison of the red frequency shift of the stretching symmetrical vibration of the nitrito-group in the course of the isomerization (Fig. 7.4a) (Dulepov 1992; Dulepov and Boldyreva 1992, 1994; Boldyreva 1994), with the blue shift of the same band induced by pressure (Fig. 7.4b) (Boldyreva *et al.* 1992) seem to be in good agreement with an assumption that during the nitrito–nitro isomerization—the crystal expands, and during the reverse nitro–nitrito transformation the crystal contracts; the values of the effective pressures achieved in the course of the reaction being estimated as 1 GPa.

A common method to evaluate quantitatively the expansion (contraction) of a crystal is X-ray powder diffraction. We have measured the lattice parameters of nitro-isomers of the series $[Co(NH_3)_5NO_2]XY$ ($XY = Cl_2$, Br_2, I_2, $Cl(NO_3)$, $(NO_3)_2$) and the lattice parameters of the corresponding nitrito-isomers formed as a result of the irradiation of the solid samples of nitro-isomers. Using these data, we have calculated the correspond-

Table 7.3 Lattice strain induced by solid state nitro–nitrito isomerization in five Co(III)–nitropenta-ammine complexes

	$[Co(NH_3)_5NO_2]XY$				
	Cl_2	Br_2	I_2 (in Pnma)	$Cl(NO_3)$	$(NO_3)_2$
$\Delta a/a$ (%)	+ 0.13(4)	+ 0.94(7)	≈ 0.0	- 2.34(5)	- 1.47(8)
$\Delta b/b$ (%)	+ 2.0(1)	+ 1.37(5)	- 1.2(1)	- 1.8(1)	- 1.47(8)
$\Delta c/c$ (%)	- 3.42(5)	- 3.47(5)	- 0.92(8)	+ 3.51(7)	+ 2.56(5)
$\Delta\beta$ (°)	- 4.78(2)	- 4.52(3)	—	—	—
$\Delta V/V$ (%)	- 0.95(6)	- 0.84(6)	- 2.1(2)	- 0.7(1)	- 0.4(1)
$\Delta l_1/l_1$ (%)	+ 3.01(2)	+ 3.38(4)	≈ 0.0	+ 3.51(7)	+ 2.56(5)
$\Delta l_2/l_2$ (%)	+ 1.99(2)	+ 1.37(2)	- 0.92(8)	- 1.8(1)	- 1.47(8)
$\Delta l_3/l_3$ (%)	- 5.96(2)	- 5.59(3)	- 1.2(1)	- 2.34(5)	- 1.47(8)

a, b, c, β are lattice parameters; Δa, Δb, Δc, $\Delta\beta$ are their changes in the course of isomerization; V is molar volume; gDV is its change in the course of isomerization; $\Delta l_1/l_1$, $\Delta l_2/l_2$, $\Delta l_3/l_3$ are linear strain along the principal axes of strain ellipsoid.

ing relative changes in the elementary cell volumes. A more careful analysis of the data on the changes in the lattice parameters revealed also a strong anisotropy of lattice distortion (Table 7.3).

Not taking into account the anisotropy of lattice distortion may lead to erroneous conclusions concerning the stresses arising in the crystal. For example, when irradiated, the single crystals of $[Co(NH_3)_5NO_2]Cl(NO_3)$ bend in such a way that their convex side faces the light source (see Fig. 7.3b) (Boldyreva *et al.* 1984), as if the molar volume increased. However, the molar volume of starting nitro-form is larger than that of the nitrito-one, but it is the anisotropy of lattice distortion (strong expansion along the axis of the needle and compression in perpendicular directions) that is responsible for the effect.

Because of the anisotropy of lattice distortion, it turned out to be too large an oversimplification to treat the mechanical stresses arising in the crystals during linkage isomerization in terms of internal pressure, despite any reasonably good correlations of spectroscopic data. Moreover, for bromide and chloride, monoclinic in nitro-form and orthorhombic after solid state nitro–nitrito photoisomerization (Börtin 1968; Cotton and Edwards 1968; Boldyreva *et al.* 1993; Masciocchi *et al.* 1994), it is not very intelligible to compare the changes in the linear lattice parameters, a, b and c, since the values of the angle parameter, β, also change. A correct analysis of the anisotropy of lattice distortion requires a calculation of the strain tensors based on the measured changes in the lattice parameters (Nye 1957; Hazen and Finger 1982). A comparison of the distortion of different structures should be based on the comparison of linear strain in the directions of the principal axes of strain tensor. These are the three directions which remain mutually perpendicular during distortion (Nye 1957). Table 7.3 (the last three lines) compares linear strain in the directions of principal axes of the strain tensor during nitro–nitrito photoisomerizations in five complexes with different anions (calculated on the basis of the experimental data of Boldyreva *et al.* 1993). One can also compare linear strain in some particular directions, for example with respect to the orientation of bonds in the complex cation (Fig. 7.5a; Fig. 7.5b will be discussed later).

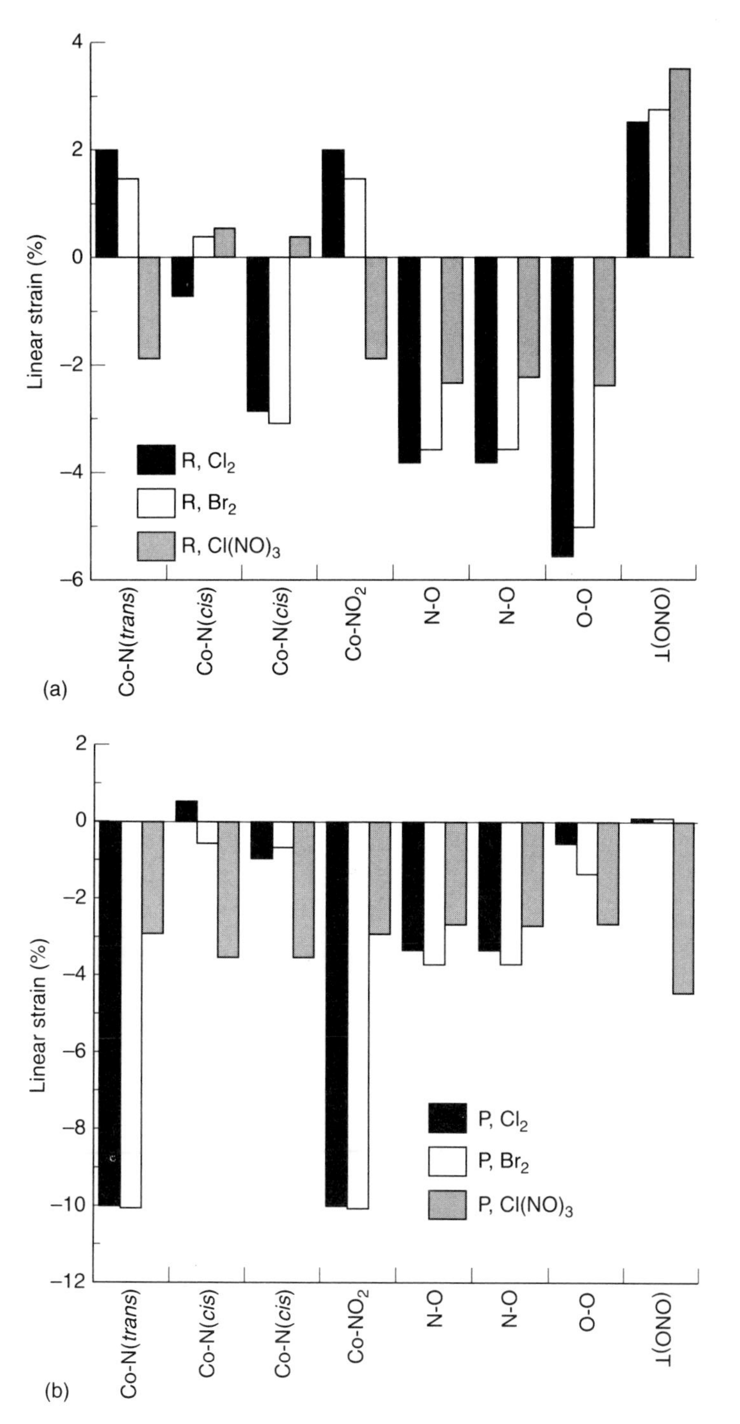

Figure 7.5 Linear strain in selected directions (directions of bonds in a complex cation) for structural distortion of $[Co(NH_3)_5NO_2]XY$ ($XY = Cl_2, Br_2, Cl(NO_3)$). (a) Lattice distortion in the course of linkage nitro–nitrito isomerization, (b) lattice distortion under hydrostatic pressure ($P = 2$ GPa).

From Table 7.3 and Fig. 7.5a one can clearly see some similarities in the lattice distortion, induced by nitro–nitrito isomerization in $[Co(NH_3)_5NO_2]XY$ complexes with different anions, X and Y. The most striking one is the fact that in all complexes the lattice expands in some directions, although the integral molar volume decreases as a result of nitro–nitrito isomerization. For example, it is obvious from Fig. 7.5a that in chloride, bromide and chloride-nitrate the lattice expands in the direction perpendicular to the plane of the nitro-group in the starting nitro-isomer. This seems to be in good agreement with the hypothesis that the terminal oxygen atom goes out of the original nitro-plane as nitro–nitrito isomerization proceeds. For the solid state photoisomerization in the bromide this was confirmed by a Rietveld solution and refinement of the structure of the nitrito-isomer, which gave fractional coordinates of the atoms in the product structure, in addition to the measured changes in lattice parameters (Masciocchi *et al.* 1994).

One can consider the environment of a complex cation in a structure and calculate linear strain in the directions of the shortest contacts between the atoms of the cation and the atoms of the environment. If lattice distortion in the course of the reaction is measured and the crystal structure of the product is known, one can consider as well the actual changes in the lengths of these contacts.* We have done this for nitro–nitrito isomerization in $[Co(NH_3)_5NO_2]Br_2$, based on our experimental data published in the paper of Masciocchi *et al.* (1994) (Fig. 7.6). One can clearly see from Fig. 7.6 that the reaction cavity is not at all rigid; the environment of complex cation does relax, the relaxation being essentially anisotropic.

One can also see from Fig. 7.5a and Table 7.3 that, despite the above-mentioned similarities, the crystal structure distortion of different complexes resulting from the isomerization is not identical. This proves that different structures respond differently to the same changes in complex cations. Different relaxation of the environment may be at least one of the factors responsible for the effect of the anion on the kinetics of the reaction in the solid state. Relaxation of mechanical stresses in the reacting crystals is known to be one of the main factors affecting kinetics and spatial propagation of solid state reactions in general (Chupakhin *et al.* 1987).

In order to get a better understanding of the role of the lattice relaxation in the solid state isomerization, it is necessary to know the intermolecular interactions in the crystals. Then, one could model the intramolecular reaction taking the relaxation of the crystal structure into account directly. One can hope to understand the interrelation between the intramolecular transformation and the response of the environment, taking into consideration electrostatic interactions between the nitro-ligands and the negatively charged anions, the hydrogen bonds between the anions and the ammine-ligands of complex cations or between the nitro-ligands of one cation and the ammine-ligands of the other.

One of the ways to achieve a better understanding of intermolecular interactions is to induce the distortion in the structure by different means other than the reaction (low

*Linear lattice strain in a particular direction, in general, is not necessarily equal to (and is not even necessarily proportional to) the relative changes in the interatomic distances between atoms which are joined in the starting structure by a vector pointing in this direction. This is what one can actually see in Fig. 7.6. There are two main reasons of this discrepancy. First, the distances between different pairs of atoms can change differently even if the vectors joining the atoms in these pairs are colinear (for example, obviously enough, intramolecular bonds and intermolecular distances usually change differently). Second, the orientation of the intramolecular bonds and the intermolecular contacts with respect to crystallographic axes and to the principal axes of strain tensor is, in general, changed as a result of distortion.

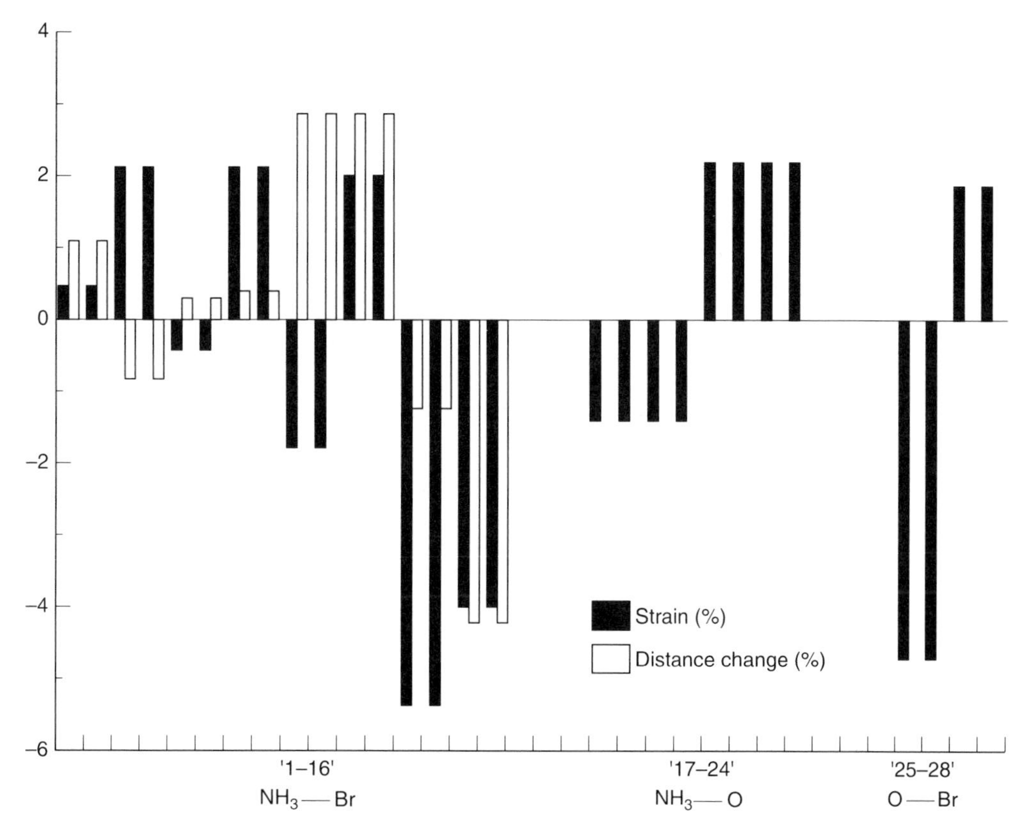

Figure 7.6 Structural distortion of $[Co(NH_3)_5NO_2]Br_2$ resulting from solid state nitro–nitrito linkage isomerization. Changes in the distances between the atoms of a complex cation and the neighbouring atoms, compared with lattice strain in the directions of these contacts in the starting structure of $[Co(NH_3)_5NO_2]Br_2$. Changes in the distances are shown only for ammine-Br contacts. Since nitro-ligand changes its coordination, it is not possible to correlate directly lattice strain and distance changes in the directions of ammine–O and O–Br contacts in the two isomeric forms. (For additional explanations see the Appendix)

temperature, high pressure) and to compare the results (Boldyreva 1994; Boldyreva *et al.* 1994, 1996).

• What is different in the crystal structure distortion of complexes with different anions, or in the distortion of different polymorphs, when distortion is induced by the same 'tool' (low temperature, high pressure, isomerization)?

• What is different in crystal structure distortion of the same compound when the 'tool' of distorting is varied?

• What are the reasons for different structural distortion? Which interactions in the crystal are mainly responsible for the observed anisotropy of structural distortion?

These questions are essential for understanding the role that the relaxation of the crystalline environment plays in the reaction. They are also important if we try to interpret the effect of some external actions, for example of hydrostatic pressure, on the reaction. We are presently at the very beginning of these studies and I would like to limit myself by giving only a few illustrative examples.

Anisotropy of structural distortion can be visualized most easily if a strain ellipsoid is plotted (Nye 1957). The strain ellipsoid shows how a sphere would be distorted if it were made of a material with the same anisotropy as the crystal structure under study

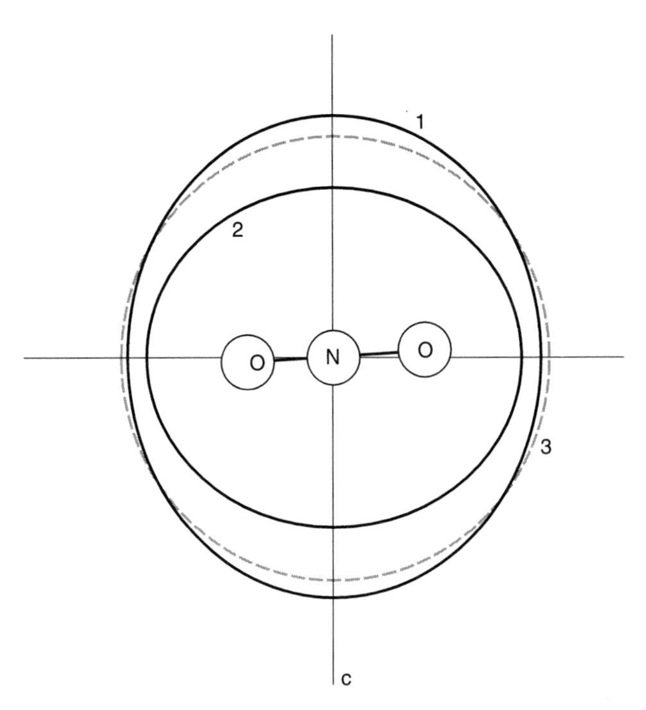

Figure 7.7 Principal sections of strain ellipsoids for the distortion of crystal structure of $[Co(NH_3)_5NO_2]Cl(NO_3)$: 1, nitro–nitrito isomerization (100% nitro→100% nitrito); 2, hydrostatic pressure ($P = 4$ GPa); 3, reference sphere (dashed line). a, c, crystallographic axes. Orientation of nitro-ligand is shown. Strain is exaggerated. (From Boldyreva *et al.* 1994.)

(Nye 1957). Figure 7.7 shows principal sections of pressure and reaction-induced strain ellipsoids calculated from the corresponding changes in the lattice parameters of $[Co(NH_3)_5NO_2]Cl(NO_3)$ (Boldyreva *et al.* 1993; Boldyreva 1994; Boldyreva *et al.* 1994). One can see clearly that the direction of major compression under pressure (*c* axis) is the direction of major expansion during the reaction. It can mean that in this direction the lattice is 'soft', and, depending upon the direction of the applied force, either contracts or expands most. It is interesting to note that the *c*-direction in $[Co(NH_3)_5NO_2]Cl(NO_3)$ also remains the 'softest' when the structure compression results from a decrease in temperature, although, in general, the anisotropies of compression under pressure and at low temperature are different.

The situation, however, is not always so simple. For example, in the structures of $[Co(NH_3)_5NO_2]Cl_2$ and $[Co(NH_3)_5NO_2]Br_2$ the 'softest' direction for distortions induced by pressure and by linkage isomerization do not coincide (Boldyreva *et al.* 1993, 1994, 1996). Moreover, even if distortion of the same structure is induced by two different scalar influences, such as low temperature and hydrostatic pressure, the softest and the most rigid directions do not coincide, and anisotropy of distortion is essentially different (Boldyreva *et al.* 1996). As an example, Fig. 7.8 compares linear strain in the directions of the principal axes of the strain tensor when the structure of $[Co(NH_3)_5NO_2]Br_2$ is distorted due to nitro–nitrito isomerization (Fig. 7.8a) (Masciocchi *et al.* 1994), with increasing pressure (Fig. 7.8b) (Boldyreva *et al.* 1994) and with a decrease in temperature ((Fig. 7.8c) Boldyreva *et al.* 1996). Fig. 7.9 shows the orientation of principal axes of strain tensors for all three types of distortion with respect to crystallographic axes.

The difference in the anisotropy of lattice distortion induced by different influences is obviously a manifestation of different intermolecular interactions, responsible for the compression in each case. In general, one can consider the following interactions in the crystals of $[Co(NH_3)_5NO_2]XY$ complexes: (i) Coulombic interactions between charged ions; (ii) van der Waals interactions; (iii) hydrogen bonds (N–H--O–N, N–H--Hal); and (iv) specific weak non-bonding interactions between nitro-groups and such anions as, for example, I. It is not yet possible to make a definite conclusion as to which of these interactions play the main role in each type of anisotropic compressions. Further experimental studies are required, first of all, including studies of the changes of atomic coordinates during compression induced by pressure. Based on these experimental data, one could undertake a computer simulation of structural distortion and try to estimate the relative changes in the contributions of each type of interaction to the total crystal energy. Common sense suggests that hydrogen bonds and specific interactions must be more important for distortion at low temperature, whilst distortion at high pressure can be expected to be determined first of all by optimal packing considerations. Future studies will show if common sense gives us a good prompt. There is evidence in the literature that hydrogen bonds do determine the anisotropy of thermal expansion (compression) of some other crystal structures (see, as examples, the early papers of Ubbelohde and co-workers (Ubbelohde 1939; Robertson and Ubbelohde 1939a, b; McKeown *et al.* 1951; Gallagher *et al.* 1955), as well as more recent publications (Nitta 1973; Whuler *et al.* 1978; Krishnan *et al.* 1979; Tanaka *et al.* 1982; Filhol and Thomas 1984; Swaminatham *et al.* 1984; Wang *et al.* 1985). Analysis of the data on the lattice strain induced in the crystals of $[Co(NH_3)_5NO_2]XY$ ($XY = Cl_2$, Br_2, $Cl(NO_3)$) by a decrease in temperature has shown that the directions of principal axes of strain ellipsoids seem to correlate with the directions of hydrogen-bonded 'bands' in these crystal structures. As an example, see Fig. 7.10 (Boldyreva *et al.* 1996).

The most difficult case is of course that of structural distortion induced by reaction. When a distortion of crystal structure results from the intramolecular isomerization, it reflects, first of all, the *intra*molecular reorganization, i.e. the way in which atoms change their positions within the complex cation. The lattice distortion, the changes in *inter*molecular juxtapositions, respond to this reorganization, and are secondary. One can expect, for example, that after the crystals of nitro-isomers are irradiated and the nitro–nitrito isomerization proceeds, the orientation of nitrito-ligand with respect to the surrounding atoms will be determined by both spatial distribution of free space in the structure and by the possibility of formation of hydrogen bonds between the terminal oxygen-atom and the ammine-ligands of the surrounding cations. Analysis of the structure of photochemically obtained $[Co(NH_3)_5ONO]Br_2$ (see the environment of a complex cation in the structure in Fig. 7.11) allows one to conclude that, at least for this particular complex, the possibility of formation of hydrogen bonds seems to be no less important than packing considerations. Formation of N–H--O bonds with the terminal oxygen of the nitrito-ligand in nitrito-coordination seems not only to direct the orientation of this group in the crystal structure, but also affects the Co–O–N linkage, making the Co–O–N close to (or even equal to) 180°, as if oxygen were a bridging atom. The anisotropy of lattice distortion observed in the course of nitro–nitrito isomerization may result from the tendency of the ammine-ligands of the neighbouring cations to approach the terminal oxygen of the nitrito-ligand, to facilitate the formation of

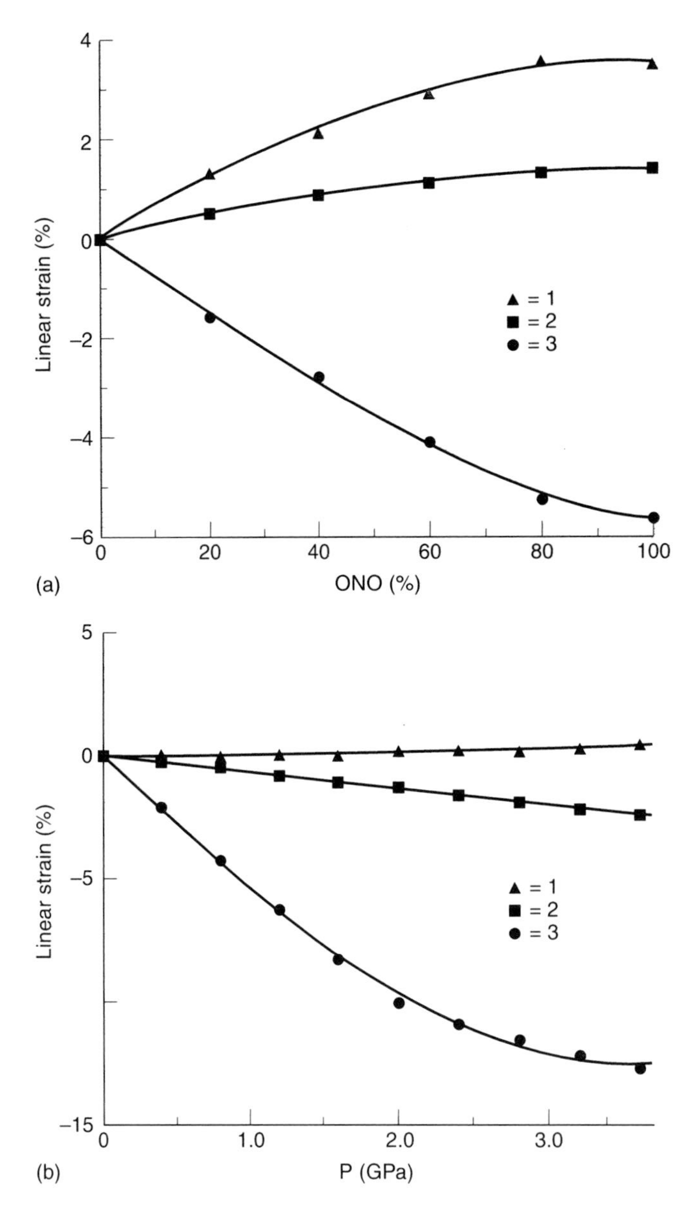

Figure 7.8 Distortion of the structure of $[Co(NH_3)_5NO_2]Br_2$ induced by, (a) nitro–nitrito isomerization, (b) high pressure and (c) (*opposite*) low temperature. Linear strain in the directions of principal axes 1, 2 and 3 of strain tensors. In accordance with the generally accepted system of notations (Nye 1957), 1, 2 and 3 are the principal axes, along which the strain ellipsoid has maximum, medium and minimum linear dimensions.

O--HN hydrogen bonds. One should also take into account that when the nitro-ligand changes its nitro-coordination for nitrito-coordination, the interactions between this group and the neighbouring halide anions in the structure are changed as well. This may induce the motion of the halide anions with respect to the cation. Since the halide anions are hydrogen bonded to *trans*-ammine-ligands of the neighbouring cations, this may also contribute to the observed changes in the relative juxtapositions of the cations. If one assumes that the interactions via the network of hydrogen bonds play an important role in the relaxation of crystal structure in the course of the intramolecular

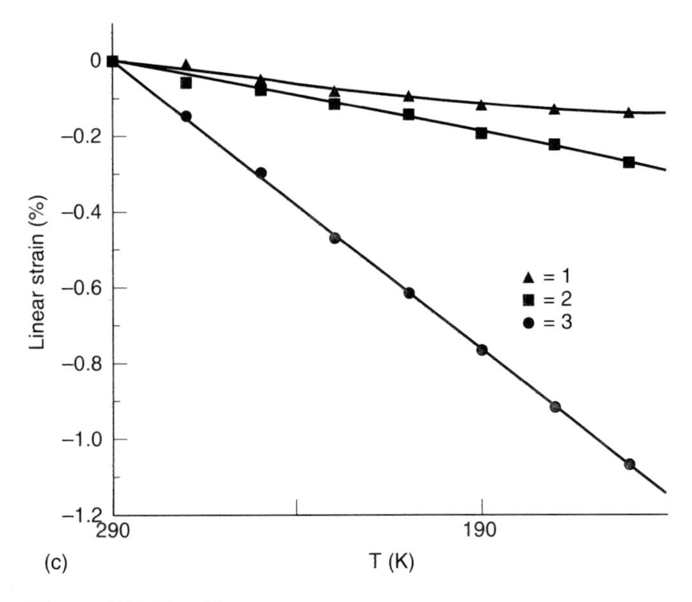

Figure. 7.8 *Cont'd.*

isomerization, one can find explanation for both the strong anisotropy of lattice distortion and of its reversibility as the back reaction proceeds.

A detailed comparison of distortion of the same structure in the course of the reaction and under high pressure is also important when considering the effect of pressure on this reaction in the solid state. As an example, compare in Fig. 7.5 and Fig. 7.7 lattice strain in $[Co(NH_3)_5NO_2]XY$ ($XY = Cl_2$, Br_2, $Cl(NO_3)$) induced by

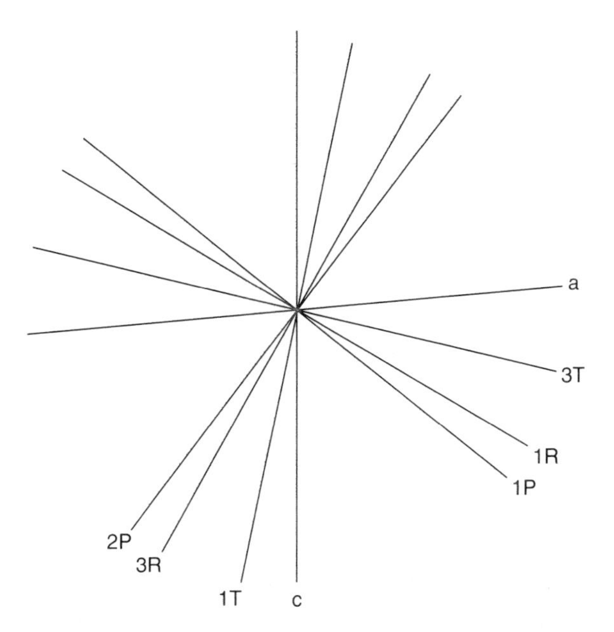

Figure 7.9 Distortion of the structure of $[Co(NH_3)_5NO_2]Br_2$ induced by nitro–nitrito isomerization (R), high pressure (P) and low temperature (T). Orientation of the principal axes of the corresponding strain tensors with respect to crystallographic axes. All the shown axes are in the crystallographic plane (a × c). The absent axes (2T, 3P and 2R) are perpendicular to this plane (parallel to the crystallographic axis b).

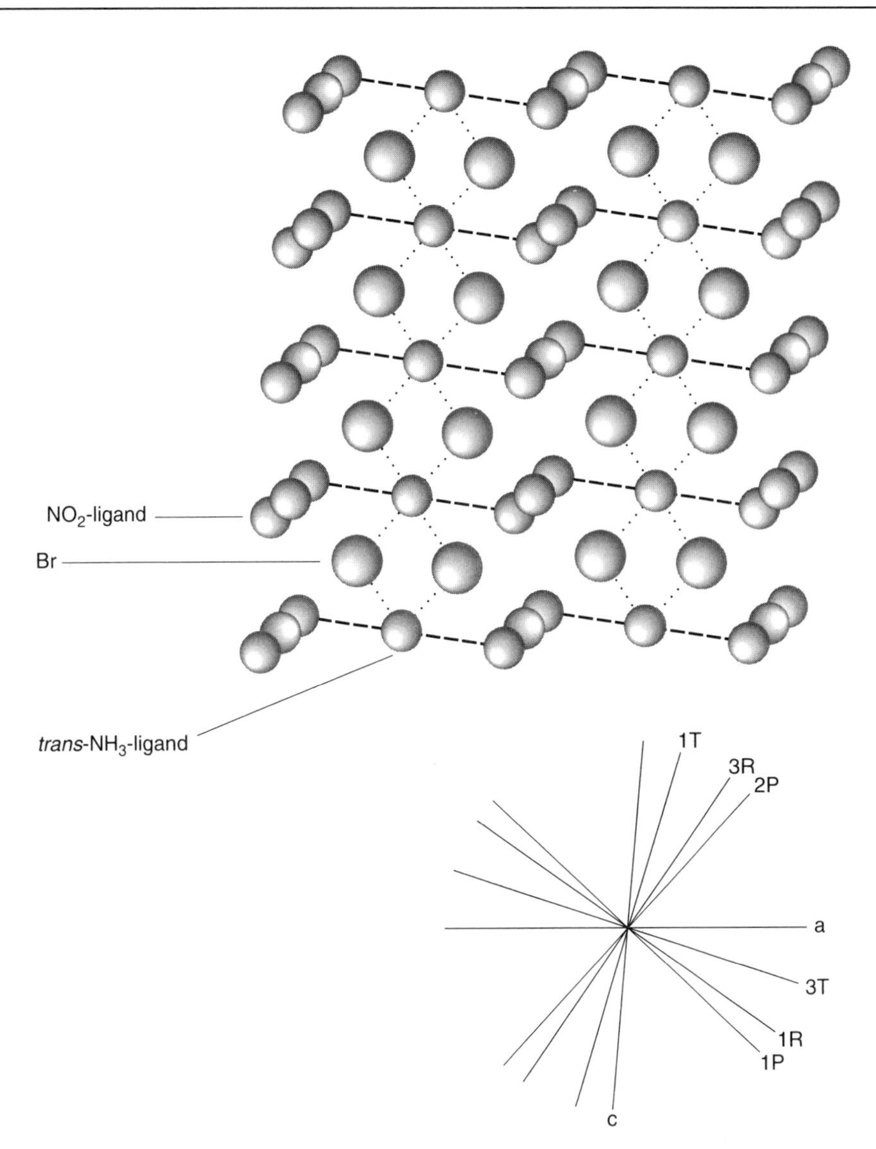

Figure 7.10 A fragment of the crystal structure of $[Co(NH_3)_5NO_2]Br_2$. Plane $(a \times c)$, atoms in a layer centred at the level $y = 0$. Dotted lines show hydrogen bonds *trans*-ammine-Br; dashed lines show long (4.572 Å at 290 K and 4.511 Å at 150 K) contacts between nitro-groups and ammine-ligands. Distances from the plane at 290 K: Br: ±0.01544b (±0.1365 Å); *trans*-ammine: ±0.0631b (±0.5577/Å); O: ±0.0765b (±0.6761 Å); N (nitro): ±0.0069b (±0.061 Å).

pressure, with lattice strain as a result of nitro–nitrito photoisomerization. This comparison helps one to understand why high pressure seems to inhibit the nitro–nitrito isomerization, and to facilitate the reverse nitrito–nitro transformation in the solid state (E.V. Boldyreva and H. Ahsbahs, unpublished results), similar to how it does in solution (Mares *et al.* 1978), although molar volume of nitrito-isomer in the solid state is smaller than that of nitro-isomer (Boldyreva *et al.* 1993). Despite the overall decrease in the molar volume in the course of nitro–nitrito isomerization, some of the directions in the structure expand considerably, and they seem to be most important for the reaction. Pressure results in contraction of the lattice in these directions, and thus inhibits the nitro–nitrito transformation. Local anisotropic changes in free volume at

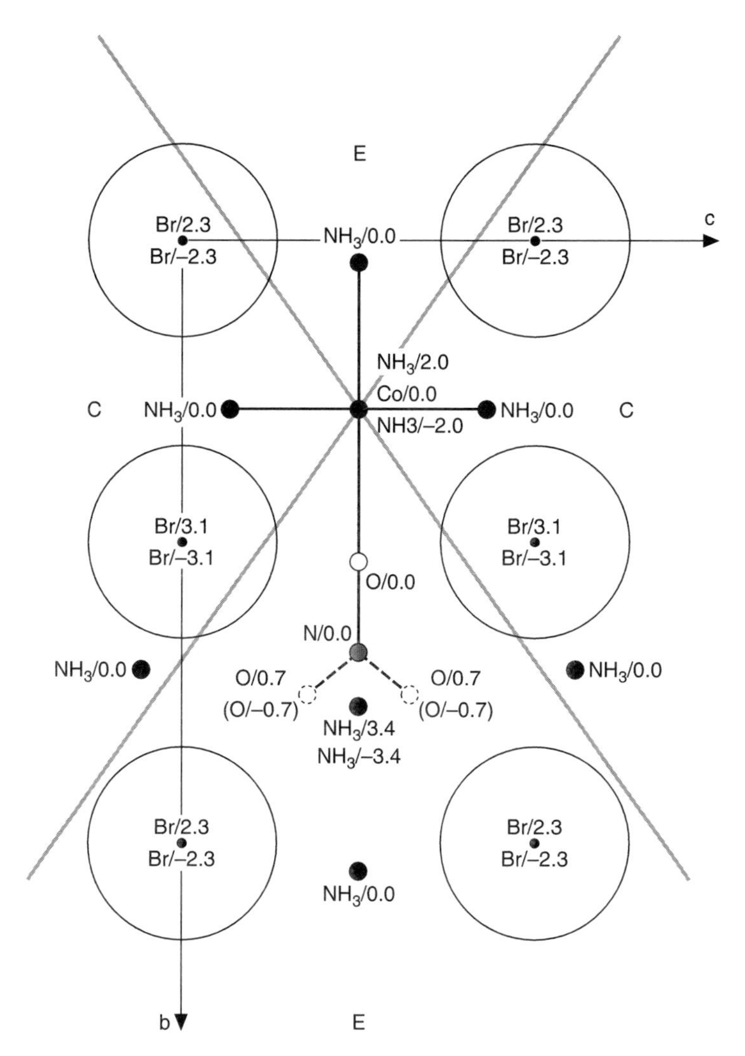

Figure 7.11 Environment of a complex cation in the structure of $[Co(NH_3)_5ONO]Br_2$ synthesized by irradiation of solid $[Co(NH_3)_5NO_2]Br_2$. The numbers show the distance of the atoms from the projection plane (Å); b and c are crystallographic axes; C and E are cones of compressed and expanded (as compared with the starting structure of nitro-isomer) directions in the structure of nitrito-isomer. The terminal oxygen atom is disordered over four equivalent positions (shown by dashed lines); hydrogen atoms are not shown.

the site of reaction turn out to have a larger effect on the reaction course than an averaged integral value of the change in the molar volume.

8 Summary and outlook

I hope to have succeeded, at least to some extent, in showing that homogeneous solid state reactions are both interesting and complicated and, therefore, are worthy of a detailed study. They give some challenges to a researcher as compared with the heterogeneous processes. One of the main challenges is that more 'routine' experimental techniques can be applied for their study, giving detailed information on the chemical and structural changes in the crystal during the reaction. A combination of complementary techniques, such as, for example, spectroscopy, X-ray diffraction, low-temperature and high-pressure experiments, etc., can give a really deep insight into the mechanism of the reaction.

Until now, the attention to homogeneous reactions was drawn mainly because of the practical interest in single crystal to single crystal solid state transformations. I am sure that the studies in this direction will continue, and more interesting materials and practically important processes will be found. At the same time, I would like to mention once again that studying homogeneous reactions can contribute considerably to solving some of the fundamental problems of chemistry in general and of solid state chemistry in particular.

- Comparing heterogeneous and homogeneous reactions, and trying to understand why the interface is formed in one case and is not formed in another, one can hope to be able finally to find the conditions under which the same solid state reaction will proceed either in a homogeneous or, alternatively, in a heterogeneous mode. This would mean a considerable progress in solving the problem of spatial control of solid state reactions (Boldyrev *et al.* 1979).
- Homogeneous solid state reactions can be good models for fundamental mechanochemistry.* One can follow experimentally at a quantitative level both the strain induced in the crystals *by* the reaction and the effect of stresses and strain *on* the reaction. One can also study, comparatively, strain in the same crystals that results from different ways of stressing the crystal. In such a way one can learn much of the interactions mainly responsible for a particular strain, including strain induced by the reaction itself. One can also get a better understanding of the mechanism of the effect of strain on the reaction (Boldyreva 1992b).
- This book is published within the series *Chemistry for the 21st Century*. There are numerous predictions that the main emphasis in the 21st century will be shifted from the chemistry of individual molecules to the chemistry of molecules either incorporated in some external environment or organized in different types of assemblies (see, as examples, Ramamurthy 1991; Balzani and Scandola 1991; Schneider and Durr 1991; Ramamurthy and Turro 1993; Weiss *et al.* 1993; Behr 1994; Lehn 1995). Solid state homogeneous reactions deserve finding their proper place in these studies. Concluding this chapter, I would like to give some more comments on this last statement.

The first models describing the effect of the environment on the intramolecular reactions were suggested for liquids. *Steric effects* were taken into account, for example, by studying correlations between the viscosity of the solution and the reaction rate. A steric model (Rabinowitch and Wood 1936) treated the environment as a 'cage' preventing a free escape of the reaction products, increasing the probability of their recombination and, in this way, effecting the reaction yield. At present, this model is often applied to radical reactions (for more recent references see, Bamford and Tipper 1969; Gurman and Pergushov 1981). Alternatively, an emphasis on the *electronic effects* is made when the reaction rate is correlated with the ionic force of solution or with its dielectric properties. One of the more general models takes into account both steric and electronic factors and considers the species of interest (solutes) to be

* *Mechanochemistry*—a branch of solid state chemistry dealing with the interrelation between the mechanical stresses and strain and the transformations of solids.

embedded in a 'cavity' of a continuous dielectric (solvent), the size, shape and surface charge of the cavity being the factors affecting the reactivity of the solute (see the review of Bishop 1994 for more details and for the references to earlier publications related to this model). Finally, environment can be considered not as a continuous medium, but in a more 'discrete' way, and formation of various intermolecular outer-sphere complexes or 'intermolecular associates' can be supposed to be responsible for the effect of the environment on the intramolecular properties and reactivity.

One of the main difficulties of studying the effect of the environment on the properties and reactions in liquid state is that the cavities (as well as intermolecular associates) in liquids are not very well defined and are transitory. Similar problems arise when the effect of environment is studied in glasses, liquid crystals or micelles (Holt 1982, Grätzel 1987; Ramamurthy 1991; Ramamurthy *et al.* 1993). On the contrary, the mutual juxtaposition of atoms in crystals is regular and much more stationary than that in liquid. It can be studied experimentally more easily, and any conclusions concerning the details of the structure of the environment of a molecule in a crystal are more definite and unambiguous than those made when studying liquids. It is also important that in a homogeneous solid the number of different types of environment is essentially limited, and is usually quite small, whilst in liquids there can be many different types of environment; their structure being subjected to permanent fluctuational changes. Therefore, the effect of solid host environments on the properties and reactivity of guest molecules attracts more and more attention (as examples, see, Turro 1986; Kalyanasundaram 1987; Ramamurthy 1991; Ramamurthy *et al.* 1993; Weiss *et al.* 1993). The first studies of solid state reactions took the models developed earlier for liquid state processes as a starting point (Salikhov *et al.* 1967; Boldyrev 1971). Nowadays, when so many efficient tools are available to study solid state structure and reactivity, solid state chemistry and, in particular chemistry of intramolecular homogeneous reactions, could themselves contribute to gaining new knowledge important for liquid state processes. One can say that using solids allows one to put the studies of the effect of the environment on a more solid basis.

Finally, a few words concerning possible contribution of studies of homogeneous solid state reactions to the problem of intermolecular interactions in molecular assemblies. Subtle intermolecular interactions (hydrogen bonds, specific non-covalent interactions such as, for example, I–nitro) are responsible for the processes of 'molecular recognition'. Biological activity of compounds is largely determined by specific intermolecular interactions. The intermolecular interactions determine the way in which the molecules 'self-organize' to form not only unusual large 'supramolecules' and 'self-assembled structures', but crystals—giant supermolecules which are really amazing in their regularity over enormous (as compared with atomic size) distances. Knowledge of intermolecular interactions forms the basis of 'supramolecular chemistry' (Lehn 1995) and 'crystal engineering' (Desiraju 1989, 1995). Crystal structures can be 'designed' in order to get materials with interesting physical properties (for example, non-linear optical materials), or to simulate some of the interactions of biological importance ('biomimetic chemistry'). Examples can be found in Desiraju (1989, 1995). Presently, the main information concerning the intermolecular interactions is derived from the comparative analysis of packing patterns in crystals of series of compounds or, alternatively, from physical (mainly spectroscopic) studies of individual solids in their stationary state. Studies of solid state homogeneous reactions could give important complementary knowledge of how the structures of solids affects intramolecular chemical transformations, how the crystal structures are distorted in response to intramolecular reactions and what is the role of intermolecular interactions in the observed effects. This would be not only important for studying solid state reactivity, but also would be helpful for a better understanding of intermolecular interactions.

9 Acknowledgements

The major part of this chapter was written during my stay as a Visiting Fellow at the Chemistry Department of Durham University (Great Britain). I thank the Royal Society for the financial support and Professor J.A.K. Howard for her warm hospitality.

10 References

Adell, B. (1952) *Z anorg allg Chem* **271**, 49–64.

Adell, B. (1955) *Z anorg allg Chem* **279**, 219–224.

Anderson, J. (1972) *Proceedings of the 7th Intern Symposium on Reactivity of Solids, Bristol,* (eds J. Anderson, M.W. Roberts & F. Stone), pp. 1–12. Chapman & Hall, London.

Ariel, S., Askari, S., Evans, S.V. *et al.* (1987) *Tetrahedron,* **43**, 7, 1253–1272.

Atkins, P.W. & Beran, J.A. (1992) *General Chemistry,* 2nd edn, p. B–10. Scientific American Books, New York.

Balzani, V. & Scandola, F. (1991) *Supramolecular Photochemistry.* Ellis Horwood, Chichester, England.

Bamford, C. & Tipper, C. (eds) (1969) *Comprehensive Chemical Kinetics,* Vol. 2, Ch. 4, pp. 302–376. Elsevier, Amsterdam.

Barret, P. (ed) (1975) *Reaction Kinetics in Heterogeneous Chemical Systems.* Elsevier, Amsterdam.

Basilevsky, M.V., Gerasimov, G.N. & Petrochenko, S.I. (1985) *Chem Phys* **97**, 331–343.

Basolo, F. & Pearson, R. (1958) *Mechanisms of Inorganic Reactions. A Study of Metal Complexes in Solution.* Wiley, New York.

Baughman, R.H. (1978) *J Chem Phys* **68 (7)**, 3110–3121.

Baughman, R.H. & Chance, R.R. (1978) *Ann NY Acad Sci* **313**, 705–725.

Baughman, R.H. & Chance, R.R. (1980) *J Chem Phys* **73 (8)**, 4113–4125.

Baughman, R.H. & Yee, K.C. (1978) *J Polym Sci Macromol Rev* **13**, 219–239.

Beattie, I.R. & Satchell, D.P.N. (1952) *Trans Farad Soc* **52**, 1590–1593.

Behr, J.-P. (ed.) (1994) *Perspectives in Supramolecular Chemistry,* Vol. 1. Wiley, New York.

Ben Amor, M. & Mutin, J.C. (1982) *CR Seances Acad Sci Ser 2* **295 (6)**, 665–668.

Ben Amor, M., Mutin, J.C., Aubry, A. & Courtois, A. (1983a) *J Solid State Chem* **48 (2)**, 215–230.

Ben Amor, M., Mutin, J.C., Aubry, A. & Courtois, A. (1983b) *J Solid State Chem* **48 (2)**, 231–241.

Berlyand, L.V., Chukanov, N. & Korsunski, B. (1989) *Proceedings of the Xth All-Union Conference on the Kinetics and Mechanisms of Chemical Reactions in Solids, Alma-Ata,* Vol. 1, p. 208. Institute of Chemical Physics, Chernogolovko, Nauka.

Bernal, J.D. (1960) *Schweiz Arch* **26**, 69–75.

Berrehar J., Lapersonne-Meyer, C., Schott, M. & Villain, J. (1989) *J Phys France* **50**, 923–935.

Bertrand, G., Sanfeld, A., Watelle, G. & Larpin, J.-P. (1986) *J Chim Phys* **83 (11/12)**, 695–702.

Bishop, D.M. (1994) *Int Rev Phys Chem* **13 (1)**, 21–39.

Bloor, D. (1983) *Mol Cryst Liq Cryst* **93 (1–4)**, 183–199.

Bloor, D. (1985) *NATO ASI Ser Ser E,* **102**, 1–24.

Bloor, D., Day, R.J., Ando, D.J. & Motevalli, M. (1985) *Br Polym J* **17 (3)**, 287–293.
Boldyrev, V.V. (1971) *Int J Rad Phys Chem* **3**, 155–169.
Boldyrev, V.V. (1973) *Russ Chem Rev (Uspekhi Khim)* **42**, 1161–1183.
Boldyrev, V.V. (1977) *Izv Sib Otd Akad Nauk SSSR Ser Khim Nauk* **3**, 48–58.
Boldyrev, V.V., Bulens, M. & Delmon, B. (1979) *Control of the Reactivity of Solids.* Elsevier, Amsterdam.
Boldyrev, V.V., Lyakhov, N.Z., Pavlyukhin, Yu.T., Boldyreva, E.V., Ivanov, E.Yu. & Avvakumov, E.G. (1990) *Sov Sci Rev B Chem* **14**, 105–161.
Boldyreva, E.V. (1982) *Izv Sib Otd Akad Nauk SSSR Ser Khim Nauk* **5**, 18–27.
Boldyreva, E.V. (1987a) *React Solids* **(3)**, 185–203.
Boldyreva, E.V. (1987b) *React Solids* **(3)**, 205–225.
Boldyreva, E.V. (1988) *Feed-back and its effect on the kinetics and spatial propagation of 'intramolecular' solid state reactions.* PhD. Thesis. Novosibirsk Institute of Solid State Chemistry, Russia.
Boldyreva, E.V. (1990) *React Solids* **8**, 269–282.
Boldyreva, E.V. (1992a) *J Thermal Anal* **38**, 89–97.
Boldyreva, E.V. (1992b) *Proceed IV-th Jpn–Russ Symp Mechanochem.* 125–135.
Boldyreva, E.V. (1994) *Mol Cryst Liq Cryst Inc Nonlin Opt* **242** (2), 17–52.
Boldyreva, E.V. & Naumov, D.Yu. (1996) To be published.
Boldyreva, E.V. & Salikhov, K.M. (1985) *React Solids* **1** (1), 3–17.
Boldyreva, E.V. & Sidel'nikov, A.A. (1987) *Izv Sib Otd Akad Nauk SSSR Ser Khim Nauk* **5**, 139–145.
Boldyreva, E.V., Ahsbahs, H. & Uchtmann, H. (1994) *Ber Bunsenges Phys Chem* **98 (5)**, 738–745.
Boldyreva, E.V., Kivikoski, Ju. & Howard, J.A.K. (1996) Submitted to *Acta Cryst B.*
Boldyreva, E.V., Sidel'nikov, A.A., Chupakhin, A.P., Lyakhov, N.Z. & Boldyrev, V.V. (1984) *Dokl Akad Nauk SSSR* **277**, 893–896.
Boldyreva, E.V., Sidel'nikov, A.A., Rukosuev, N.I., Chupakhin, A.P. & Lyakhov, N.Z. (1988) *Photometer.* Russian patent N.1368654 Al.
Boldyreva, E.V., Virovets, A.V., Burleva, L.P., Dulepov, V.E. & Podberezskaya, N.V. (1993) *Russ J Struct Chem Russ Ed (Z Strukt Khim)* **34**, 128–138.
Boldyreva, E.V., Burgina, E.B., Baltakhinov, V.P. *et al.* (1992) *Ber Bunsenges Phys Chem* **96 (7)**, 931–937.
Börtin, O (1968) *Acta Chem Scand* **22**, 2890–2898.
Braga, D., Grepioni, F., Johnson, B., Lewis, J. & Martinelli, M. (1990) *J Chem Soc Dalton Trans* 1847–1852.
Braun, H. & Wegner, G. (1983a) *Makromol Chem* **184 (5)**, 1103–1119.
Braun, H. & Wegner, G. (1983b) *Mol Cryst Liq Cryst* **96 (1–4)**, 121–139.
Braunschweig, F. & Bässler, H. (1980) *Ber Bunsenges Phys Chem* **84**, 177–181.
Burstein, A.I., Klyuchikhin, V.L. & Trakhtenberg, L.I. (1984) *Khim Phyz (Russ J Chem Phys)* **2**, 155–161.
Cahn, J.W. (1962) *Acta Metallurg* **10**, 907–913.
Cahn, J.W. (1968) *Trans Met Soc American Institute of Mechanical Engineers* **242**, 166–180.
Chang, H., Popovitz-Biro, R., Lahav, M. & Leiserovitz, L. (1982) *J Am Chem Soc* **104**, 614–616.
Cheng, K. & Foxman, B. (1977) *J Am Chem Soc* **99**, 8102–8103.
Chernyaev, I. (1926) *Izvest Inst Izuchen Platiny* **4**, 213.
Chukanov, N., Golovanova, O. & Korsunski, B. (1986) *Proceedings of the IX-th All-Union Conference on the Kinetics and Mechanisms of Chemical Reactions in Solids, Alma-Ata*, p. 38. Institute of Chemical Physics, Chernogolovka, Nauka.
Chupakhin, A.P., Sidel'nikov, A.A. & Boldyrev, V.V. (1987) *React Solids* **3**, 1–19.
Cohen, M.D. (1975) *Angew Chem Int Ed Engl* **14**, 386–393.
Cohen, M.D. (1979) *Mol Cryst Liq Cryst* **50**, 1–10.
Cohen, M.D. (1987) *Tetrahedron* **43 (7)**, 1211–1225.

Collins, M.A. & Craig, D.P. (1981) *Chem Phys* **54**, 305–321.
Cotton, F.A. & Edwards, W. (1968) *Acta Cryst* **B24**, 474–477.
Craig, D.P. & Mallett, C.P. (1982) *Chem Phys* **65**, 129–142.
Craig, D.P., Lindsay, R.N. & Mallett, C.P. (1984) *Chem Phys* **89**, 187–197.
Desiraju, G.R. (1989) *Crystal Engineering. The Design of Organic Solids*. Elsevier, Amsterdam.
Desiraju, G.R. (1995) *Angew Chem* **107**, 2541–2558.
Doba, T., Ingold, K. & Siberand, W. (1984) *Chem Phys Lett* **103**, 339–342.
Doron, V.E. (1968) *Inorg Nucl Chem Lett* **4**, 601–606.
Dulepov, V.E. (1992) *Effect of crystal structure and outer-sphere anion on the intra-sphere linkage isomerization in nitrito-penta-ammines of Co(III)*. Graduation paper. Novosibirsk State University, Russia.
Dulepov, V.E. & Boldyreva, E.V. (1992) *Siberian Chem J* **5**, 109–117.
Dulepov, V.E. & Boldyreva, E.V. (1994) *React Kinet Catal Lett* **53 (2)**, 289–296.
Enkelmann, V., Wegner, G., Novak, K. & Wagener, K.B. (1993) *J Am Chem Soc* **115**, 10 390–391.
Enkelmann, V., Wegner, G., Novak, K. & Wagener, K.B. (1994) *Mol Cryst Liq Cryst Inc Nonlin Opt* **242**, 121–127.
Feitknecht, W. (1964) *Pure Appl Chem* **9 (3)**, 423–440.
Filhol, A. & Thomas, M. (1984) *Acta Cryst* **B40**, 44–59.
Freund, F. (1968) *Fortsch Chem Forsch* **10**, 347–373.
Gallagher, K., Ubbelohde, A.R. & Woodward, I. (1955) *Acta Cryst* **8**, 561–566.
Galwey, A.K., Laverty, G.M., Baranov, N.A. & Okhotnikov, V.B. (1994) *Phil Trans Roy Soc Lond A* **347**, 139–156, 157–184.
Garner, W.E. (1955) *Chemistry of the Solid State*, Ch. 8. Butterworth, London.
Gavezzotti, A. (1987a) *Acta Cryst* **43B**, 559–562.
Gavezzotti, A. (1987b) *Tetrahedron* **43**, 1241–1251.
Gavezzotti, A. (1988) *Mol Cryst Liq Cryst Inc Nonlin Optics* **156A**, 25–34.
Gavezzotti, A. & Bianchi, R. (1986) *Chem Phys Lett.* **128**, 295–299.
Gavezzotti, A. & Simonetta, M. (1982) *Chem Rev* **82 (1)**, 1–13.
Gillot, B. & Rousset, A. (1994) *Heterog Chem Rev* **1 (1)**, 69–98.
Gillot, B., Rousset, A. & Dupre, G. (1978) *J Solid State Chem*, **25**, 263–271.
Giovanoli, R. & Brutsch, R. (1974) *Chimia* **28 (4)**, 188–191.
Giovanoli, R. & Brutsch, R. (1975) *Thermochem Acta* **13**, 15–36.
Grätzel, M. (1987) *Tetrahedron* **43 (7)**, 1679–1688.
Grenthe, I. & Nordin, E. (1979a) *Inorg Chem* **18**, 1109–1116.
Grenthe, I. & Nordin, E. (1979b) *Inorg Chem* **18**, 1869–1874.
Guarini, G.G.T. & Dei, L. (1983) *J Chem Soc Farad Trans I* **79**, 1599–1604.
Guarini, G.T. & Magnani, A. (1988) *React Solids* **6**, 277–280.
Gurman, V.S. & Pergushov, V.I. (1981) *Chem Phys* **55**, 131–135.
Hasegawa, M. (1983) *Chem Rev* **83 (4)**, 507–534.
Hasegawa, M., Arioka, H., Harashina, H., Nohara, M., Kubo, M. & Nishikubo, T. (1985) *Israel J Chem* **25**, 302–305.
Hazen, R. & Finger, L. (1982) *Comparative Crystal Chemistry Temperature, Pressure, Composition, and the Variation of the Crystal Structure*. Wiley, New York.
Hillert, M. (1961) *Acta Metallurg* **9**, 525–535.
Hirshfeld, F.L. & Schmidt, G.M.J. (1964) *J Polym Sci* **A2**, 2181–2190.
Hitchman, M.A. & Rowbottom, G.L. (1982) *Coord Chem Rev* **42**, 55–132.
Hollingsworth, M. & McBride, J.M. (1990) In: *Advances in Photochemistry*, Vol. 15, (eds D. Vollman, G. Hammond & K. Gollnick), pp. 279–379 Wiley, New York.
Holt, S. (ed.) (1982) *Inorganic Reactions in Organized Media*, p. 177. American Chemistry Society Symposium Series.
Jackson, W.G., Lawrance, G.A., Lay, P.A. & Sargeson, A.M. (1982) *Austr J Chem* **35**, 1561–1580.
Jessen, S.M. & Küppers, H. (1991) *J Appl Cryst* **24**, 239–242.

Jorgensen, S. (1893) *Z Anorg Chem* **5**, 168–174.
Kaiser, J. & Wegner, G. (1972) *Reactivity of Solids. Proceedings of the ISRS 7th, 1972* (ed. J.S. Anderson). Chapman & Hall, London, 506–514.
Kaiser, J., Wegner, G. & Fischer, E.W. (1972a) *Israel J Chem* **10**, 157–171.
Kaiser, J., Wegner, G. & Fischer, E.W (1972b) *Kolloid-Z Z Polym* **250 (11–12)**, 1158–1161.
Kalyanasundaram, K. (1987) *Photochemistry in Microheterogeneous Systems.* Academic Press, New York.
Kamiya, N., Tanaka, I. & Iwasaki, H. (1993) In: *Reactivity in Molecular Crystals* (ed. Y. Ohashi), pp. 25–44. VCH, Tokyo.
Kearsley, S.K. & Desiraju, G.R. (1985) *Proc Roy Soc Lond* **A 397**, 157–181.
Kravtsova, E.A., Boldyreva, E.V., Mazalov, L.N. *et al.* (1996) *Zh Strukt Khim (Russ J Struct Chem)* In press.
Krishnan, R.S., Srinivasan, R. & Devanarayanan, S. (1979) *Thermal Expansion of Crystals, International Series in the Science of the Solid State*, Vol. 12. Pergamon Press, Oxford.
Kubota, M. & Ohba, S. (1992) *Acta Cryst* **B48**, 627–632.
Kutyrkin, V.A., Madarashvili, I.R., Karpukhin, O.N. & Anisimov, V.M. (1984) *Kinet Katal (Russ J Kinet Catal)* **25 (6)**, 1310–1314.
Lamartine, R., Decoret, C., Royer, J. & Vicens, J. (1986) *Mol Cryst Liq Cryst* **134**, 197–218.
Langmuir, I. (1916) *J Am Chem Soc* **38 (11)**, 2221–2295.
Lehn, J.-U. (1995) *Supramolecular Chemistry.* VCH, Weinheim.
Lebedev, Ya.S. (1978) *Kinet Katal (Russ J Kinet Catal)* **19 (6)**, 1367–1376.
McBride, J.M. (1983) *Acc Chem Res* **16**, 304–312.
McBride, J.M., Segmuller, B., Hollingsworth, M., Mills, D. & Weber, B. (1986) *Science* **234**, 830–835.
McKeown, P.J.A., Ubbelohde, A.R. & Woodward, I. (1951) *Acta Cryst* **4**, 391–395.
Mares, M., Palmer, D.A. & Kelm, H. (1978) *Inorg Chim Acta* **27**, 153–156.
Martin, M. (1991) *Mat Sci Rep* **7 (1/2)**, 1–86.
Masciocchi, N., Kolyshev, A., Dulepov, V., Boldyreva, E. & Sironi, A. (1994) *Inorg Chem* **33**, 2579–2585.
Misra, T.N. & Prasad, P.N. (1982) *Chem Phys Lett* **85 (4)**, 381–386.
Murmann, R. & Taube, H. (1956) *J Am Chem Soc* **78**, 4886–4890.
Murthy, G.S., Arjunan, P., Venkatesan, K. & Ramamurthy, V. (1987) *Tetrahedron* **43 (7)**, 1225–1240.
Mutin, J.C. & Dusausoy, Y. (1981) *J Solid State Chem* **38 (3)**, 394–405.
Mutin, J.C. & Watelle-Marion, G. (1972) *Bull Soc Chem Fr* **12**, 4488–4492.
Mutin, J.C. & Watelle-Marion, G. (1977) *J Phys (Paris) Colloq* **7**, 123–127.
Mutin, J.C. & Watelle-Marion, G. (1979) *J Solid State Chim* **28 (1)**, 1–12.
Mutin, J.C., Watelle-Marion, G. & Dusausoy, Y. (1979) *J Solid State Chem* **27 (3)**, 407–421.
Nakanishi, H., Hasegawa, M. & Sasada, Y. (1977a) *J Polym Sci Polym Phys Ed* **15 (1)**, 173–191.
Nakanishi, H., Jones, W. & Thomas, J. (1980a) *J Chem Soc Chem Commun* **N13** 611–612.
Nakanishi, H., Jones, W. & Thomas, J. (1980b) *Chem Phys Lett* **71**, 44–48.
Nakanishi, H., Hasegawa, M., Kirihara, H. & Yurugi, T. (1977b) *Nippon Kagaku Kaishi* **7**, 1046–1050.
Nakanishi, H., Jones, W., Thomas, J., Hasegawa, M. & Rees, W.L. (1980c) *Proc Roy Soc* **A369** (1738), 307–325.
Nakanishi, H., Jones, W., Thomas, J., Hursthouse, M. & Motevalli, M. (1981) *J Phys Chem* **85**, 3636–3642.
Niepce, J.C. & Watelle-Marion, G. (1973) *CR Acad Sci Ser C* **276 (8)**, 627–630.
Niepce, J.C., Dumas, P. & Watelle-Marion, G. (1973) *Fine particles. Proceedings of the 2nd International Conference of the Electrochemical Society, Boston,* (ed. W.E. Kuhn), pp. 256–268. Princeton.
Nitta, I. (1973) *Acta Cryst* **A29**, 317.
Norris, K., Gray, P., Craig, D.P., Mallett, C.P. & Markey, B.R. (1983) *Chem Phys* **79**, 9–19.

Novak, K., Enkelmann, V., Wegner, G. & Wagener, K.B. (1993) *Angew Chem Int Ed Engl* **32 (11)**, 1614–1616.

Novak, K., Enkelmann, V., Köhler, W., Wegner, G. & Wagener, K.B. (1994) *Mol Cryst Liq Cryst Inc Nonlin Opt* **242**, 1–8.

Nye, J. (1957) *Physical Properties of Crystals: Their Representation by Tensors and Matrices*, 1st edn. Oxford University Press, Oxford.

Ohashi, Y. (1988) *Acc Chem Res* **21**, 268–274.

Ohashi, Y., Uchida, A. & Sekine, A. (1993) In: *Reactivity in Molecular Crystals* (ed Y. Ohashi), pp. 115–153. VCH, Tokyo.

Ohashi, Y., Uchida, A., Sasada, Y. & Ohgo, Y. (1983) *Acta Cryst* **B39**, 54–61.

Ohashi, Y., Yanagi, K., Kurihara, T., Sasada, Y. & Ohgo, Y. (1982) *J Am Chem Soc* **104**, 6353–6359.

Ohgo, Y. & Arai, Y. (1993) In: *Reactivity in Molecular Crystals* (ed. Y. Ohashi), pp. 263–275. VCH, Tokyo.

Panunto, T.W., Urbanczyk-Lipkowska, Z., Johnson, R. & Etter, M. (1987) *J Am Chem Soc* **109**, 7786–7797.

Penland, R.B., Lane, T.J. & Quagliano, J.V. (1956) *J Am Chem Soc* **78**, 887–889.

Phillips, W.M., Choi, S. & Larrabee, J.A. (1990) *J Chem Educ* **67**, 267–269.

Podberezskaya, N.V., Virovets, A.V. & Boldyreva, E.V. (1990) *Proceedings of the VIII-th All-Union Symposium on Intermolecular Interactions and Molecular Conformations*, Vol. 1, p. 73. Novosibirsk.

Podberezskaya, N.V., Virovets, A.V. & Boldyreva, E.V. (1991) *Russ J Struct Chem* **32**, 693–704.

Prodan, E.A., Pavlyuchenko, M.M. & Prodan, S.A. (1976) *Topochemical Reactions*. Nauka i Tekhnika, Minsk.

Rabinowitch, E. & Wood, W. (1936) *Trans Farad Soc* **32**, 1381–1387.

Ramamurthy, V. (ed.) (1991) *Photochemistry in Organized and Constrained Media*. VCH, New York.

Ramamurthy, V. & Turro, N. (1993) *Chem Rev* **93 (1–2)**, Introduction by guest editors.

Ramamurthy, V. & Venkatesan, K. (1987) *Chem Rev* **87 (2)**, 433–481.

Ramamurthy, V., Weiss, R.G. & Hammond, G.S. (1993) In: *Advances in Photochemistry* Vol. 18 (ed.) D.H. Volman, D. Neckers & G.S. Hammond, pp. 67–234. Wiley-Interscience, New York.

Robertson, J.M. & Ubbelohde, A.R. (1939a) *Proc Roy Soc (Lond)* **170A**, 222–240.

Robertson, J.M. & Ubbelohde, A.R. (1939b) *Proc Roy Soc (Lond)* **170A**, 241–251.

Roginskii, S.Z. (1948) *Adsorbtion and Catalysis at Inhomogeneous Surfaces*. Izd-vo AN SSSR, Moscow.

Rose, E.J. & McClure, D.S. (1981) *J Photochem* **17**, 171.

Salikhov, K.M., Medvinskii, A.A. & Boldyrev, V.V. (1967) *Khim Vys Energ* **1**, 381–386.

Scandola, F., Bartocci, C. & Scandola, M.A. (1974) *J Phys Chem* **78 (6)**, 572–575.

Scheffer, J.R. (1980) *Acc Chem Res* **13**, 283–290.

Schklover, V.E. & Timofeeva, T.V. (1985) *Russ Chem Rev (Uspekhi Khim)* **54 (7)**, 1057–1099.

Schklover, V.E., Timofeeva, T.V. & Struchkov, Yu.T. (1986) *Russ Chem Rev (Uspekhi Khim)* **55 (8)**, 1282–1318.

Schmalzried, H. (1990) *J Chem Soc Farad Trans* **86 (8)**, 1273–1280.

Schmidt, G.M.J. (1967) *Photochemistry of the Solid State*. Wiley-Interscience, New York.

Schneider, H.-J. & Durr, H. (eds) (1991) *Frontiers in Supramolecular Organic Chemistry and Photochemisry*. VCH, Weinheim.

Sidel'nikov, A.A., Mitrofanova, R.P. & Boldyrev, V.V. (1994) *Thermochem Acta* **234**, 269–274.

Sieberand, W. & Widman, T. (1986) *Acc Chem Res* **19**, 238–243.

Siebert, H. (1958) *Z Anorg Allg Chem* **298**, 51–63.

Swaminathan, S., Craven, B.M. & McMillan, R.K. (1984) *Acta Cryst* **B40**, 300–306.

Swiatkiewicz, J., Eisenhardt, G., Prasad, P.N., Thomas, J.M., Jones, W. & Theocharis, C.R. (1982) *J Phys Chem* **86**, 1764–1767.

Takenaka, Y., Ohashi, Y., Tamura, T. *et al.* (1993) *Acta Cryst* **B49 (2)**, 272–277.
Tanaka, M., Tsujikawa, I., Toriumi, K. & Ito, T. (1982) *Acta Cryst* **B38**, 2793–2797.
Taylor, R., Mullaley, A. & Mullier, G. (1990) *Pestic Sci* **29 (2)**, 197–213.
Timonova, I.N., Kravtsova, E.A., Boldyreva, E.V. & Mazalov, L.N. (1987) *Russ J Struct Chem Russ Ed (Zh Strukt Khim)* **28 (4)**, 69–76.
Tolkatchev, V.A. (1996) In: *Reactivity of Solids: Past, Present and Future. A Chemistry for the 21st Century Monograph*, (ed. V.V. Boldyrev), pp. 185–221. Blackwell Science Limited, Oxford.
Turro, N. (1986) *Pure Appl Chem* **58**, 1219–1228.
Ubbelohde, A.R. (1939) *Proc Roy Soc (Lond)* **173A**, 417–427.
Uchida, A. & Dunitz, J. (1990) *Acta Cryst* **B46**, 45–54.
Uchida, A., Ohashi, Y. & Ohgo, Y. (1991) *Acta Cryst* **C47**, 1177–1180.
Uchida, A., Danno, M., Sasada, Y. & Ohashi, Y. (1987) *Acta Cryst* **B43**, 528–532.
Uchida, A., Ohashi, Y., Sasada, Y., Ohgo, Y. & Baba, S. (1984) *Acta Cryst* **B40**, 473–478.
Virovets, A.V. & Podberezskaya, N.V. (1992) *Kristallografia (Russ J Crystallog)* **37 (4)**, 1017–1019.
Virovets, A.V., Podberezskaya, N.V., Boldyreva, E.V., Burleva, L.P. & Gromilov, S.A. (1992) *Russ J Struct Chem Russ Ed (Zh Strukt Khim)* **33**, 146–156.
Vorobiev, A.Kh. & Gurman, V.S. (1982) *J Photochem* **20**, 123–137.
Wang, W. & Jones, W. (1987) *Tetrahedron* **43**, 1273–1279.
Wang, Y., Tsai, C.J., Liu, W.L. & Calvert, L.D. (1985) *Acta Cryst* **B41**, 131–135.
Wegner, G. (1969) *Z Naturforsh* **24b (7)**, 824–832.
Wegner, G. (1971) *Makromol Chem* **145**, 85–94.
Wegner, G. (1977) *Pure Appl Chem* **49**, 443–454.
Wegner, G. & Fischer, E.W. (1970) *Ber Bunsenges Phys Chem* **74 (8,9)**, 909–912.
Weiss, R.G., Ramamurthy, V. & Hammond, G.S. (1993) *Acc Chem Res* **26**, 530–536.
Wendlandt, W.W. & Woodlock, J.H. (1965) *J. Inorg Nucl Chem* **27**, 259–260.
Whuler, P.A., Spinat, P. & Brouty, C. (1978) *Acta Cryst* **B34**, 793–799.
Wilkinson, G., Gillard, R. & McCleverty, J. (eds) (1987) *Comprehensive Coordination Chemistry*, Vol. 1. Pergamon Press, Oxford.
Zamaraev, K.I. (1994) *New J Chem* **18**, 3–18.
Zotov, N. (1990) *Acta Cryst* **A46**, 627–628.
Zotov, N. & Petrov, K. (1991) *J Appl Cryst* **24**, 227–231.

Appendix: Some additional explanations to the legends of Fig 7.2 and 7.6

Generation of symmetry equivalent atoms:

(a) **$[Co(NH_3)_5NO_2]Cl_2$:**

N(2A), N(3A), O(1A), Hal(1B): $-x$, y, $0.5\text{-}z$;
N(2B), Hal(1E): $0.5-x$, $0.5+y$, $0.5-z$;
N(2C), Hal(1F): $-0.5+x$, $0.5+y$, z;
Hal(1A): $-x$, $-y$, $-z$;
Hal(1C): x, $-y$, $0.5+z$;
Hal(1D): $0.5-x$, $0.5-y$, $-z$;
Hal(1G): $-0.5+x$, $0.5-y$, $0.5+z$;
O(1B): $0.5-x$, $-0.5+y$, $0.5-z$;
O(1C): x, $1-y$, $0.5+z$;
O(1D): $-0.5+x$, $-0.5+y$, z;
O(1E): $-x$, $1-y$, $-z$;
N(3B): x, $1-y$, $-0.5+z$;
N(3C): $-x$, $1-y$, $1-z$.

N(1), nitrogen of *trans*-ammine; N(2) and N(3), nitrogens of *cis*-ammine; N(4), nitrogen of nitro-ligand

(b) **[Co(NH_3)$_5$$NO_2$]Cl($NO_3$):**
N(2A), N(3A), O(3A): $x, y, -0.5-z$;
O(2A), N(3B), C1(1B): $-x, -y, -z$;
O(4A), C1(1A), O(3B), O(1A): $0.5-x, 0.5+y, -z$;
O(1B), O(4B): $0.5-x, 0.5+y, -1-z$;
N(2B): $0.5-x, -0.5+y, -z$;
N(2C): $0.5-x, -0.5+y, -0.5+z$;
N(3C): $-x, -y, -0.5+z$;
O(3C), N(5A): $-0.5+x, 0.5-y, z$;
O(3D): $-0.5+x, 0.5-y, -0.5-z$;
O(3E): $0.5-x, 0.5+y, -0.5+z$;
O(2B): $-x, -y, -0.5+z$;
C1(1C): $x, y, -1+z$.

N(1), nitrogen of *trans*-ammine; N(2) and N(3), nitrogens of *cis*-ammine; N(4), nitrogen of nitro-ligand; N(5), nitrogen of NO_3-anion; O(1) and O(2), oxygens of nitro-ligand; O(3) and O(4), oxygens of NO_3-anion. The lengths of the contacts of complex cations with the nearest neighbours (at 290 K and ambient pressure) are the following (Å):

[Co(NH_3)$_5$$NO_2$]$Cl_2$: 1, N(2)–Cl(1E) and **2**, N(2A)–Cl(1F): 3.351 (2); **3**, N(3)–Cl(1C) and **4**, N(3A)-Cl(1A): 3.357(2); **5**, N(3)–Cl(1B) and **6**, N(3A)–Cl(1): 3.408(2); **7**, N(1)–Cl(1A) and **8**, N(1)–Cl(1C): 3.459(2); **9**, N(2)–Cl(1) and **10**, N(2A)–Cl(1B): 3.485(2); **11**, N(3)–Cl(1E) and **12**, N(3A)–Cl(1F): 3.463(2); **13**, N(1)–Cl(1) and **14**, N(1)–Cl(1B): 3.485(2); **15**, N(2)–Cl(1D) and **16**, N(2A)–Cl(1G): 3.525(2); **17**, N(3)–O(1C), **18**, O(1)–N(3B); **19**, N(3A)–O(1E) and **20**, O(1A)–N(3C): 2.970(2); **21**, N(2)–O(1B), **22**, O(1)–N(2B), **23**, N(2A)–O(1D) and **24**, O(1A)–N(2C): 3.083(2); **25**, O(1)–Cl(1D) and **26**, O(1A)–Cl(1G): 3.360(2); **27**, O(1)–Cl(1E) and **28**, O(1A)–Cl(1F): 3.817(2).

[Co(NH_3)$_5$$NO_2$]Cl($NO_3$): 1, N(3)–O(4A) and **2**, N(3A)–O(4B): 2.982(2); **3**, N(1)–O(3B) and **4**, N(1)–O(3E): 3.016(2); **5**, N(3)–O(3C): 3.078(2); **6**, N(2)–O(3) and **7**, N(2A)–O(3A): 3.132(2); **8**, N(1)–O(3C) and **9**, N(1)–O(3D): 3.348(2); **10**, N(3)–O(2A); **11**, O(2)–N(3B), **12**, O(2)–N(3C) and **13**, N(3A)–O(2B): 3.092(2); **14**, N(2)–O(1A), **15**, O(1)–N(2B), **16**, O(1)–N(2C) and **17**, N(2A)–O(1B): 3.279(2); **18**, N(2)–Cl(1) and **19**, N(2A)–Cl(1): 3.386(2); **20**, N(1)–Cl(1A): 3.391(2); **21**, N(2)–Cl(1A) and **22**, N(2A)–Cl(1A): 3.414(2); **23**, N(3)–Cl(1) and **24**, N(3A)–Cl(1C): 3.448(2); **25**, N(3)–Cl(1B) and **26**, N(3A)–Cl(1B): 3.536(2); **27**, N(1)–N(5A): 3.274(3); **28**, O(1)–N(5): 2.811(3); **29**, O(1)–O(4): 2.908(4); **30**, O(1)–O(3) and **31**, O(1)–O(3A): 3.161 (2); **32**, O(2)–Cl(1B): 3.542(2): **33**, N(4)–Cl(1) and **34**, N(4)–Cl(1C): 3.661(0).

8 Kinetics of the Simplest Radical Reactions in Solids

V.A. TOLKATCHEV

Institute of Chemical Kinetics and Combustion, Siberian Branch of the Russian Academy of Sciences, Institutskaya, 3 Novosibirsk-90, 630090, Russia

1 Introduction

The development of electron spin resonance (ESR) technique has allowed an active study of the chemical transformations of radicals in solids. It has been established that in a solid phase the kinetic behaviour of the simplest reactions differs much from those known for gas and liquid phases. Therefore, in the gas and liquid phases the study of the kinetics of chemical transformations has given a great body of information on the mechanism of various reactions; in the solid phase such an approach has met with difficulties in interpreting the kinetic rules. These problems still exist, which is evident from the following discussion. Nevertheless, much is clear now and the kinetic investigations can be used and are used in chemical applications.

Professor V.V. Voevodsky is known as one of the most active propagandists of the ESR technique at the initial stage of its application (Blumenfeld *et al.* 1962). He also stimulated the study of the simplest reaction kinetics in solids. Although he took part in only some studies on the subject, he was always very interested in these problems and tried to promote them. A lot of data have been obtained by the scientists of his school, to which the author also belongs.

Now, there are many monographs devoted to the study of elementary reactions in solids, for example, those on electron tunnelling (Zamaraev *et al.* 1985), the reactions of polymeric systems (Emmanuel and Buchachenko 1988) and the different tunnelling processes in chemical physics (Goldanskii *et al.* 1989). In these monographs many problems on the kinetics of elementary solid phase reactions are under discussion. The aim of this chapter is not to give in detail the material already available but rather to pay more attention to the less known data on the study of the kinetics of elementary reactions.

2 Formal kinetics laws

2.1 *Kinetics of a simple cage reaction*

The cage reaction is a chemical transformation that occurs due to the reaction with the nearest neighbours. In this case, the rate constant of the process is equal to the sum of the probabilities of the reaction with all possible partners, $k = \Sigma W_i$. If its rate constant does not vary with time, the process is a classical example of the first-order reaction and is described by an exponent. Thus, for radical reactions the concentration of radicals due to the reaction with neighbours, is of the form:

$$[R] = [R(0)]\exp(-kt) = [R(0)]\exp(-\Sigma W_i t) \tag{8.1}$$

where [R] is radical concentration, k is rate constant, t is time. The problem in the description of this process is if the radical environment can include partners with different reactivity.

The problem has been studied using the reactions of atomic H abstraction:

$$R + R'H \rightarrow RH + R' \tag{8.2}$$

in the zink acetate and acetic acid matrices consisting of the mixture of H-containing molecules and the molecules for which the H atom is substituted by a deuterium (D) atom (Syutkin and Tolkatchev 1985a, 1989). Since these molecules substitute each other without changing the lattice geometry, and the isotopic effect of substitution is actually absent, these systems appear to be rather convenient for experimental studies of the above problem. In zinc acetate dihydrate the kinetics of H-atom abstraction by methyl radicals forming under γ-irradiation is known to be exponential (Tolles *et al.* 1968).

Consider first the simplest theoretical variant that explains the experimental idea. In the substance consisting of H- and D-containing molecules, let all the molecules occupy, with equal probability, all sites in a crystal or a glassy matrix. Then, the probability of H-containing molecules resident in a given site will equal the mole fraction of H-containing molecules. Denote it by c. If the reaction of radical in the H-containing matrix is exponential, then in the matrix of a mixed isotopic composition the reaction kinetics in the simplest variant must be of the form (Syutkin and Tolkatchev 1985a):

$$[R(c,t)] = [R(c,0)] \prod_{i=1}^{N} (c\exp(-W_i^{H}t) + (1-c)\exp(-W_i^{D}t)) \tag{8.3}$$

Where W_i^H and W_i^D are the probabilities of H- and D-atom abstraction, respectively, from the molecule occupying the ith position in the cage; N is the number of accessible molecules; $[R(c,t)]$ is the concentration of radicals at time t in the matrix with the composition $c/(1-c)$. Indeed, let the radical react with a molecule only in one position. The position is denoted by index i. Then, the probability that the radical will not react by t is either $c\exp(-W_i^{H}t)$ or $(1-c)\exp(-W_i^{D}t)$ for the reaction with the H- or D-containing molecule, respectively. The total probability of radical survival by t is:

$$c\exp(-W_i^{H}t) + (1-c)\exp(-W_i^{D}t) \tag{8.4}$$

Assuming that the probability of the reaction with the molecule occupying a certain position in the cage is independent of the composition of other positions, Equation (8.4) gives (8.3).

Generally speaking, assumptions given in Equation (8.3) are not evident. The mutual effect of the different sites is quite permissible. Besides, the change in the isotopic composition of matrix affects the intermolecular vibrations of the lattice. According to the up to date theoretical concepts (see, for examples, Goldanskii *et al.* 1989; Benderskii *et al.* 1994) this can have an effect on the rate of atom transfer. Note also that Equation (8.3) can directly be applied only to the equiprobable stabilization of radicals in cages with different isotopic compositions. If this is not the case, the changes should be introduced that correspond to a statistic composition of the cages.

Therefore, one of the aims of Syutkin and Tolkatchev (1985a, 1989) was an attempt to understand whether this simple approach is suitable for describing real kinetics or not. The problem of the initial radical distribution in cages with different isotopic compositions has also been solved for zinc acetate dihydrate and acetic acid. Under radiation both of the systems give rise to a methyl radical. In both of the systems the radical environment contains a site that participates in the reaction at the instant of methyl stabilization, providing this site is occupied by a H-containing molecule. This reaction is likely to be determined by hot methyl during its cooling after formation. The probability if the stabilization of methyl radical with a H-containing molecule at this site is 0.33 and 0 for zinc acetate and acetic acid, respectively. It is unity in both of the cases of this site is occupied by a deuterated molecule. The magnitude of this probability has been determined by analysing the dependence of the amount of methyl radicals and radical products of Equation (8.2) on the isotopic composition of the matrix. Furthermore, this site is denoted by index 1.

The method for determining the N value (the number of molecules with which the radical can react) is obvious from Equation (8.3). For large transformation depths, when the H-containing cages burn out, the decay of radicals can be described by the asymptotic law:

$$[\mathrm{R}(c,t)] = [\mathrm{R}(c,0)](1-c)^N \exp(-\Sigma W_i^{\mathrm{D}} t) \tag{8.5}$$

Comparing this behaviour to that in the D-containing matrix:

$$[\mathrm{R}(c,t) = [\mathrm{R}(c,0)]\exp(-\Sigma W_i^{\mathrm{D}} t) \tag{8.6}$$

it is readily understood that N can be estimated by comparing the asymptotic behaviour of the kinetics for large contents of deuterated molecules with the decay of radicals in a pure D-matrix.

A correction of Equation (8.3) for the non-equiprobability in the stabilization of methyl radicals with regard to molecule 1 in the case of zinc acetate, gives the following form for the kinetics of methyl radical decay (Syutkin and Tolkatchev 1985a):

$$[\mathrm{R}(c,t)]/[\mathrm{R}(c,0)] = \frac{(0.33c\exp(-W_1^{\mathrm{H}}t)+1)(1-c)\exp(-W_1^{\mathrm{D}}t)}{0.33c+(1-c)} \times \prod_{i=2}^{N} (c\exp(-W_i^{\mathrm{H}}t)+(1-c)\exp(-W_i^{\mathrm{D}}t)) \tag{8.7}$$

The structure of the first factor in Equation (8.7) is the following. In the numerator the numeric factors (0.33 and 1) are equal to the probabilities of methyl radical formation for the H and D substitutions of site 1. The denominator coincides with the numerator at $t = 0$ and normalizes the expression.

Figure 8.1 exemplifies the kinetic curves of the reaction of methyl radicals in zinc acetate for some compositions. The asymptotic behaviour of radical concentration, in this case:

$$[\mathrm{R}(c,t)] = [\mathrm{R}(c,0)](1-c)^N \exp(-\Sigma W_i^{\mathrm{D}} t)/(1-0.67c) \tag{8.8}$$

is denoted by a dashed line. A segment of the crossing of the dashed line with the ordinate can be used to calculate N. It is 3 for this case. Solid curves in Figure 8.1 have been calculated using Equation (8.7); the parameters were obtained using the least-squares fit in the treatment of all kinetic curves with different c (Syutkin and Tolkatchev 1985a). The results are given in Table 8.1. They merit a detailed consideration, which is

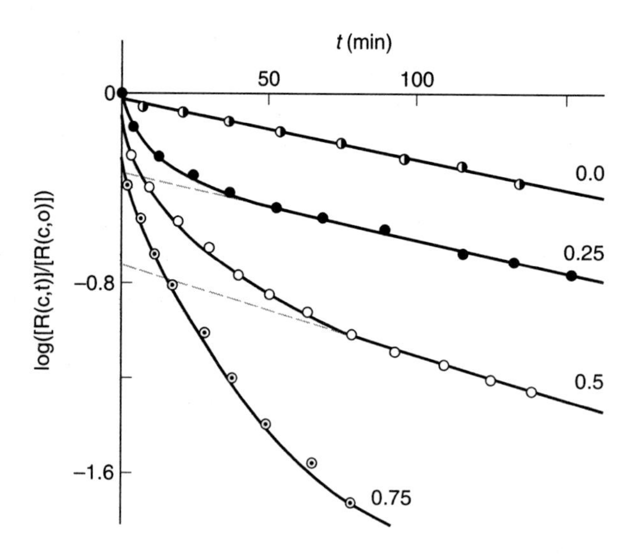

Figure 8.1 Kinetics of methyl radical decay in the crystals of zinc acetate dihydrate $(CH_3CO_2)_2Zn/(CD_3CO_2)_2Zn$ according to the reaction $CH_3 + (CH_3CO_2)_2Zn \rightarrow CH_4 + CH_2CO_2Zn\ CH_3CO_2$. (From Syutkin and Tolkatchev 1985a.) Numbers near the curves correspond to a mole fraction of H-containing molecules (c). $T = 160$ K.

Table 8.1 The probabilities of H-atom transfer

T (K)	135	160
W_1 (min^{-1})	$(8.5 \pm 0.5) \times 10^{-2}$	1.3 ± 0.4
W_2, W_3 (min^{-1})	$(2 \pm 1) \times 10^{-3}$	$(5.5 \pm 1.5) \times 10^{-2}$

performed in Section 3.1. Now, note that a similar method for extracting the individual probabilities of H-atom transfer has appeared to be rather efficient. The similar results have also been obtained for acetic acid (Syutkin and Tolkatchev 1989).

Thus, Equation (8.3) can be used to describe the kinetic behaviour of a simple exponential cage reaction. Unfortunately, such studies are quite few. Therefore, it is now difficult to estimate the commonness of the description. It is clear that a strong matrix isotopic effect (see Section 3.2) can restrict the applicability of Equation (8.3) in the isotopic dilution of systems. However, the effectiveness and simplicity of the result provide certain foundations for its usage in analysing more complex kinetic processes.

2.2 *Formal kinetic laws of retarded reactions*

2.2.1 THE OVERALL PICTURE

Many examples exist when the kinetics of the simplest reactions in solids is described by simple equations for gas phase or liquid phase reactions. In the previous section we have already mentioned the exponentiality of the reaction of H-atom abstraction by methyl radicals in zinc acetate dihydrate. For other examples of a simple kinetic behaviour see Tsvetkov *et al.* (1961); Kuzminskii *et al.* (1962); Bresler *et al.* (1963);

Neiman *et al.* (1963); Buben *et al.* (1964); Butiagin (1965); Griffiths and Sutcliffe (1966); Nara *et al.* (1967, 1968); French and Willard (1968); Mikhailov and Lebedev (1968); Thryion and Baijai (1968); Ermolaev *et al.* (1968); Ivantchev *et al.* (1968); Lebedev (1969); Cutten and Ericson (1970); Horan *et al.* (1970); McGhie *et al.* (1970); Wang and Williams (1972); Grinberg *et al.* (1972); Stentz *et al.* (1972); Serdobov *et al.* (1973); Jakimchenko *et al.* (1974, 1980); Klinshpont and Milintchuk (1975); Jakimchenko and Lebedev (1978); Sprague (1979); Jakimchenko and Degtjarev (1980); Smith *et al.* (1981); Syutkin and Tolkatchev (1982, 1985b); McBride (1983); Nilsson and Lund (1984); Lazarev *et al.* (1990, 1992).

There are many processes in solids that must obey the first- or second-reaction orders. However, these are more retarded with transformations than is required by simple laws (see, for examples, Campbell and Looney 1962; Ermolaev *et al.* 1962; Sullivan and Koski 1963, 1964; Bresler *et al.* 1963; Schelimov *et al.* 1963; Tsvetkov and Falaleev 1964; Judeikis and Siegel 1965; Zhuzhgov *et al.* 1965; Aditya and Willard 1966; Belevskii *et al.* 1966; French and Willard 1968; Shirom and Willard 1968; Vacek and Schulte-Frohlinde 1968; Butiagin *et al.* 1968; Cutten and Ericson 1970; Butiagin 1972, 1974; Johnson *et al.* 1973; Burshtein and Tsvetkov 1974; Neiss and Willard 1975; Fuks *et al.* 1975; Bol'shakov and Tolkatchev 1976; Dole and Salik 1977; Shimada *et al.* 1977, 1981; Tria and Johnsen 1977a,b, 1978; Stepanov *et al.* 1978; Zaskulnikov and Tolkatchev 1979, 1980; Plonka *et al.* 1979; Tria *et al.* 1979; Levina and Tolkatchev 1980; Senthilnathan and Platz 1980; Bol'shakov *et al.* 1980; Myagkikh *et al.* 1980; Syutkin and Tolkatchev 1982; Platz *et al.* 1982; McBride 1983; Plonka 1983, 1986, 1991; Plonka and Pietrucha 1983; Vyazovkin *et al.* 1983; Plonka and Kevan 1984; Doba *et al.* 1984a,b,c; 1985; Syutkin *et al.* 1984; Emmanuel and Buchachenko 1988; Lazarev *et al.* 1991). In a formal description, the approximation functions are convenient because they allow a quantitative description of the experiment. For monomolecular and bimolecular processes occurring in the excesses of one of the reagents, i.e. in quasi-monomolecular conditions, the Kohlrausch law is a good approximation:

$$[R(t)] = [R(0)]\exp(-\mathrm{k}t^{\alpha}) \tag{8.9}$$

where k and α are the empirical constants with $0 < \alpha \leq 1$ (Fuks *et al.* 1975; Bol'shakov and Tolkatchev 1976; Stepanov *et al.* 1978; Zaskulnikov and Tolkatchev 1979, 1980; Plonka *et al.* 1979; Levina and Tolkatchev 1980; Senthilnathan and Platz 1980; Bol'shakov *et al.* 1980; Platz *et al.* 1982; Plonka 1983, 1986, 1991; Plonka and Pietrucha 1983; Vyazovkin *et al.* 1983; Plonka and Kevan 1984; Doba *et al.* 1984a,b,c, 1985; Syutkin and Tolkatchev 1985b,c). This can be exemplified by the kinetics of methyl radical reaction with methanol in the methanol matrix in which this law successfully describes the experiment from 20 to 105 K, i.e. within 4kT (Vyazovkin *et al.* 1983). This is depicted in Fig. 8.2 by some kinetic curves of this reaction. Another successful approximation is the dependence proposed for monomolecular processes (Kutyrkin *et al.* 1984; Mardaleishvily *et al.* 1984):

$$[c(t)]/[c(0)] = (1 + W_0 nt)^{-1/n} \tag{8.10}$$

where W_0 and n are the constants characterizing the kinetic process: W_0 is its initial rate; n characterizes the degree of retardation. A normal (exponential) process corresponds to $n \Rightarrow 0$. An increase in n corresponds to the increase of retardation. Figure 8.3

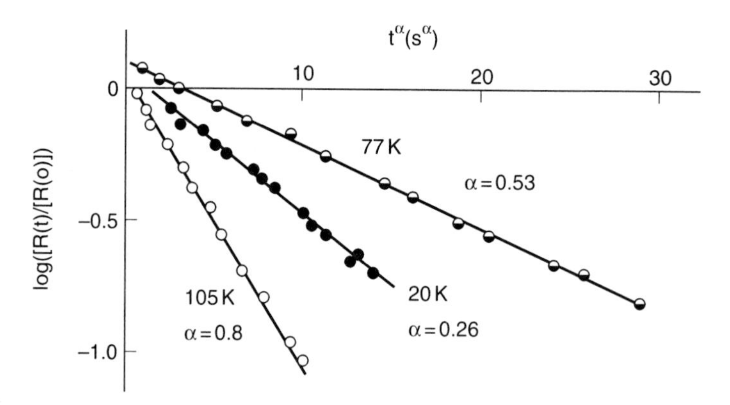

Figure 8.2 Kinetics of methyl radical decay in methanol glass at different temperatures (Vyazovkin *et al.* 1983) according to the reaction $CH_3 + CH_3OH \rightarrow CH_4 + CH_2OH$. The straight lines are shifted along the ordinate.

depicts the application of this approximation. In this case, the kinetics of rhodamine 6G photodissociation depends on the wavelength of mercury lamplight (Mardaleishvily and Anisimov 1986). Under full-range illumination, the kinetics are exponential. The figure shows two kinetic curves. One of these has been obtained under the light, the intensity of which was attenuated at wavelengths $\lambda \leqslant 260$ nm, and the other has been obtained using a lightfilter that passes light at $\lambda \geqslant 350$ nm. In the second case, the kinetics are more retarded (Fig. 8.3a). As follows from Fig. 8.3b, both of these can be well approximated by the dependence of Equation (8.10).

The limiting case of retarded reactions is the step kinetic process in which the reaction first is very quick and at the final transformation stage actually stops and runs slowly (Bamford *et al.* 1960; Loy 1960, 1961; Cracco *et al.* 1962; Ermolaev *et al.* 1962; Flournoy *et al.* 1962; Smith and Jacobs 1962; Bensasson *et al.* 1963; Mikhailov *et al.* 1964, 1965, 1972a,b; Fedoseeva and Kuzminskii 1967; Kozlov 1967; Bol'shakov *et al.* 1971; Grinberg *et al.* 1972; Stentz *et al.* 1972; Serdobov *et al.* 1973; Jakimchenko

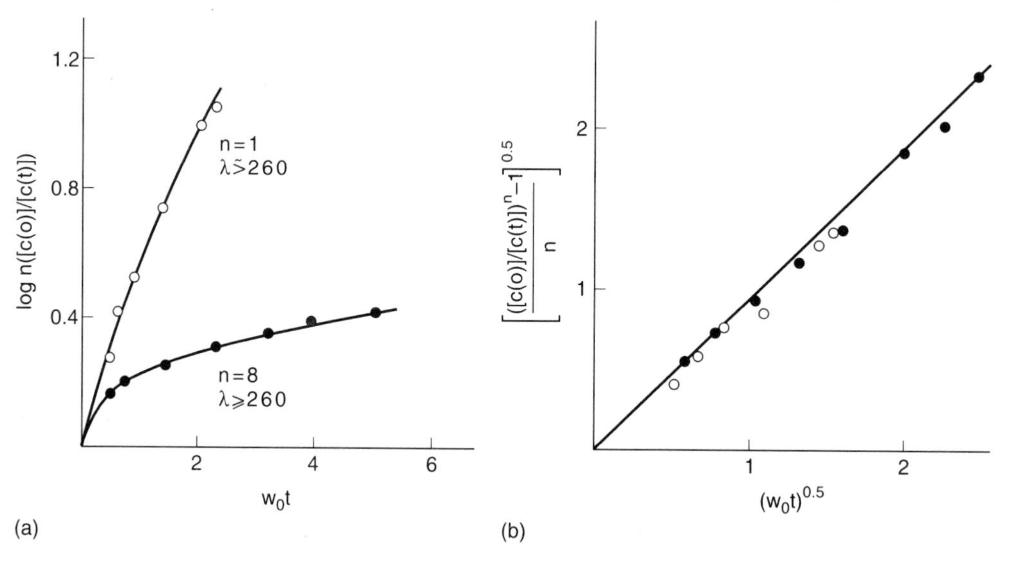

Figure 8.3 Kinetics of rhodamine 6G photodissociation in the films of polymethylmethacrylate under the light of a mercury lamp with different spectra compositions. (From Mardaleishvily and Anisimov 1986.)

et al. 1974, 1975, 1978; Kuzina and Mikhailov 1976; Aleksandrov *et al.* 1976; Radtsig 1976; Radtsig and Rajnov 1976; Davydov *et al.* 1977; Plotnikov *et al.* 1977; Jakimchenko and Lebedev 1978, 1979; Vorotnikov *et al.* 1983, 1986; Emmanuel and Buchachenko 1988). At the latter stage the concentration varying with time is often described by the logarithmic time function, $R \propto -\log t$.

2.2.2 RATE CONSTANT DISTRIBUTION

The retardation of chemical processes in solids is usually attributed to their inhomogeneity. A model of rate constant distribution is in wide use. The model assumes that at different sites in solids the reactivity of particles is different and the above anomalies are caused by the inhomogeneities of the reaction rate. For a quantitative description of the model the authors Mikhailov *et al.* (1965) and Lebedev (1978) assume that a solid can be divided into fairly homogeneous regions each having its own rate constant. Consequently, a usual kinetic equation can be applied to each region. It is evident that for the reactions in which the reagents are far apart, this automatically means that the sizes of regions must be large enough to involve a great amount of reacting particles. In this case, a rate constant can be introduced to characterize the region. Then, if within the identical regions the reaction kinetics is described as $f = f(\mathrm{k}t)$ where k is the rate constant, the change in the total concentration will be described by the convolution over all regions:

$$[\mathrm{R}(t)] = [\mathrm{R}(0)] \int_{\mathrm{k}_{min}}^{\mathrm{k}_{max}} g(\mathrm{k}) f(\mathrm{k}t)\, \mathrm{dk} \qquad (8.11)$$

f is the first- or second-order law; $g(\mathrm{k})$ is the weight of regions with a given k.

When the spectrum of constant k is wide enough, for example, due to a wide distribution in the activation energies of elementary processes E, the total kinetic process will be sufficiently inhomogeneous. When $E_{max} - E_{min} \approx E_{max}$ the important kinetic laws can be explained in the frame of the simple rectangular distribution in activation energy. These include the step character of the kinetics, the logarithmic character of the reactant decay with time and the wide temperature range of chemical process (Mikhailov *et al.* 1965; Lebedev 1978).

This approach has been well developed and the means have been found to treat the experiment in order to extract the kinetic reaction parameters for wide k distributions (Mikhailov *et al.* 1972a; Kuzina and Mikhailov 1976; Plotnikov *et al.* 1977). The questions of both the mixing constants and the recovery of distribution functions by thermal and mechano-activation processes have been considered (Radtsig and Rajnov 1976; Mardaleishvily *et al.* 1984). A very successful review of this approach, including numerous examples of application, is given in Emmanuel and Buchachenko (1988).

Equation (8.11) adequately describes the experiment for retarded kinetic processes that proceed to completion and do not degenerate into step ones. This was first mentioned by Jakimchenko *et al.* (1980) and later shown by computer calculation (Burshtein *et al.* 1984; Spath and Raff 1992). In this case, the kinetic parameters can be extracted by computer processing of the experiments (Bagryansky *et al.* 1988; Syutkin *et al.* 1992; Kondratenko *et al.* 1993). When using approximation dependencies

(Equation 8.10) it is convenient to assume that in Equation (8.11) $k_{min} = 0$ and $k_{max} = \infty$. Then, the initial rate of the process is:

$$W_0 = \int_0^{\infty} k\, g(k) dk \tag{8.12}$$

Following reverse Laplace transformation, Equation (8.10) can be used to reproduce analytically the distribution function $g(k)$ (Kutyrkin *et al.* 1984) via the parameters W_0 and n:

$$g(k) = (W_0 n)^{-1/n}\, \Gamma^{-1}(1/n) \exp(-k/(W_0 n)) k^{-1+1/n} \tag{8.13}$$

where Γ is the γ-function.

Bagryansky (1987) has demonstrated that for the diffusion-controlled reactions occurring in the excess of a diffusing reagent, two concentration regimes are possible when the kinetics can be described in terms of Equation (8.11)-type rate constant distribution. The first regime has already been described and in this case it corresponds to the criterion $[B]\, l^3 \gg 1$, where [B] is the concentration of excess diffusing reagent and l is the characteristic space size of the ordered microregion characterized by the rate constant k. The second regime corresponds to the opposite criterion $[B]\, l^3 \ll 1$, when the B reagent covers, on average, a few microregions before colliding with the partner. In this case the spectrum of constant is averaged, but not to the end because the averaging by motion is interrupted by the reaction. As a result, distribution (Equation 8.11) becomes more narrow. These calculations have experimentally been verified by Bagryansky *et al.* (1988).

2.2.3 GENERAL REMARKS ON THE TIME-DEPENDENT RATE CONSTANT

The pioneering papers describing the retarded kinetic processes from the other point of view in Section 2.2.2 are the papers of Campbell and Looney (1962), Butiagin (1972, 1974) and Burshtein and Tsvetkov (1974). As reported in these papers, radical recombination in a series of polymers and dicarboxylic acids can be described either by the equation:

$$[R(t)] = [R(0)] \exp(-k\sqrt{t}) \tag{8.14}$$

or by:

$$1/[R(t)] - 1/[R(0)] = k_1 t + k_2 \sqrt{t} \tag{8.15}$$

The authors have tried to account for the results in the frame of the first- or second-order reactions with the help of the Smoluchowski diffusion model for a rate constant. The rate constant in this case is of the form $k = 4\pi\rho D(1 + \rho/\sqrt{\pi D t})$ where ρ is the reaction radius, D is the diffusion coefficient and t is the time. Assuming that the second (unstationary) term in the expression for the rate constant dominates in the first case, and that in the second case both of the terms are comparable, then, the recombination kinetics can be described formally.

This explanation is rather contradictory as indicated in the last of these papers (Burshtein and Tsvetkov 1974). The meaning of the first-order reaction in radical recombination is rather vague. Also unclear is the origin of the following paradox. The second time-dependent term in the expression for the Smoluchowski constant reflects the decay of particles initially situated at the distance of the order of the reaction radius. It can be observed in the curve of radical decay, providing $[R(0)]\rho^3 \gg 1$ (Burshtein and Tsvetkov 1974). This criterion fails for any particle concentration in the papers mentioned above. As a result, the processing of experimental data causes abnormally large values of the reaction radius that are of the order of the mean distance between the radicals, i.e. 70–100 Å. Such a reaction radius seems to be incredible. Considering such a treatment as an approximation, the rate constant of reactions in solids can be assumed to be time dependent. This assumption allows a formal description of these simple reactions.

So far as we know, Hamill and Funabashi (1977) were the first to introduce $k(t)$ (for electron reactions). More recently, such an assumption has been made by Zaskulnikov and Tolkatchev (1979, 1980, 1982); Bol'shakov *et al.* (1979); Plonka *et al.* (1979); Zaskulnikov *et al.* (1981); Plonka (1983); Plonka and Pietrucha (1983); Bagryansky *et al.* (1983); Plonka and Kevan (1984); Vyazovkin *et al.* (1985a). The recently published reviews of Plonka (1986, 1991) contain the other references.

For bimolecular reactions the kinetics of such a process is of the form:

$$\frac{d[A]}{dt} = -k(t)[A][B] \tag{8.16}$$

The expression for $k(t)$ can readily be deduced by differentiating the approximation form of the experimental kinetic curve of reagent disappearance if the concentrations of reacting particles can be derived in a factor form.

Before analysing, take note of the two principle differences between this description and that performed in terms of rate constant distribution.

1 As follows from Equation (8.16), chemical transformation can be described as a process occurring in a homogeneous matrix. Its retardation is assigned to the time-dependent reaction probability.

2 Unlike conventional concepts, such a process possesses a unique memory about the moment of its onset, i.e. it is not Markovian. This should be taken into account in any multistage reaction because, due to the memory, the complex transformations cannot be described by a usual set of differential equations.

Two simple examples have been analysed: (i) two successive monomolecular reactions in γ-irradiated malonic acid (Levina and Tolkatchev 1980); and (ii) the reaction of H-atom abstraction by methyl radicals in methanol glass that occurs during a simultaneous generation of methyl radicals (Bol'shakov *et al.* 1979).

The reasons for the appearance of $k(t)$ are the following.

1 If the reaction is controlled by diffusion in the homogeneous matrix and reagent concentrations are high enough, the unstationary term of the Smoluchowski rate constant is of importance. This means that the rate constant depends on time.

2 A solid matrix in which the chemical process occurs is short of being in a thermodynamically equilibrium state, for example because it was too quickly frozen upon preparation compared to the rates of the relaxation processes that can stabilize the structure.

3 The active particles in a solid are often obtained under external high-energy actions:

radiation–chemical, photochemical, mechanochemical. Therefore, a reaction particle can find itself in the non-equilibrium environment (see, for example, McBride 1983). The relaxation of matrix structure as well as of local environment can change the chemical transformation rate (Ermolaev *et al.* 1962; Scwarc 1962, 1966; Neiss and Willard 1975; Myagkikh *et al.* 1980; Zaskulnikov and Tolkatchev 1982; McBride 1983; Bagryansky *et al.* 1983; Grebenkin and Bol'shakov 1991). If these changes occur at reaction temperature, k(t) forms.

In Plonka *et al.* (1979) and Plonka (1986), another variant of k(t) formation has been assumed. It is evident that when the particles diffuse in the inhomogeneous medium, their residence time in deeper traps is longer. The instantaneous start of such particles, which can be caused by temperature rise, can be followed by a decrease in their jump rate due to the recapture of particles by deep traps. Thereafter, the rate can acquire a stationary value. It is assumed (Plonka *et al.* 1979; Plonka 1986) that the time of attainment of this value is large enough in many systems and that the time-dependent rate constant, reflecting the decrease of reagent collisions with time, can be introduced. A similar idea (called the dispersion transport) is well known for transport of electrons and holes in dielectric disordered media (see reviews of Scher and Montrol 1975; Pfister and Scher 1978; Marschall 1983). In certain cases, when mobile particles can react and their concentration substantially exceeds that of the second reagent, the time-dependent rate constant can be introduced (Hamill and Funabashi 1977; Helman and Funabashi 1979; Zumofen *et al.* 1983; and references therein). However, the space–inhomogeneous process cannot always be converted into the space–homogeneous one retarded with time (Pollak 1977; Goldanskii *et al.* 1989). Strictly speaking, this is possible when the length of a jump for a random walker is much larger than the size of the ordered microregion. It means that the jump length must be much larger than the distance between the neighbouring sites of the lattice. This often holds for electron transfer. Only in this case can the matrix properties be averaged so that the particle flow will uniformly decrease with time over the whole matrix. This allows a time-dependent rate constant to be introduced. If the jump length is less than the size of quasi-homogeneous regions, the introduction of k(t) is known to be incorrect. When the jump length is of the order of intermolecular distances, which is likely to be observed in radical reactions, the question of the correctness of k(t) introduction becomes problematic.

2.2.4 THE DIFFERENCES IN THE REACTIONS DESCRIBED BY RATE CONSTANT DISTRIBUTION AND TIME-DEPENDENT RATE CONSTANT

For monomolecular reactions no kinetic criterion is available that could allow one to determine whether the retardation is due to distribution in rate constants or caused by the variation of rate constant with time. For bimolecular processes such a criterion has been formulated by Zaskulnikov and Tolkatchev (1979, 1980). Its qualitative idea is the following. For rate constant distribution, provided each constant is independent of time, the transformation kinetics in a separate region can be described as usual according to the law of mass action. In this case, the change in concentration must change the reaction time according to the same law. Thus, a double decrease in concentration must double the reaction time because the same number of reagent collisions will take twice as much time. If the rate constant varies with time the case is quite different. The decrease in reagent concentration will lead to the increase of collision time and, as a result, to the decrease of rate constant during the process. Consequently, the reaction time will further increase.

Consider this criterion quantitatively using the reaction:

$$R + O_2 \rightarrow RO_2 \qquad (8.17)$$

which occurs in excess oxygen. Let the kinetics be determined by rate constant distribution. In a separate region:

$$\frac{d[R]}{dt} = -k_i[R][O_2] \qquad (8.18)$$

Integrating in excess oxygen we obtain:

$$[R_i] = [R_i(0)]\exp(-k_i[O_2]t) \qquad (8.19)$$

Multiplying by region volume, we derive a similar relation for the number of particles:

$$N_i = N_i(0)\exp(-k_i[O_2]t) \qquad (8.20)$$

Summing up over all regions and taking into account that $N_i(0) = N(0)g_i$, where g_i is the fraction of the ith-type regions, we obtain:

$$N(t) = N(0)\,\Sigma\, g_i \exp(-k_i[O_2]t) \qquad (8.21)$$

Dividing by sample volume, we deduce a similar relation for concentration that, in the integral form, has the Equation 8.11-type form:

$$[R(t)] = [R(0)] \int_{k_{min}}^{k_{max}} g(k)\exp(-k[O_2]t)dk \qquad (8.22)$$

As follows from Equation (8.22) $[O_2]t$ is the universal variable unambiguously determining the transformation depth. A double decrease in concentration will double the period, during which the reaction reaches the same depth.

Assume now that the oxidation kinetics can be described by Equation (8.16). In our case it is of the form:

$$\frac{d[R]}{dt} = -k(t)[R][O_2] \qquad (8.23)$$

which in excess oxygen after integration gives:

$$[R(t)] = [R(0)]\exp(-[O_2] \int_0^t k(t)\, dt) \qquad (8.24)$$

From Equation (8.24) it follows that the transformation depth can be given unambiguously by the $[O_2] \int_0^t k(t)\, dt$ value, which automatically allows for the decrease of rate constant with time. Thus, if $k(t) = \beta/(2\sqrt{t})$, oxidation kinetics in Equation (8.24) will obey the Kohlrausch law:

$$[R(t)] = [R(0)]\exp(-[O_2]\beta\sqrt{t}) = [R(0)]\exp(-\beta\sqrt{[O_2]^2 t}) \qquad (8.25)$$

One can see when the oxygen concentration decreases by a factor of two the reaction time must be increased fourfold for the reaction to reach the same transformation depth. For a more general case, if the kinetics of oxidation in excess oxygen is well approximated by the Kohlrausch law (Equation (8.6)), the coincidence between kinetic curves in the coordinates:

$$\frac{\log([R(t)]/[R(0)])}{([O_2]t)^{\alpha}} \tag{8.26}$$

corresponds to rate constant distribution, and in the coordinates:

$$\frac{\log([R(t)]/[R(0)])}{[O_2]t^{\alpha}} \tag{8.27}$$

corresponds to the dependence of rate constant on time.

Figure 8.4 exemplifies the kinetic data on the oxidation of ethyl radicals in glassy methanol at 87 K and Fig. 8.5 depicts the same for matrix radicals in the annealed glassy 3-methylheptane glass at 90 K. The coordinates in the figures correspond to the above reaction models. It is seen that for methanol the kinetics corresponds to k(t) and in 3-methylheptane it corresponds to rate constant distribution.

It is important that the approximation accuracy of experimental curves has no effect on the analysis of the kinetics based on Equations (8.22) and (8.26). For Equation (8.27) the case is quite different because this formula results from integration of Equation (8.24) with a certain form of approximation formulae. Therefore, the simplest analysis of experiments based on Equation (8.27) is possible most often with relatively small changes in the concentration of excess reagent. In these cases the chemical processes run at comparable times. Thus, the approximation parameters do not change much from one kinetic curve to another, which allows this analysis. Strictly speaking, if the rate constant varies with time, the kinetics must be analysed in terms of dependence (Equation 8.24). It should be verified that $(1/[O_2])\ \ln([R(t)]/[R(0)])$ is the non-linear time function within a wide range of oxygen concentrations (Bagryansky *et al.* 1983).

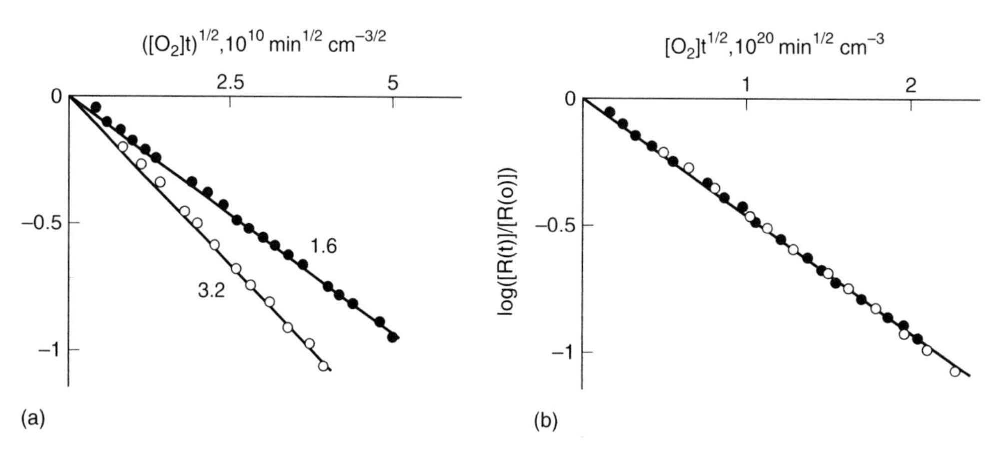

Figure 8.4 Kinetics of the decay of ethyl radicals in methanol-d_4 glass with 9% D_2O for different oxygen concentrations in the reaction $C_2H_5 + O_2 \rightarrow C_2H_5O_2$ (Zaskulnikov *et al.* 1981) in the coordinates corresponding to: (a) rate constant distribution and (b) time-dependent rate constant. The numbers denote oxygen concentrations (10^{19} cm^{-3}). $T = 87$ K. $[O_2] \gg [C_2H_5]$.

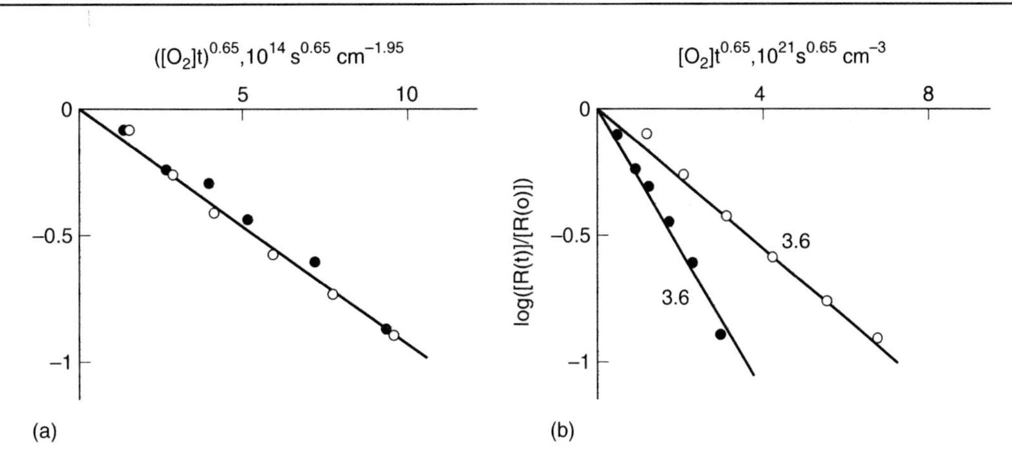

Figure 8.5 Kinetics of 3-methylheptyl radicals in 3-methylheptane glass preliminarily annealed at reaction temperature at different oxygen concentrations according to $C_8H_{17} + O_2 \rightarrow C_8H_{17}O_2$ (Vasenkov and Tolkatchev 1992) in the coordinates corresponding to: (a) rate constant distribution; (b) time-dependent rate constant. The numbers denote oxygen concentrations (10^{18} cm^{-3}). $T = 90$ K. Seven days is the time of the glass annealing. $[O_2] \gg [C_8H_{17}]$.

Figure 8.6 gives such an analysis on the data of Zaskulnikov *et al.* (1981) for the above-mentioned reaction of ethyl radical oxidation in methanol glass. Note that when the Kohlrausch law is an exact equation for kinetics, the kinetic data must be described by a straight line by the coordinates used in Fig. 8.6. Actually, a slightly bent curve has been observed. As follows from its slope, when its region is approximated by Equation (8.9) is used for approximation, the α-value varies from 0.45 to 0.6 at large and small oxygen concentrations, respectively.

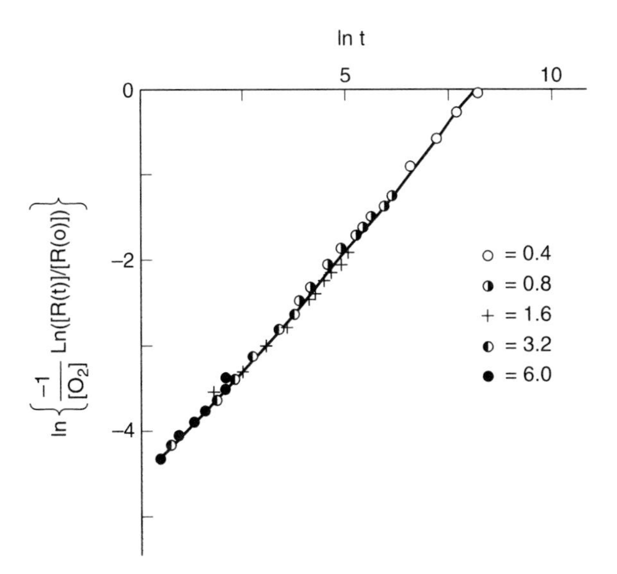

Figure 8.6 Kinetics of the decay of ethyl radicals in methanol-d_4 glass with 9% D_2O within a wide range of oxygen concentrations according to $C_2H_5 + O_2 \rightarrow C_2H_5O_2$, calculated from data of Zaskulnikov *et al.* (1981) in the coordinates corresponding to time-dependent rate constant. The numbers denote oxygen concentration (10^{19} cm^{-3}). $T = 87$ K. $[O_2] \gg [C_2H_5]$.

Papers (Zaskulnikov and Tolkatchev 1980; Bagryansky *et al.* 1989; Bagryansky and Tolkatchev 1991) report on the dependence of the oxidation kinetics of a series of alkyl radicals on oxygen concentration in toluene-d_8 (87 K), 2,4-dimethylpentane (87 K), ethanol-d_6 (77 K), isopropanol-d_8 (87 K), *n*-butanol-d_{10} (87 K) and of matrix radicals in polymethylmethacrylate oligomers (130–135 K). It is concluded that in low-molecular glasses the reaction rate constant is time dependent and in oligomers, as molecular weight increases, the language of description must transform to rate constant distribution. According to later studies (Vasenkov and Tolkatchev 1991, 1992; Vasenkov *et al.* 1993) the concentration dependence can be distorted by the capture of a part of oxygen by deep traps forming in the unannealed glassy matrices. Moreover, a correct consideration of the influence of traps has led Vasenkov and Tolkatchev (1991) to change the above conclusion for the *n*-butanol-d_{10} matrix. Therefore, the data given in papers (Zaskulnikov and Tolkatchev 1980; Bagryansky *et al.* 1989; Bagryansky and Tolkatchev 1991) should be considered as preliminary.

A few papers (Zaskulnikov and Tolkatchev 1979, 1982; Zaskulnikov *et al.* 1981; Bagryansky *et al.* 1983; Zapadinsky *et al.* 1985, 1986; Korolev *et al.* 1991) are devoted to the study of the reasons for rate constant time dependence in methanol glass. $k(t)$ has been demonstrated to be independent of radicals (Bagryansky *et al.* 1983). It divides into two factors, one of which is dependent on the time of preliminary matrix annealing at reaction temperature (before radical formation) and the second is independent of annealing (Zaskulnikov and Tolkatchev 1982; Bagryansky *et al.* 1983). It has been shown that the first factor is determined by the decrease of oxygen diffusion coefficient with time due to matrix relaxation (Zapadinsky *et al.* 1986; Korolev *et al.* 1991). The origin of the other is unclear. For *n*-butanol-d_{10} glass the case is quite different. The kinetics of the annealed matrix have been described in terms of rate constant distribution (Vasenkov and Tolkatchev 1991). The unannealed matrix relaxes at reaction temperatures (90 K, 98 K) and the kinetics are described in terms of the specturm of constants varying according to the same time law (Vasenkov and Tolkatchev 1991).

In polymethylmethacrylate, polystyrene, polycarbonate and polysulphone the reactions of radical oxidation (Equation 8.17) are described by rate constant distribution (Bagryansky and Tolkatchev 1987a, 1991; Bagryansky *et al.* 1988, 1989; Kondratenko *et al.* 1993). The high-temperature reactions of radiation defects with Ag^+ ions in phosphate glasses controlled by Ag^+ ion diffusion, are described in the same terms (Syutkin *et al.* 1991, 1992, 1995).

2.2.5 ANALYSIS OF THE KINETIC BEHAVIOUR OF H-ATOM ABSTRACTION REACTIONS

The reactions of abstraction are interesting with regard to kinetics because radicals can react with their environment without diffusion shifts. Matrices have been diluted with deuterated analogue in order to analyse their kinetic behaviour. As already noted, this procedure has no effect on the geometry of the systems and has a minor effect on its macroscopic physical properties (Rabinovitch 1968). When reaction is slowed down greatly by isotopic effect, the isotopic dilution plays the same role as a decrease in reagent concentration in the analysis of the kinetic behaviour of the process involving diffusion. The simplest case has been considered in Section 2.1, whereas here a more complicated behaviour is given.

Consider first Equation (8.3). Expanding it for minor t and differentiating log [R], we obtain:

$$d[R]/dt = (-c\Sigma W_i^H - (1-c)\Sigma W_i^D)[R] = -(ck^H + (1-c)k^D)[R] \quad (8.28)$$

Consequently, the kinetics of cage reaction correspond to the normal kinetic liquid phase laws at the initial stage, even in the absence of reagent motion. According to analysis carried out by Syutkin and Tolkatchev (1985c), this stage is the larger the greater the number of radical-accessible molecules. It corresponds to the cage composition that is not distorted by the reaction. As soon as the most reactive sites in cages begin to burn out, the reaction is retarded and gradually reaches asymptotic behaviour (Equation 8.5). This means that even in the absence of radical motion in the matrix, the initial stage of transformation can be analysed according to the principles given in the previous section. If the radical or matrix molecules begin to move over the lattice, then this stage is sure to increase with transformation depth.

Figure 8.7 presents the data on methyl radical transformation in the lattice of crystalline sodium acetate trihydrate (Syutkin and Tolkatchev 1985c; Syutkin *et al.* 1985). The $(ct)^{0.4}$ is plotted on the x-axis. The transformation depth is observed to depend only on the product of ct. Consequently, the non-exponentiality of the kinetic behaviour in a pure H-matrix ($c = 1$) is related to rate constant distribution.

When similar processes have been analysed in low-molecular glasses, in all cases the kinetics of these reactions in H-containing matrices appeared to be well approximated by the Kohlrausch law (Equation 8.9) (Bol'shakov and Tolkatchev 1976; Levina and Tolkatchev 1980; Senthilnathan and Platz 1980; Bol'shakov *et al.* 1980; Zaskulnikov *et*

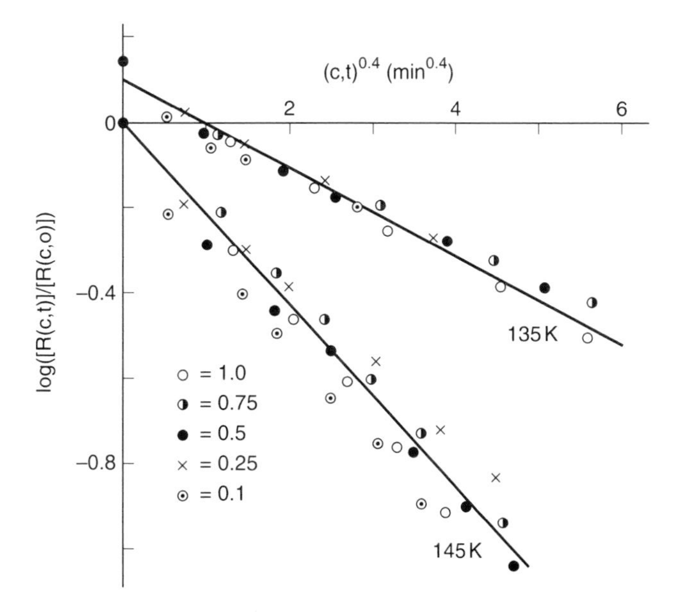

Figure 8.7 Kinetics of the methyl radical decay in the crystals of sodium acetate trihydrate CH_3CO_2Na/CD_3CO_2Na according to $CH_3 + CH_3CO_2Na \rightarrow CH_4 + CH_2CO_2Na$. (From Syutkin and Tolkatchev 1985c.) The kinetic curve at 135 K is arbitrarily shifted along the ordinate. The figures are equal to the mole fraction of CH_3CO_2Na.

al. 1981; Platz *et al.* 1982; Vyazovkin *et al.* 1983; Doba *et al.* 1984a,b,c, 1985; Vyazovkin *et al.* 1985a; for example see Fig. 8.2). When glasses are diluted with deuterated analogues, the reaction retardation cannot be assigned to rate constant distribution, whereas the experiment can be described well by introducing $k(t)$ (Zaskulnikov *et al.* 1981; Vyazovkin *et al.* 1985a). This is exemplified by Fig. 8.8 for the same reaction as in Fig. 8.2. It is seen that the ct coordinates are not universal and do not determine the transformation depth compared to ct^{α}. Probably in this case, a time-dependent rate constant should be introduced that can be approximated by the law $k(t) = \beta/t^{0.68}$.

At present the reasons for such a kinetic behaviour are not clear. The most likely hypotheses are the following. One of the reasons can be the fact that a local non-equilibrium zone forms at a radical localization site. The non-equilibrium can be caused by both the radiation damage of substance upon radical formation and the peculiarities of their stabilization under irradiation (Vyazovkin *et al.* 1985a). It is known that the relaxation of non-equilibrium parameters in different physical processes in the solid phase can be described by the Kohlrausch law, which in its form is similar to Equation (8.9). Therefore, it is likely that in low-molecular glasses the chemical reaction follows the relaxation of local radical environment, thus leading to the dependence of reaction rate constant on time. In the frame of a rough model this result can be obtained theoretically (Tolkatchev 1987).

The second possible cause of such a kinetic behaviour is the following. In a number of papers (Burshtein *et al.* 1984; Doba *et al.* 1984a,b,c, 1985) an attempt has been made to relate the Kohlrausch law for the kinetics of H–atom transfer to the peculiarities of H–atom tunnelling using both a sharp dependence of transfer probability on distance and the changes of these distances in disordered media. In the simplest variants this idea leads to rate constant distribution. To this end one should assume that all the probabilities of the reaction between the radical and the individual sites (W_i, see Equation (8.28) synchronously change with changing radical positions in the matrix. Then, the total probability, which is the rate constant for the reaction with either H or D surrounding,

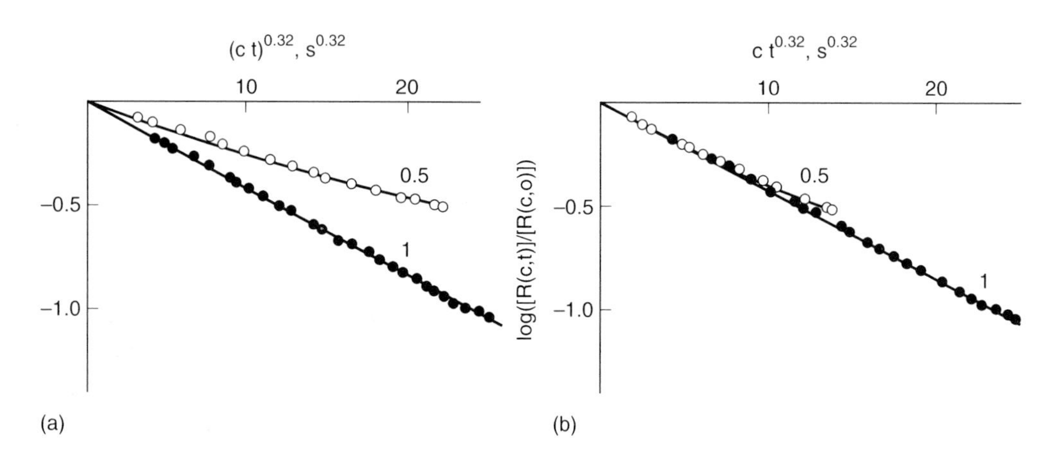

Figure 8.8 Kinetics of the decay of methyl radicals in methanol glass CH_3OH/CD_3OH according to $CH_3 + CH_3OH \rightarrow CH_4 + CH_2OH$. (From Vyazovkin *et al.* 1985a.) The numbers are equal to the mole fraction of CH_3OH. $T = 45$ K. (a) and (b) are the different experimental treatments.

must be averaged. The averaging reduces to Equation (8.11). However, as mentioned above for glassy matrices, this model fails for the experiments on the change of the initial process rate with changing concentration of H-containing molecules.

An interesting variant was proposed by Burshtein *et al.* (1984). The idea is that within the limits of the cage, the reaction kinetics in separate sites should be averaged independently of the rest. For example, in Equation (8.3) each cofactor must be averaged independently. This uncorrelated averaging is possible providing the difference in the position of any molecule of the cage is independent of the positions of other molecules in the glass matrices. Then, each cofactor in Equation (8.3) can yield the term of the following form:

$$\langle c\exp(-W_i^{\rm H}t) + (1-c)\exp(-W_i^{\rm D}t)\rangle = c\exp(-\gamma_i^{\rm H}t^{\alpha_i}) + (1-c)\exp(-\gamma_i^{\rm D}t^{\beta_i}) \quad (8.29)$$

and the total kinetics can be written as:

$$\begin{aligned}[\mathrm{R}(c,t)] &= [\mathrm{R}(c,0)]\langle\prod_{i=1}^{N}(c\exp(-W_i^{\rm H}t) + (1-c)\exp(-W_i^{\rm D}t))\rangle \\ &= \prod_{i=1}^{N}\langle(c\exp(-W_i^{\rm H}t) + (1-c)\exp(-W_i^{\rm D}t))\rangle \\ &= \prod_{i=1}^{N}(c\exp(-\gamma_i^{\rm H}t^{\alpha_i}) + (1-c)\exp(-\gamma_i^{\rm D}t^{\beta_i}))\end{aligned} \quad (8.30)$$

If α_i and β_i are almost the same at all sites ($\alpha_i \cong \beta_i \cong \alpha$), the procedure used to deduce Equation (8.28) can be employed to obtain the following:

$$\mathrm{d[R]}/\mathrm{d}t = (-c\Sigma\gamma_i^{\rm H}\alpha/t^{1-\alpha} - (1-c)\Sigma\gamma_i^{\rm D}\alpha/t^{1-\alpha})[\mathrm{R}] \quad (8.31)$$

This corresponds to the experiment on studying the dependence of the reaction rate on the concentration of H-containing molecules. Indeed, the requirement for the closeness between α and β imposes rather rigid limitations on the model. The other limitations follow from the next section. Here we are going to discuss the following. In this case the real probabilities of reaction are time independent. Nevertheless, the concentration dependence of the real kinetics at the initial stage imitates the time dependence of the constant. This will always hold for the uncorrelated averaging over the separate sites of the lattice providing, in a separate site, the averaging gives the necessary kinetic law. Thus, if the matrix-averaged kinetics with one site gives the behaviour of the Kohlrausch law type, for example due to relaxation of particle local environment (Tolkatchev 1987), the concentration dependence of the initial rate will correspond accurately to Equation (8.16).

An attempt has been made by Doba *et al.* (1993) to average the probabilities over the individual sites of the lattice for the reaction of methyl radicals with methanol in the methanol matrix using the experimental data on the structure of the nearest environment (Doba *et al.* 1987). However, Doba *et al.* (1993) have failed to reach an agreement between calculation and experiment for mixed isotopic matrices. The reason is not clear. Probably, this is related to a simplified character of averaging, neglecting the role of intermolecular vibrations (Goldanskii *et al.* 1989; Benderskii *et al.* 1994), and the influence of the collective effects both discussed in the monograph of Benderskii *et al.* (1994) and given below.

3 Collective effects

3.1 *A form of the barrier of the elementary solid phase reaction*

When considering chemical reactions in the gas phase, the profile of potential surface along the reaction path is commonly depicted as a smoothly changing barrier separating the initial and final states. This simple situation is drastically complicated in solids. It is mainly complicated because a few molecules must be shifted for reagents to move by more or less optimal reaction paths. This must give rise to many maxima and minima on the surface of the potential energy of the interaction of a molecular cluster, the shifts in which are substantial for chemical transformations. Consequently, many maxima and minima must appear in the reaction path.

A problem of the analysis of potential energy surface during molecule shifts has repeatedly been discussed in papers on the theory of the crystal packing of molecules (see the monograph of Kitajgorodskii 1971). It has appeared that due to the multiplicity of interactions in the crystal lattice of organic compounds the system can pass through a set of energy minima even upon rotation of one molecule. As an example this can be confirmed by calculating the potential surface by the method of atom–atom potentials for naphthalene crystals (Kitajgorodskii 1971). In this case, within the limits of the rotation through 8° about the main equilibrium position with the depth of the minimum of about 18 kcal mol^{-1}, one can observe two additional minima with the depth of about 13 kcal mol^{-1}. The number of the minima sharply increases with increasing number of the degrees of freedom for movements (Kitajgorodskii 1971).

From studying the physical properties of a glassy state it is evident that these cannot be described assuming only one potential minimum for molecules in the equilibrium state. The analysis of the question has become possible with the appearance of studies on the anomalous behaviour of heat capacity and heat conductivity in different glasses at helium temperatures (Zeller and Pohl 1971) and by the anomalous hypersonic absorption in them (Heinicke *et al.* 1971). The experiments performed at temperatures near 4 K can be explained assuming that the particle states should be described by a double asymmetrical potential well with almost equivalent positions and a wide spectrum of localized tunnelling levels in these wells (Phillips 1972; Anderson *et al.* 1972). It is agreed that the spectrum is determined by disorders in the glass structure.

The situation can easily be explained using heat capacity as an example. It has been found experimentally that the heat capacity as a temperature function behaves itself as $C = \alpha T + \beta T$. The second term corresponds to the Debay theory. The first term that dominates at $T \cong 0.2$ K calls for additional assumptions. It can exist only with a sufficient density of levels near the well bottoms. Only this can provide the absence of heat capacity freezing out at superlow temperatures, which is achieved in the frame of the above hypotheses. Indeed, for a sufficient density of levels near the well bottoms and a smooth change of this density $n(E)$, the width of the excited zone at temperature T is proportional to kT. The mean energy of the excited level is also $\cong$ kT. As a result, the energy absorbed during heating is proportional to $k^2T^2n(0)$ and the heat capacity is $k^2Tn(0)$. Thus, the linear temperature term in the expression for heat capacity can be explained.

Evidently the hypothesis that the potential well is only double is not necessary. It can, for example, be triple and this also satisfies the experiment but complicates the description. Therefore, a double well is quite enough for a qualitative description of the phenomenon. Moreover, at very low temperatures the most essential wells are the two lowermost near the potential minima.

Now an idea of a two-well system of equilibrium positions is widely used to interpret the data on other physical studies. These include the effects of acoustic and dielectric saturation, the observing of the signals of acoustic echo, etc. (see the review of Smolyakov and Khaimovich 1982). Recently, this concept has been used in the theory of low-temperature nuclear spin relaxation (Kanert *et al.* 1991) and in spin echo studies (Dzuba and Tsvetkov 1988) of the reorientation of stable radicals in glasses.

The conformation transitions observed by the ESR method in radical pairs are the direct experiments that can record transitions between the close potential minima of a potential surface. We are dealing with the pairs $R_1...R_2$, which are the free radicals situated in the neighbouring sites of either crystal lattice or glass, and that form under irradiation of organic compounds. Due to a strong dipole–dipole interaction the ESR spectrum of such a pair is highly sensitive to the distance between R_1 and R_2 and allows one to distinguish distances of about 0.1 Å.

As an example consider the transitions of the B-type radical pairs in γ-irradiated single crystals of potassium deuteromalonate (Knopp and Muller 1981, 1983; Syutkin *et al.* 1984; Syutkin and Tolkatchev 1986a). The pairs represent the particles:

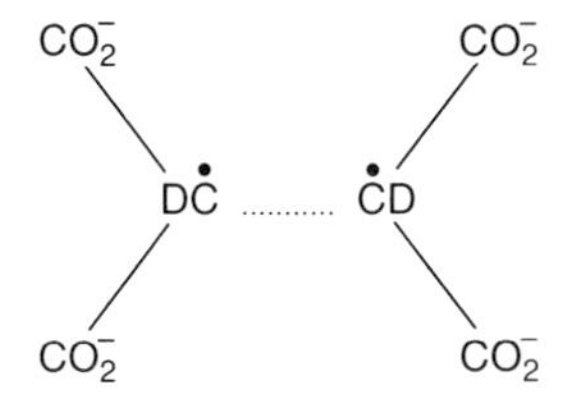

which occupy the neighbouring sites of the lattice along the crystallographic axis *b*. Figure 8.9 exemplifies the ESR spectra for pairs B_2, B_3 and B_4. As follows from the spectra, the different pair states can be resolved well, which has allowed a detailed quantitative study of transitions in this system. A scheme of transitions and the distances between unpaired electrons for each stable pair state are shown in Fig. 8.10. Near 150 K the irreversible transition of the B_1–B_3 pairs has been observed. The reversible transitions occur between the B_2, B_3 and B_4 pairs in the range 220–270 K. Near 270 K these three pairs gradually transform into the B_5 state. As follows from the figure, in all transitions the distances in pairs change by about 0.2 Å. This value is close to the mean-square amplitude of the intermolecular heat vibrations of molecules in crystals (Currie and Speakman 1970; Sime *et al.* 1970; Kitajgorodskii 1971). The intensities of the ESR lines of the B_2, B_3 and B_4 pairs have been used to measure their relative equilibrium concentrations and to calculate the thermodynamic parameters of transitions in the temperature range 225–275 K (Table 8.2; Syutkin and Tolkatchev 1986a).

These variations in thermodynamic parameters can hardly be explained by such a minor change in the distance between pair radicals when assuming only the change in the distance between two particles. The case is that for the given differences in the distance of the pair the changes in enthalpy may not be so large. For example, when

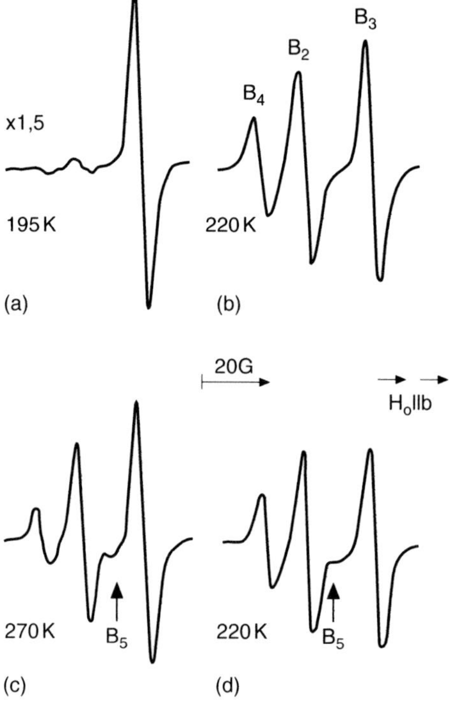

Figure 8.9 ESR spectra of high-field components of the B-type radical pairs in γ-irradiated single crystals of potassium deuteromalonate (Syutkin *et al.* 1984) after successive annealing at: (a) 195 K for 1 hour; (b) 220 K for 8.5 hours; (c) 270 K for 20 min; (d) 220 K for 2 hours. The arrow denotes the B_5 pairs. The spectra have been recorded at 77 K.

crystalline benzene shifts from the equilibrium position, the sharpness of potential is about 0.1 kcal mol in a 0.1 Å shift (Kitajgorodskii 1971). Besides, for molecular organic crystals the standard values of entropy at 298 K are 30–40 cal mol grad (Mischenko and Ravdel 1983). Somewhat lower values correspond to 250 K. Thus, the changes in entropy with changing distance in the pair can be compared to the value of entropy of the crystal itself. It is assumed, then, that the different pair configurations differ not only in the mutual distribution of radicals but also in the mutual distribution of the neighbouring molecules. This means that each state of a radical pair corresponds to a certain packing of the cluster consisting of surrounding molecules and a pair. The packing corresponds to the energy minimum in configuration space. Transitions should be considered as the rearrangement of the structure of this cluster. The Electron Nuclear Double Resonance (ENDOR) study of this system (Knopp and Muller 1983) is in fair agreement with the above conclusion.

Let us turn to studying the chemical properties that must be sensitive to a potential

Table 8.2 Thermodynamic parameters of transitions for the B_2, B_3 and B_4 pairs in the temperature range 225–275 K

	Transition	
Parameter	B_3–B_2	B_3–B_4
$-\Delta H$ (kcal mol)	1.15	3.1
$-\Delta S$/R	2.8	7.3

ΔH, change in enthalpy; ΔS, change in entropy.

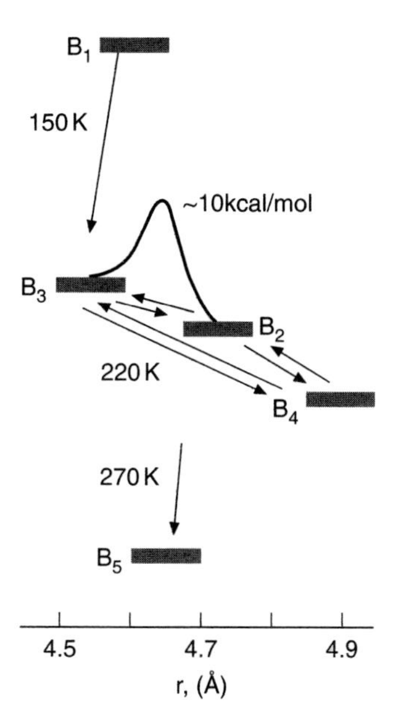

Figure 8.10 Scheme of the transitions between the state of B-type radical pairs in γ-irradiated potassium-d_3 hydromalonate (Knopp and Muller 1981; Syutkin *et al.* 1984; Syutkin and Tolkatchev 1986a).

barrier form. These include the isotopic effects in the reaction of H-atom abstraction by free radicals from matrix molecules (see Equation 8.2) and the probabilities of H-atom transfer depending on the distance between reagents in the same process.

A theoretical consideration of the classical isotopic effect of the reaction in Equation (8.2) is usually based on the theory of a transitional state. Its temperature dependence is commonly represented by the relation $k_H/k_D = A\exp(-\Delta E/kT)$, where k_H and k_D are the rate constants of H-atom abstraction reactions, A is a constant, (Melander and Saunders 1980; Kwart 1982). Experiments performed in the gas phase for similar reactions (Kondrat'ev 1970) correspond to the linear symmetrical transitional state with $A \cong 1$. Difference in activation energy ΔE, is determined by the main frequencies of the stretching vibrations of C–H and C–D and amounts to $h(\nu_H - \nu_D)$. The theoretical estimations of the classical isotopic effect with such a difference in activation energies (about 1.5 kcal mol^{-1}) and at room temperature, give the value of about 10. A temperature drop as well as the consideration of reaction tunnelling must increase the isotopic effect.

The isotopic effects in the reaction in Equation (8.2) have been studied repeatedly. At low temperatures the isotopic effects are large enough in many systems, which is usually attributed to the tunnelling character of the reaction (Goldanskii *et al.* 1989). However, at both low and high temperatures the isotopic effects are sometimes very low or even unavailable (Dainton *et al.* 1976; Senthilnathan and Platz 1980; Toriyama *et al.* 1980; Zanin *et al.* 1980; Platz *et al.* 1982; Zaitsev *et al.* 1983; Wright and Platz 1984; Syutkin and Tolkatchev 1985b, 1986b). When reagent R diffuses this can be explained readily assuming diffusion to be the limiting reaction stage. Especially interesting are the cases when the isotopic effect is anomalously small and the reaction involves the neighbouring molecule, i.e. the diffusion stage is lacking. Consider now such an example.

Figure 8.11 depicts the kinetics of the reaction of $CH_2CO_2^-$ radicals with matrix molecules in γ-irradiated glycine. The matrix molecules are a mixture of the isotopes

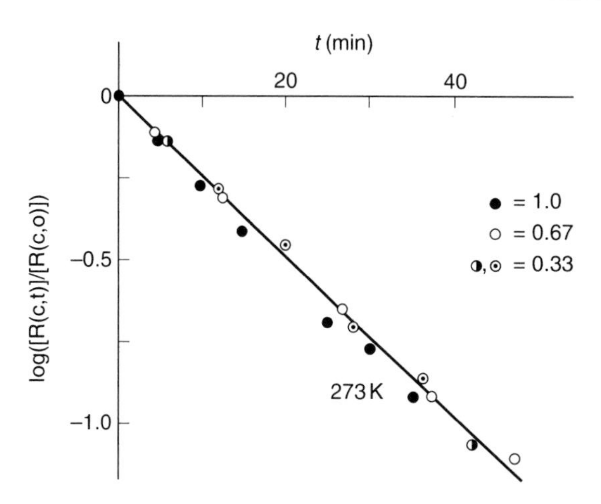

Figure 8.11 The decay of $CH_2CO_2^-$ radicals in γ-irradiated glycine-d_3/d_5 according to $CH_2CO_2^- + ND_3^+CH_2CO_2^- \rightarrow CH_3CO_2^- + ND_3^+CHCO_2^-$ and $CH_2CO_2^- + ND_3^+CD_2CO_2^- \rightarrow CH_2DCO_2^- + ND_3^+CDCO_2^-$ (Syutkin and Tolkatchev 1985b). The numbers denote the mole fraction of $ND_3^+CH_2CO_2^-$ molecules in the matrix.

$ND_3^+CH_2CO_2^-$ and $ND_3^+CD_2CO_2^-$. Therefore, in this case the reaction (Equation 8.2) follows the paths:

$$CH_2CO_2^- + ND_3^+CH_2CO_2^- \rightarrow CH_3CO_2^- + ND_3^+CHCO_2^-$$
$$CH_2CO_2^- + ND_3^+CD_2CO_2^- \rightarrow CH_2DCO_2^- + ND_3^+CDCO_2^- \qquad (8.32)$$

The different points in Fig. 8.11 correspond to the different isotopic compositions of the matrix, which has no effect on the abstraction rate. Thus, the isotopic effect is considered to be unity, which contradicts any theoretical concepts, but only if the potential barrier is assumed to be smooth and to have one maximum. When many maxima are assumed the contradiction vanishes. Indeed, let us suggest that in this case one of the minima is separated from the neighbouring ones by fairly high barriers and corresponds to the reactive state. Then, its formation can limit the process by destroying the isotopic effect. This is obvious from the simple two-stage reaction scheme:

$$R + R'H \underset{W_-}{\overset{W_+}{\rightleftarrows}} R...R'H \overset{k_h}{\rightarrow} RH + R'$$
$$R + R'D \underset{W_-}{\overset{W_+}{\rightleftarrows}} R...R'D \overset{k_d}{\rightarrow} RD + R' \qquad (8.33)$$

where R...R′H and R...R′D denote this reaction state; W_+ and W_- are the probabilities of its formation and decay. Then, from the principle of quasi-stationary concentration it is easy to state that the isotopic effect equals:

$$k_H/k_D = \frac{W_+k_h/(W_- + k_h)}{W_+k_d/(W_- + k_d)} \qquad (8.34)$$

It is evident that if $W_- \ll k_h$, k_d then $k_H/k_D = 1$.

Now analyse the second question of the dependence of the probability of H-atom tunnelling on the distance between reagents. These values can be measured experimentally in the systems in which the isotopic effect is large enough and the number of molecules with which a free radical can react is not large. These measurements have been performed by Syutkin and Tolkatchev (1985a, 1989) when studying the kinetics of the reaction of methyl radicals in two crystalline matrices (zinc acetate dihydrate and acetic acid). The kinetic description of these reactions has been given above. Now we are going to discuss in detail the data obtained in zinc acetate. In this system, forming under γ-irradiation, the methyl radical can react with three molecules at the temperatures studied. The corresponding probabilities are given in Table 8.1. Before we comment on them, we discuss the position of reagents in the crystal lattice.

According to the literature data, in monomolecular decays or molecules under irradiation the fragments remain almost in the same positions as a parent molecule (for details see Syutkin and Tolkatchev 1985a; Syutkin *et al.* 1985). A scale of distortions is related to the changes of molecular radii and it amounts to about 0.2 Å. Therefore, we compare the obtained values of transfer probabilities to the positions of molecules in the undamaged lattice according to crystallographic data that, for zinc acetate, are given in Niekerk *et al.* (1953).

Figure 8.12 shows the relative location of both the R molecule yielding a methyl radical, and the four nearest molecules. The distance between the C_1 atoms of other molecules and those of molecule R is no less than 5.3 Å. Figure 8.12 shows that the molecules are non-equivalent in their relative orientation and in the distance to radical, even assuming a 0.2 Å shift and rotations through 10°. The difference in the distances is 0.5–1.0 Å.

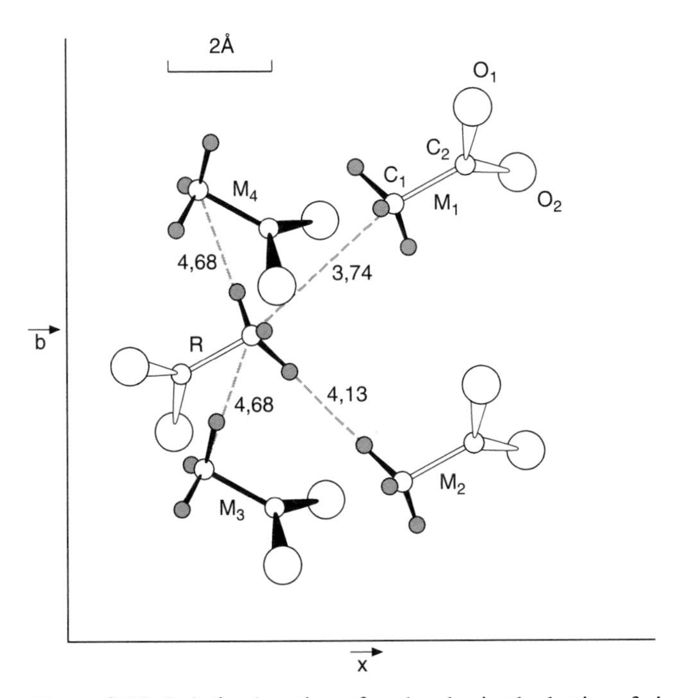

Figure 8.12 Relative location of molecules in the lattice of zinc acetate dihydrate (Niekerk *et al.* 1953). Projection along axis *z* produces an angle of 40.5° relative to axis *c*. All the C–C bonds are in the figure plane. The *Z* coordinates are: $Z_R = 0$, $Z_{M_1} = Z_{M_2} = -0.4$ Å, $Z_{M_3} = Z_{M_4} = 3.78$ Å. The distances between the carbon atoms of methyl groups are given in ångströms. Symbol R denotes a molecule producing a methyl radical.

Regarding the correspondence between the probabilities of the transfer and the mutual location of molecules and a radical, the following may be concluded. The probability W_1 can be compared with that of H-atom abstraction from the molecule at site M_1 (Fig. 8.12) because its equilibrium position is most optimal. As the methyl radical is free to rotate only about its own third-order axis, for the reaction with the molecules in positions M_2 and M_3 to occur, the p_z-orbital of the methyl radical must be orientated on these molecules. The W_2 and W_3 values can be assigned to the reaction with these two molecules. The molecules in position M_4 are likely to be less accessible for H-atom abstraction than the others due to the disadvantageous location of the methyl-group protons of this molecule relative to radical.

The magnitude of the activation energy of the probability of W_1 is equal to 5 kcal mol^{-1} (Syutkin and Tolkatchev 1985a). A lower estimate of the value of isotopic effect obtained from position one is about 600 (Syutkin and Tolkatchev 1985a). A low value of the activation energy and a large value of the isotopic effect correspond to tunnelling. However, in this case an increase in the distance of transfer by 0.5–1 Å must lead to a 10^9–10^{17}-fold decrease of the transfer probability according to the Gamov formula. The intermolecular vibrations that modulate the barrier must cause no large changes in the effect because the estimations with allowance for vibration give the same figures (Burshtein *et al.* 1984). Experimentally, however, a decrease in the probability by an order of two orders of magnitude has been observed. For acetic acid, a similar result has been obtained (Syutkin and Tolkatchev 1989). Thus, in these cases the discrepancy between the up-to-date theoretical concepts and experiment is cardinal.

An explicit discrepancy between the tunnelling character of reaction and the weak dependence of transfer probability on distance can be explained assuming the existence of a barrier with a great number of potential minima. It is believed, then, that the behaviour of such a cluster in the reaction does not only reduce to a complex vibrational process, but one should consider the final lifetime of the cluster in the potential minima. The existence of the minima can change the dependence of transfer probability on distance first of all because the transfer begins from the distances having no direct connection with the equilibrium ones. In terms of the above two-stage reaction scheme, at least the $W_- \gg k_d$ relation must hold for the existence of a large isotopic effect. This gives the value of the isotopic effect:

$$k_H/k_D = \frac{W_+ k_h/(W_- + k_h)}{W_+ k_d/W_-} = \frac{k_h W_-}{k_d(W_- + k_h)} \tag{8.35}$$

which may be large because k_h/k_d is large. At the same time, from the same two-stage scheme, it follows that the equilibrium states of reagents do not determine the isotopic effect.

An assumption that, with the changes of the reaction coordinates for a cage process, the reaction has in its way not only two main minima but also the additional stable one of potential energy corresponding to the optimal position of reagents, is used in other cases as well. This idea has allowed Lebedev (1985) to account for the long-performed experiments (Karpukhin 1978), in which the relationship between the reaction rates of different types in the liquid and solid phases have been compared. All the processes in the liquid phase have been controlled by chemical transformations. It has experimentally been verified that the faster reactions are retarded to a greater extent upon conversion into the solid phase. In this case, the $k_s \propto k_l^{1/2}$ relationship holds where k_s and k_l are the reaction constants in the solid and liquid phases, respectively. An introduction of a double potential minimum and some additional assumptions on the barrier form allows one to explain these results (Lebedev 1985).

Thus, both the physical and chemical studies permit one and the same conclusion, i.e. that a form of the potential barrier in solid phase reactions substantially differs from those in the gas phase. This determines a series of peculiarities of elementary reactions in solids.

3.2 *Matrix isotopic effect*

We have discussed the isotopic effect in the transfer of H atom upon its substitution by D isotope. It seems that in the substitution of atoms in organic compounds in the same parts of the molecule that are not involved in the reaction, the minor isotopic effects are expected. The case is that such a procedure actually has no effect on a mass of molecules, which thus causes the minor changes in macrophysical properties (Rabinovitch 1968). The influence on the reaction of H–D changeover in the non-reacting molecule parts or non-reacting molecules (Bockmann and Moan 1976; Bol'shakov *et al.* 1977; Jakimchenko and Lebedev 1978; Levina and Tolkatchev 1980; Bol'shakov *et al.* 1980; Zaskulnikov *et al.* 1981; Syutkin and Tolkatchev 1982; Wright and Platz 1984; Bagryansky and Tolkatchev 1987b; Zapadinsky and Tolkatchev 1988) is called the matrix isotopic effect.

As a rule, the matrix isotopic effects are small. However, sometimes these are rather large, which is exemplified by Fig. 8.13. Since the reaction of radical oxidation in methanol is diffusion controlled (Bagryansky *et al.* 1983), it is concluded that the substitution of H by D in a hydroxyl group reduces almost twice the diffusion coefficient of molecular oxygen. A strong slowing down of the process has been observed upon H-atom abstraction in the reaction of methyl radicals with sodium acetate trihydrate when substituting the H_2O molecules by D_2O (Bol'shakov *et al.*

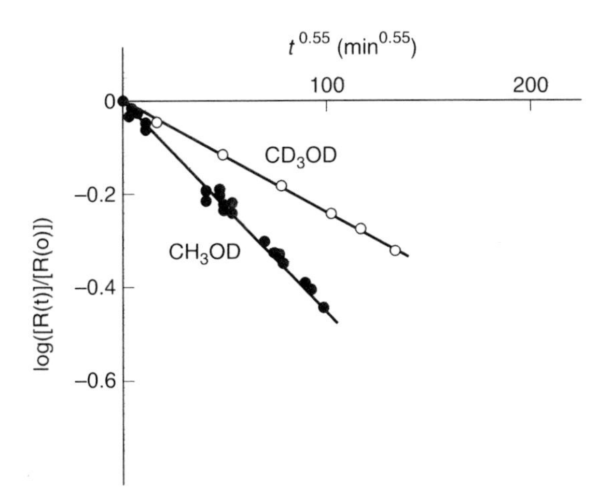

Figure 8.13 Kinetics of the oxidation of hydroxymethyl radicals by molecular oxygen in excess in glassy methanol with different isotopic compositions with 9% D_2O (Zapadinsky and Tolkatchev 1988). $T = 77$ K. Oxygen concentration 2.4×10^{19} cm^{-3}.

1977), and in the reaction:

$$CH_2CO_2^- + ND_3^+CH_2CO_2^- \rightarrow CH_3CO_2^- + ND_3^+CHCO_2^- \quad (8.36)$$

in α-glycine after deuteration of the amino-group (Syutkin and Tolkatchev 1982). In all these processes the reaction is slowed down by system deuteration. An opposite precedent has been observed when studying H-atom diffusion in sulphuric acid/water glasses (Bockmann and Moan 1976), and the reaction of H-atom abstraction by methyl radicals from polymethylmethacrylate molecules (Bagryansky and Tolkatchev 1987b) in its mixtures with a deuterated analogue. Figure 8.14 depicts the kinetic curves for this reaction for some isotopic compositions of the matrix. First, the reaction rate decreases with dilution of polymethylmethacrylate with a deuterated one and then it begins to grow.

Strong matrix isotopic effects are likely to testify to the fact that the minor changes in the character of potential surface and in the frequencies of lattice vibrations upon deuteration can cardinally affect both the diffusion and the overcoming of the reaction trajectory during reagent movement.

3.3 *The size of the reaction zone in elementary reactions*

This section reports on another aspect of reaction cluster behaviour, i.e. the amount of molecules participating in the shifts leading to chemical reactions. This material is based on a number of factors observed at different times for different processes (Klochikhin and Trakhtenberg 1984; Vyazovkin *et al.* 1985b; Bagryansky and Tolkatchev 1987b, Zapadinsky and Tolkatchev 1988; Vasenkov *et al.* 1991; Syutkin *et al.* 1992, 1995; Doba *et al.* 1993; Tolkatchev *et al.* 1993; Vyazovkin and Tolkatchev 1995). This testifies to the existence of a local mobility near the reagent that does not reduce to intermolecular vibrations and involves a fairly large volume near reacting particles. In

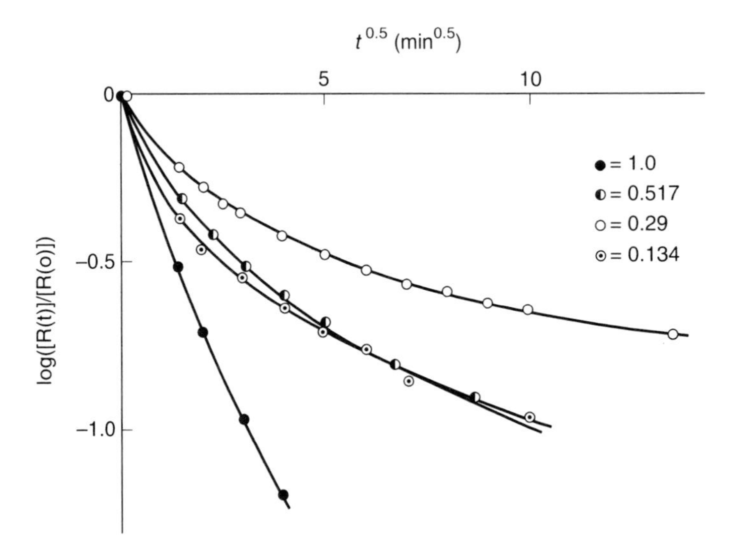

Figure 8.14 Kinetics of the decay of methyl radicals in polymethylmethacrylate-h_8/d_8 at 120 K (Bagryansky and Tolkatchev 1987b). Numbers in the figure are equal to a mole fraction of H-containing molecules.

a spherical model the radius with such a volume can reach 10 Å for quite different reactions. We shall discuss these facts in more detail.

Consider first the reaction:

$$Ag + Ag^+ \rightarrow Ag_2^+ \tag{8.37}$$

in sodium–calcium–aluminophosphate glass that runs at low temperature. It has been studied in detail by Syutkin *et al.* (1992) in excess Ag^+. The kinetics of the reaction in Equation (8.37), within the range 100–250 K displays a stepwise character (Fig. 8.15), well known for solid phase processes: after a fast rise of reaction temperature the decay of Ag first is fast (at $t < 100$ s) and then sharply decreases. The fraction of atoms decaying with a high rate depends on both the annealing temperature and Ag^+ content (Fig. 8.15). For a fixed reaction time (10^3 s) the influence of Ag^+ content on the transformation depth of Ag obeys the formula (Syutkin *et al.* 1992):

$$[Ag]/[Ag(0)] = \exp(-v(T)[Ag^+]) \tag{8.38}$$

The right-hand side of the formula equals the probability that the volume v has at least one ion, providing all Ag^+ are randomly distributed over the sample. This means that the reaction in Equation (8.37) is sure to occur in the volume limited by value v. This volume is temperature dependent and, if it is assumed to be spherical, the temperature dependence of the radius of this sphere appears to be linear. It is depicted in Fig. 8.16. For higher temperatures the kinetic character of this reaction changes drastically. The reaction goes to completion at reasonable times (Fig. 8.17) and can be described in the frame of the diffusion-controlled reaction with rate constant distribution determined by the spectrum of diffusion coefficients (Syutkin *et al.* 1992). This reaction regime has already been discussed. Note that in this case the unstationary term of the Smoluchowski diffusion rate constant has been taken into account because of the high Ag^+ concentration. Similar results have been obtained for the reaction $Ag_2^+ + Ag^+ \rightarrow Ag_3^{2+}$ in the same system (Syutkin *et al.* 1992) and for the reaction in Equation (8.37) in sodium–calcium–phosphate glass (Syutkin *et al.* 1995).

The likely results are available for the reaction of radical oxidation by molcular

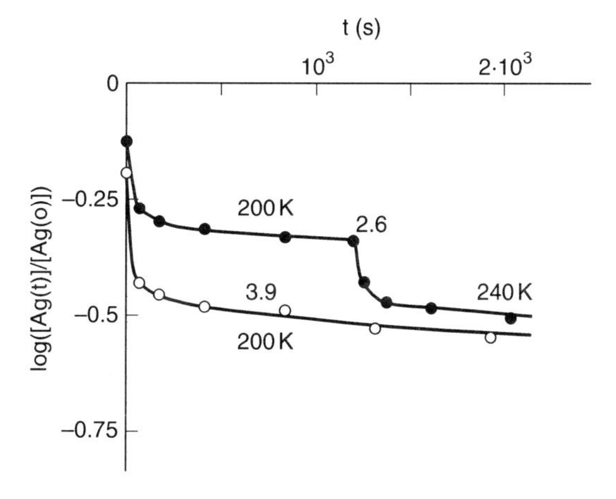

Figure 8.15 Decay of Ag atoms in phosphate glass via the reaction $Ag + Ag^+ \rightarrow Ag_2^+$ at low temperatures (Syutkin *et al.* 1992). Numbers in the figure are equal to Ag ions concentration (10^{20} cm^{-3}). $[Ag^+] \gg [Ag]$.

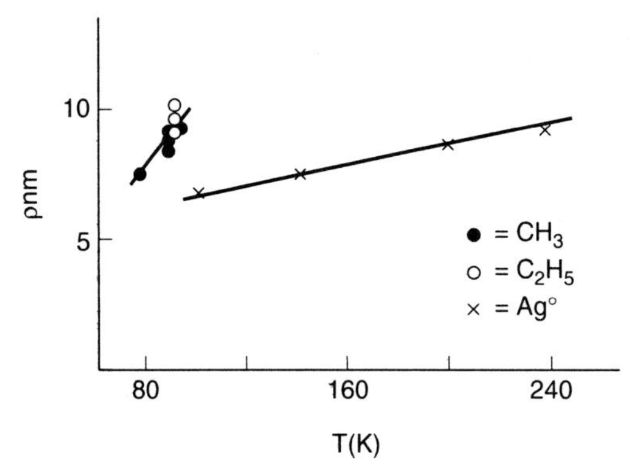

Figure 8.16 Temperature dependence of reaction zone radius (Tolkatchev *et al.* 1993) for the reaction $R + CH_3OH \rightarrow RH + CH_2OH$, where $R = CH_3$, C_2H_5, and for the reaction $Ag + Ag^+ \rightarrow Ag_2^+$ in sodium–calcium–aluminophosphate glass.

oxygen in methanol (Zapadinsky and Tolkatchev 1988), and squalane (Vasenkov *et al.* 1991). In these systems the radius of the reaction zone at 77 K is 10.1 ± 1.4 Å for methanol and 11.8 ± 0.5 Å for squalane.

The number of accessible molecules for the reaction in Equation (8.1) in alcohol glasses has been studied by Klochikhin and Trakhtenberg (1984), Vyazovkin *et al.* (1985a,b), Doba *et al.* (1993) and Vyazovkin and Tolkatchev (1995). Consider in more detail the experimental idea realized in papers by Vyazovkin *et al.* (1985a,b) and Vyazovkin and Tolkatchev (1995), which differs from others in that it has no relation to any models and is based on the experimental kinetic results only. Assume that the radical can react within a limited volume with the number of molecules N. In the matrix in which all RH molecules are substituted by the RD ones, let the kinetic law of radical decay be described by the law $[R] = [R(0)]\ f(t)$, where $f(t)$ in this case is an

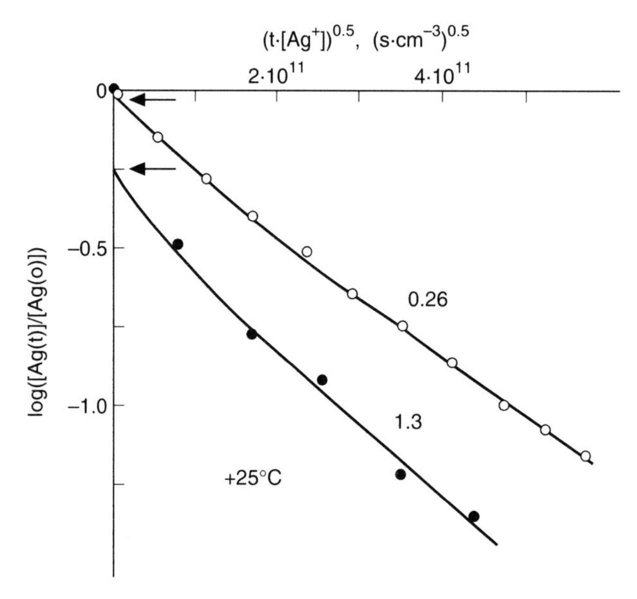

Figure 8.17 Decay of Ag atoms in phosphate glass at 298 K via the reaction $Ag + Ag^+ \rightarrow Ag_2^+$ (Syutkin *et al.* 1992). Numbers in the figure are equal to Ag ions concentration (10^{20} cm^{-3}). Arrows denote a jump in the initial concentration due to the low-temperature processes of $Ag + Ag^+ \rightarrow Ag_2^+$.

arbitrary function rather than the exponential one. Let the substance contain a mixture of the H and D molecules (RH + RD). Let all the molecules occupy the sites in the glassy matrix with equal probability. Then, as already mentioned, the probability to find the H molecule in any place equals its mole fraction (denote it with c). The probability that a given position contains no H molecules is $(1 - c)$. Consider now only the systems in which the probability of the reaction in Equation (8.2) is substantially higher than that of D-atom abstraction from RD in a similar process. Then, in the mixed matrix for large transformation depths when all the radicals in the environment, which contains at least one RH molecule, have reacted, the decay of radicals must be described by the asymptotic law:

$$[R(c,t)] = [R(c,0)](1 - c)^N f(t) \tag{8.39}$$

Note that Equation (8.39) is the same law as in Equation (8.5) for cage exponential reactions.

Such a behaviour of kinetics is shown in Fig. 8.18. In this case $f(t)$ is approximated by the $\exp(-k\sqrt{t})$ dependence. As follows from the figure, Equation (8.39) describes the experiment well. For the given experimental conditions $N = 30$. Similar values have been obtained by Doba *et al.* (1993) and Klochikhin and Trakhtenberg (1984). The temperature dependence of the radius of the reaction zone (ρ) for the reaction of H-atom abstraction by methyl radicals in methanol, and the value of ρ for ethyl radicals in the same matrix is depicted in Fig. 8.16. Thus, for quite different reactions occurring in the local environment of the reagent there are molecular motions that provide reactions between the particles, which are separated by 6–10 Å. This distance depends on temperature.

The range of molecular motions in glasses can be divided into two regions with quite different frequencies. They are commonly called the regions of the α- and β-relaxation (see the review of Johari 1985, and references therein). These regions are assumed to

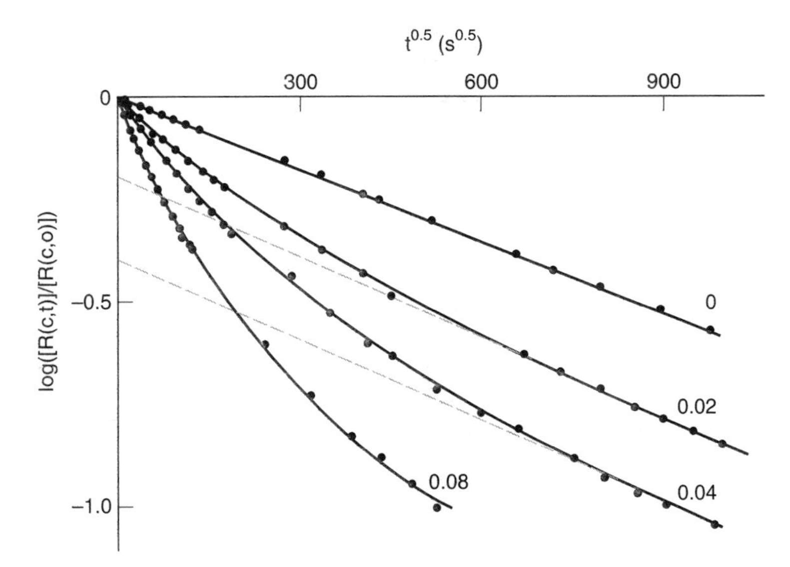

Figure 8.18 Kinetics of the decay of methyl radicals in methanol glass CH_3OH/CD_3OH via the reaction $CH_3 + CH_3OH \rightarrow CH_4 + CH_2OH$ and $CH_3 + CD_3OH \rightarrow CH_3D + CD_2OH$ (Vyazovkin *et al.* 1985b). Numbers in the figure are equal to a mole fraction of CH_3OH. $T = 77$ K.

relate to two order types in glassy matrices (Cohen and Grest 1981; Johari 1985; Dzuba and Tsvetkov 1987, and references therein): the structure consists of more dense and porous regions. The latter provides the β-relaxation with a higher frequency. In the above experiments the active particles have always been obtained under γ- or ultraviolet-action on the sample. Assuming that they are always generated in porous zones and diffuse within their boundaries, the limitations on the size of the region with high-frequency motions can be explained. One can also readily account for an increase of the reaction zone radius with temperature because at the periphery the energy barriers must be higher and the periphery becomes accessible only for reagents with increasing temperature. However, some experiments are at variance with this point of view. We shall consider these in more detail.

First of all, we shall discuss recent data of Vyazovkin and Tolkatchev (1995) on the reaction of H-atom abstraction (see Equation (8.2)), in which the methyl radicals were used as the R ones, and the molecules of ethyl alcohol CH_3CH_2OH and CH_3CD_2OH were used as the R′H ones. The deuterated methanol with a minor amount of H_2O was employed as a glassy matrix. Detailed discussion of the results is presented in the original paper of Vyazovkin and Tolkatchev (1995). Here, we are going to consider only the following. At 90 K for CH_3CH_2OH the number N of molecules accessible for the methyl radical is about 100, whereas for CH_3CD_2OH it decreases to five. When ethanol-h_6 is substituted by ethanol-d_2 the H atom abstracts from the methyl rather than methylene group. Comparing the data given in (Vyazovkin *et al.* (1985a,b) and Vyazovkin and Tolkatchev (1995) it is seen that the ratio between the rates of abstraction of H atom from ethanol and of D atom from methanol-d_4 is four orders of magnitude for ethanol-h_6 and two orders of magnitude for ethanol-d_2. Therefore, the deuteromethanol matrix in the latter case is still rather inert. If the N value in the first case has been determined by the size of the porous zone over which diffusion runs, it must not change much upon substitution of H-containing ethanol by that with a deuterated methylene group. Actually, the change has occurred. The experiments will be interpreted later. Here, we are just stating the contradiction between the idea of porous zone and the experimental results. In this case we are not discussing the fact of whether the porous zones can exist in glassy systems or not. It is quite possible that they do exist and the active particles (radicals) form in porous zones under radiation. This experiment is in conflict only with the fact that the reaction zone size coincides with that of the porous one. Other results inconsistent with the same idea are in the original paper of Vyazovkin and Tolkatchev (1995).

It is noteworthy that the anomalously large size of the reaction zone has been observed not only in glassy matrices. This conclusion follows from the data on the kinetics of the reaction in Equation (8.2) running in crystal sodium acetate crystallohydrate, given in Fig. 8.7. As the figure suggests, even for 10% of the H-containing component in the coordinates corresponding to rate constant distribution, no deviations from the kinetic law in a pure H-containing matrix and no passage to the asymptotic law have been observed. This confirms the fact (as mentioned in Section 2.2.5) that N is large enough. For this case N has been estimated to be more than 20 (Syutkin and Tolkatchev 1985a,b,c).

It may be assumed that the methyl radical in sodium acetate is free to diffuse over the matrix. However, studying the position of methyl radical by the ESR and spin-echo methods and comparing these data with those obtained by X-ray technique for a crystal cell (for details see Syutkin *et al.* (1985)), it is seen that the methyl radical preserves the methyl group orientation in the undamaged molecule. Moreover, as the spectral characteristics of the methyl radical do not vary with reaction time the orientation and position of the methyl radical are also preserved during the reaction (Syutkin *et al.* 1985). The methyl orientation can be preserved in the elementary crystal cell only if it contains the CO_2 molecules, which form upon methyl radical generation. A simultaneous diffusion of the CO_2–methyl pair is hardly probable. It is more likely that the fluctuations leading to the reaction involve a great number of molecules that shift together with the methyl radical. In the same case when no reaction is observed during these fluctuations, the crystal structure recovers with the recovering CO_2–methyl pair in the same cell.

Probably, such a reciprocating motion can also provide a contact of reagents at the distance of one or two coordination spheres in glassy matrices. These concepts make it possible to account for different N during the reaction of methyl radicals with ethanol-h_6 and -d_2 in the above experiments. Since there is no free motion in the matrix, the dependence of reaction probability on the distance between reagents should be introduced. N will be determined by the condition that the reaction rate in pure D cells with size N (without H-containing molecules) must exceed that of the reaction with H molecules being in the $N+1$ position, $\Sigma_{i=1}^{N} W_i^{\mathrm{D}} > W_{N+1}^{\mathrm{H}}$ (Vyazovkin and Tolkatchev 1995). Thus, this dependence must be different for the reactions of methyl radicals in CD_3OH matrix with C_2H_5OH (reaction with methylene group) and with CH_3CD_2OH (reaction with methyl group). This can account for the result obtained assuming a sharper distance dependence of the CH_3CD_2OH molecules than for the CH_3CH_2OH ones, because in the first case the barrier of atom transfer must be higher.

The reciprocating motion can be used to explain k(t) in the reactions of abstraction in glassy matrices as a result of the uncorrelated averaging of the transfer probabilities discussed in Section 2.2.5. The case is that free movement of radicals in the matrix will average the non-equilibrium of the lattice sites and exponentialize the kinetics. Assuming the model of reciprocating fluctuation movement, this explanation is quite possible under all other limitations of Section 2.2.5.

Finally, this model allows the interpretation of the well-known data on the stabilization of radical pairs in γ-irradiated alkanes. These include the correlation between the distance in the pair and the temperature of pair decay (Gillbro and Lund 1974; Dubinskii *et al.* 1977), the narrowing of distance–distribution radical pairs with decreasing radiation temperature (Iwasaki *et al.* 1976a,b) and the mechanism preventing the formation of new radicals at distances less than 10 Å under irradiation at 77 K (Dubinskii *et al.* 1977).

The model of reciprocating fluctuation motion makes the explanation of experimental results easier. It is, of course, not simple and evident. However, from the author's standpoint the foregoing testifies to the reality of the model. It is quite probable that the radius of such a reactive cluster may reach the critical value (about 10 Å) with temperature rise (Tolkatchev *et al.* 1993). As temperature increases, particle diffusion unfreezes.

4 Predictions for the 21st century

The author hopes that he has managed to demonstrate that little is known about the important details determining the mechanism of an elementary act in solids. First of all, it is obvious that by studying the kinetics one can solve many questions of its formal description. Nevertheless, a number of questions are still open to discussion. The case is that many of them have been studied using a small number of objects which, in the case of chemical processes often undergoing a series of effects, makes one careful in drawing conclusions. Some of these questions have just been set. These include, for example, the study of the influence of β-relaxation on the kinetics of chemical transformation. In either case, according to recent studies, this influence is complex enough (Anisimov *et al.* 1988; Vasenkov and Tolkatchev 1991). The author thinks that the study of the phenomena affecting the formal kinetics of chemical transformations in solids is sure to be continued in the future.

Although the formal kinetic rules are now more or less clear, the systematic studies of the facts determining the rate of elementary reactions in solids are actually unavailable. For example, there are no detailed studies on the reasons for the spectrum of rate constants, which is a frequent phenomenon in the chemistry of solids. The general words on why the range is caused by the inhomogeneity of a solid are very vague and cause no enthusiasm. For example, what is the relation in inhomogeneity of the facts (Vasenkov *et al.* 1991; Kondratenko *et al.* 1993) that rate constant distribution depends on the entropy factor distributions, and all the constants have one and the same activation energy? Since the investigation of both the process rates and the structure of the rate constant of elementary reactions are actually absent, and the problems on chemical transformations in solids are important and interesting, the moment is sure to come when they will be studied extensively. Since the 20th century is coming to a close, this will happen in the 21st century.

The most interesting area is the study of the collective movements of molecules leading to chemical transformations. This problem is quite general and involves not only the reactions in solids but also those occurring in condensed medium. Recently, the methods of molecular dynamics have become widespread, which allows one to analyse such details of motions in the condensed medium of which one could only dream of before. The development of computational methods is likely to afford the computation of large molecular systems. It is hoped that it will be possible to study such phenomena (scarcely probable from the point of view of statistical physics) as the realization of the reaction state. Many laws of the elementary chemical act that are difficult to verify by a physicochemical experiment will then be studied in detail. The consequences of computations performed in this direction can be so wide that the discovery of fundamental rules can be predicted which can advance science along these new lines.

All theories given in this section have no experimental basis but they will soon be verified as the 21st century is coming.

5 References

Aditya, S. & Willard, J.E. (1966) *J Am Chem Soc* **88**, 229–232.

Aleksandrov, A.I., Bubnov, N.N., Lazarev, G.G., Lebedev, Ya.S., Proklf'ev, A.I. & Serdobov, M.V. (1976) *Izvestija AN SSSR Ser Khim* **3**, 515–520.

Anderson, P.W., Halperin, B.I. & Varma, C. (1972) *Phil Mag* **25**, 1–9.

Anisimov, V.M., Mardaleishvily, I.R. & Zaitseva, N.I. (1988) *Vys Soedineniya* **30A**, 1239–1243.

Bagryansky, V.A. (1987) *Khim Fiz* **6**, 1429–1433.

Bagryansky, V.A. & Tolkatchev, V.A. (1987a) *Vys Soedineniya* **29B**, 909–911.

Bagryansky, V.A. & Tolkatchev, V.A. (1987b) *J Polym Sci* **25A** 581–593.

Bagryansky, V.A. & Tolkatchev, V.A. (1991) *Radiat Phys Chem* **37**, 517–522.

Bagryansky, V.A., Pushchaeva, L.M. & Tolkatchev, V.A. (1989) *Vys Soedineniya* **31B**, 892–896.

Bagryansky, V.A. Zaskul'nikov, V.M. & Tolkatchev, V.A. (1983) *Chem Phys* **78**, 41–48.

Bagryansky, V.A., Sokolov, A.P. & Tolkatchev, V.A. (1988) *Vys Soedineniya* **30A**, 2262–2267.

Bamford, C.H., Jenkins, A.D. & Ward, J.C. (1960), *Nature* **186**, 712–713.

Belevskii, V.I., Bugaenko, L.T. & Golubev, V.B. (1966) *Dokl AN SSSR* **168**, 122–125.

Benderskii, V.A., Makarov, D.E. & Weight, C.A. (1994) *Tunneling Dynamics at Low Temperatures*, pp. 3–200. Wiley, NY.

Bensasson, R., Durup, M., Dworkin, A., Magat, M., Marx, R. & Scwarc, H. (1963) *Disc Farad Soc* **36**, 177–185.

Blumenfeld, L.A., Voevodskii, V.V. & Semenov, A.G. (1962) *The Application of ESR in Chemistry*, pp. 3–240. Siberian Division of AN SSSR, Novosibirsk.

Bockmann, B. & Moan, J. (1976) *Chem Phys Lett* **42**, 575–577.

Bol'shakov, B.V. & Tolkatchev, V.A. (1976) *Chem Phys Lett* **40**, 468–470.

Bol'shakov, B.V., Stepanov, A.A. & Tolkatchev, V.A. (1980) *Int J Chem Kinet* **12**, 271–281.

Bol'shakov, B.V., Doctorov, A.B., Tolkatchev, V.A. & Burshtein, A.I. (1979) *Chem Phys Lett* **64**, 113–115.

Bol'shakov, B.V., Fuks, M.P., Tolkatchev, V.A. & Burstein, A.I. (1977) *Proceedings of the 4-th Tihan. Symposium on Radiation Chemistry*, p. 723. Akademiai Kiado, Budapest.

Bol'shakov, A.I., Mel'nikov, V.P., Mikhailov, A.I., Barkalov, I.M. & Goldanskii, V.I. (1971) *Khim Vis Energ* **5**, 57–62.

Bresler, S.E., Kazbekov, E.N., Fomitchev, V.N., Setch, F. & Smeitek, P. (1963) *Fiz Tverd Tela* **5**, 675–682.

Buben, N.Ja., Pristupa, A.I. & Shamshev, V.N. (1964) *Kinet Katal* **5**, 190–191.

Burshtein, A.I. & Tsvetkov, Ju.D. (1974) *Dokl AN SSSR* **214**, 369–372.

Burshtein, A.I., Klotchikhin, V.L. & Trakhtenberg, L.I. (1984) *Khim Fiz* **3**, 155–161.

Butiagin, P.Ju. (1965) *Dokl AN SSSR* **165**, 103–106.

Butiagin, P.Ju. (1972) *Pure Appl Chem* **30**, 57–76.

Butiagin, P.Ju. (1974) *Vys Soedineniya* **16A**, 63–70.

Butiagin, P.Ju., Kolbanev, I.V., Dubinskaya, A.M. & Kisluk, M.U. (1968) *Vys Soedineniya* **10A**, 2265–2277.

Campbell, I.D. & Looney, F.D. (1962) *Austral J Chem* **15**, 642–652.

Cohen, M.H. & Grest, G.T. (1981) *Phys Rev* **24B**, 4091–4094.

Cracco, F., Arvia, A.J. & Dole, M. (1962) *J Chem Phys* **37**, 2449–2457.

Currie, M. & Speakman, J.C. (1970) *J Chem Soc A* **11**, 1923–1926.

Cutten, D.R. & Ericson, L.G. (1970) *Mol Cryst Liq Cryst* **6**, 351–375.

Dainton, F.S., Holt, B.J., Philipson, N.A. & Pilling, M.J. (1976) *J Chem Soc Farad Trans I* **2**, 257–267.

Davydov, E.Ja., Pariisky, G.B. & Toptygin, D.Ja. (1977) *Vys Soedineniya* **19A**, 977–983.

Doba, T., Ingold, K.U. & Siebrand, W. (1984a) *Chem Phys Lett* **103**, 339–342.

Doba, T., Ingold, K.U., Siebrand, W. & Wildman, T.A. (1984b) *J Phys Chem* **88**, 3165–3167.

Doba, T., Ingold, K.U., Siebrand, W. & Wildman, T.A. (1984c) *Farad Disc Chem Soc* **78**, 175–181.

Doba, T., Ingold, K.U., Siebrand, W. & Wildman, T.A. (1985) *Chem Phys Lett* **115**, 51–54.

Doba, T., Ingold, K.U., Lusztyk, J., Siebrand, W. & Wildman, T.A. (1993) *J Chem Phys* **98**, 2962–2970.

Doba, T., Ingold, K.U., Reddoch, A.H., Siebrand, W. & Wildman, T.A. (1987) *J Chem Phys* **86**, 6622–6630.
Dole, M. & Salik, J. (1977) *J Am Chem Soc* **99**, 6454–6455.
Dubinskii, A.A., Grinberg, O.Ja., Tabachnik, A.A. & Lebedev, Ya.S. (1977) *Khim Vis Energ* **11**, 156–161.
Dzuba, S.A. & Tsvetkov, Yu.D. (1987) *Zhurn Strukt Khim* **28**, 15–38.
Dzuba, S.A. & Tsvetkov, Yu.D. (1988) *Chem Phys* **120**, 291–298.
Emmanuel, N.M. & Buchachenko, A.L. (1988) *Chemical Physics of Degradation and Stabilization of Polymers*, pp. 53–115. VNU, Amsterdam.
Ermolaev, V.K, Molin, Ju.N. & Buben, N.Ja. (1962) *Kinet Kat* **3**, 314–321.
Ermolaev, V.K., Milov, A.D. & Tolkatchev, V.A. (1968) *Khim Vys Energ* **3**, 224–230.
Fedoseeva, T.S. & Kuzminskii, A.S. (1967) *Khim Vis Energ* **1**, 248–252.
Flournoy, J.M., Baum, L.H. & Siegel, S. (1962) *J Chem Phys* **36**, 2229–2231.
French, W.G. & Willard, J.E. (1968) *J Phys Chem* **72**, 4604–4608.
Fuks, M.P., Bol'shakov, B.V. & Tolkatchev, V.A. (1975) *React Kinet Catal Lett* **3**, 349–354.
Gillbro, T. & Lund, A. (1974) *J Chem Phys* **61**, 1469–1474.
Goldanskii, V.I., Trakhtenberg, L.I. & Flerov, V.N. (1989) *Tunneling Phenomena in Chemical Physics*, pp. 3–294. Gordon & Bleach, New York.
Grebenkin, S.Ju. & Bol'shakov, B.V. (1991) *Vys Soedineniya* **33A**, 1859–1863.
Griffiths, W.E. & Sutcliffe L.H. (1966) *Trans Farad Soc* **62**, 2837–2842.
Grinberg, O.Ja., Dubinskii, A.A. & Lebedev, Ya.S. (1972) *Kinet Katal* **13**, 660–664.
Hamill, W.H. & Funabashi, K. (1977) *Phys Rev* **16B**, 5523–5527.
Heinicke, W. & Winterling, G. & Dransfeld, K. (1971) *J Acoust Soc Am* **49**, 954–958.
Helman, W.P. & Funabashi, K. (1979) *J Chem Phys* **71**, 2458–2463.
Horan, P.K., Henriksen, T. & Snipes, W. (1970) *J Chem Phys* **52**, 4324–4328.
Ivantchev, S.S., Konovalenko, V.V. & Gak, Ju.N. (1968a) *Dokl AN SSSR* **178**, 634–637.
Ivantchev, S.S., Jurzhenko, A.I., Lukovnikov, A.F., Gak, Ju.N. & Kwasha, S.M. (1968b) *Teor Eksper Khim* **4**, 780–787.
Iwasaki, M., Toriyama, K., Muto, H. & Nunome, K. (1976a) *Chem Phys Lett* **39**, 90–94.
Iwasaki, M., Toriyama, K., Muto, H. & Nunome, K. (1976b) *J Chem Phys* **65**, 596–606.
Jakimchenko, O.E. & Degtjarev, E.N. (1980) *Khim Vys Energ* **14**, 239–247.
Jakimchenko, O.E. & Lebedev, Ya.S. (1978) *Uspekhi Khim* **47**, 1018–1047.
Jakimchenko, O.E. & Lebedev, Ya.S. (1979) *Dokl AN SSSR* **249**, 1395–1398.
Jakimchenko, O.E., Romanova, O.M. & Lebedev, Ya.S. (1978) *Z Fiz Khim* **52**, 1199–1202.
Jakimchenko, O.E., Degtjarev, E.N., Prusakov, V.N. & Lebedev, Ya.S. (1980) *Teor Eksp Khim* **16**, 75–80.
Jakimchenko, O.E., Gaponova, I.S., Gol'dberg, V.M., Pariisky, G.B., Toptygin, D.Ja. & Lebedev, Ya.S. (1974) *Izvestija AN SSSR Ser Khim* **2**, 354–360.
Jakimchenko, O.E., Kirjuschkin, S.G., Pariisky, G.B., Toptygin, D.Ja., Schljapnikov, Ju.A. & Lebedev, Ya.S. (1975) *Izvestija AN SSSR Ser Khim* **10**, 2255–2257.
Johari, G.P. (1985) *J Chim Phys* **82**, 283–291.
Johnson, D.R., Wen, W.Y. & Dole, M. (1973) *J Phys Chem* **77**, 2174–2179.
Judeikis, H.S. & Siegel, S. (1965) *J Chem Phys* **43**, 3625–3638.
Kanert, O., Kloke, M., Kuchler, R., Ruckstein, S. & Jain, H. (1991) *Ber Bunsenges Phys Chem* **95**, 1061–1068.
Karpukhin, O.N. (1978) *Uspekhi Khim* **47**, 1119–1143.
Kitajgorodskii, A.I. (1971) *Molecular Crystalls*, pp.3–422. Nauka, Moskow.
Klinshpont, E.R. & Milintchuk, V.K. (1975) *Vys Soedineniya* **17B**, 358–361.
Klochikhin, V.L. & Trakhtenberg, L.I. (1984) *Zh Fiz Khim* **58**, 2877–2879.
Knopp, R. & Muller, A. (1981) *Mol Phys* **42**, 1245–1258.
Knopp, R. & Muller, A. (1983) *Mol Phys* **50**, 369–378.
Kondrat'ev, V.N. (1970) *Rate Constants of Gas Phase Reactions,* pp. 9–119. Nauka, Moscow.
Kondratenko, E.V., Bol'shakov, B.V. & Tolkatchev, V.A. (1993) *Vys Soedineniya* **35**, 1267–1271.

Korolev, V.V., Sushkov, D.G. & Bazhin, N.M. (1991) *Chem Phys* **158**, 129–136.
Kozlov, V.T. (1967) *Vys Soedineniya* **9A**, 515–521.
Kutyrkin, V.A., Mardaleishvily, I.R., Karpukhin, O.N. & Anisimov, V.M. (1984) *Kinet Katal* **25**, 1310–1314.
Kuzina, S.I. & Mikhailov, A.I. (1976) *Dokl AN SSSR* **231**, 1395–1398.
Kuzminskii, A.S., Neiman, M.B., Fedoseeva, T.S., Lebedev, Ya.S., Buchachenko, A.L. & Tchertkova, V.F. (1962) *Dokl AN SSSR* **146**, 611–614.
Kwart, H. (1982) *Acc Chem Res* **15**, 401–405.
Lazarev, G.G., Kuskov, V.L. & Lebedev, Ya.S. (1990) *Chem Phys Lett* **170**, 94–98.
Lazarev, G.G., Kuskov, V.L. & Lebedev, Ya.S. (1991) *Chem Phys Lett* **185**, 375–380.
Lazarev, G.G., Lara, F., Garsia, F. & Rieker, A. (1992) *Chem Phys Lett* **199**, 29–32.
Lebedev, Ya.S. (1969) *Radiat Effects* **1**, 213–227.
Lebedev, Ya.S. (1978) *Kinet Katal* **19**, 1367–1376.
Lebedev, Ya.S. (1985) *Dokl Akad Nauk SSSR* **281**, 636–640.
Levina, L.M. & Tolkatchev, V.A. (1980) *Radiat Phys Chem* **16**, 75–81.
Loy, B.R. (1960) *J Pol Sci* **44**, 341–347.
Loy, B.R. (1961) *J Pol Sci* **50**, 245–252.
McBride, J.M. (1983) *React Acc Chem Res* **16**, 304–312.
McGhie, A.R., Blum, H. & Labes, M.M. (1970) *J Chem Phys* **52**, 6141–6144.
Mardaleishvily, I.R. & Anisimov, V.M. (1986) *Z Priklad Spektrosk* **44**, 581–584.
Mardaleishvily, I.R., Kutyrkin, V.A., Karpukhin, O.N. & Anisimov, V.M. (1984) *Vys Soedineniya* **26**, 1513–1518.
Marschall, J.M. (1983) *Rep Progr Phys* **46**, 1235–1282.
Melander, L. & Saunders, W.H. (1980) *Reaction Rates of Isotopic Molecules*, pp. 12–60. Wiley, NY.
Mikhailov, A.I. & Lebedev, Ya.S. (1968) *Z Fiz Khim* **42**, 1005–1007.
Mikhailov, A.I., Lebedev, Ya.S. & Buben, N.Ja. (1964) *Kinet Katal* **5**, 1020–1027.
Mikhailov, A.I., Lebedev, Ya.S. & Buben, N.Ja. (1965) *Kinet Katal* **6**, 48–55.
Mikhailov, A.I., Bol'shakov, A.I., Lebedev, Ja.S. & Goldanskii, V.I. (1972a) *Fiz Tverd Tela* **14**, 1172–1179.
Mikhailov, A.I., Kuzina, S.I., Lukovnikov, A.F. & Goldanskii, V.I. (1972b) *Dokl AN SSSR* **204**, 383–386.
Mischenko, K.P. & Ravdel, A.A. (1983) *Handbook of Chemical and Physical Value*, pp. 46–47. Khimiya, Leningrad.
Modarelli, D.A., Lahti, P.M. & George, C. (1991) *J Am Chem Soc* **113**, 6329–6330.
Myagkikh, V.I., Zaskulnikov, V.M. & Tolkatchev, V.A. (1980) *React Kinet Catal Lett* **14**, 73–76.
Nara, S., Kashiwabara, H. & Sohma, J. (1967) *J Polym Sci A-2* **5**, 929–938.
Nara, S., Shimada, S., Kashiwabara, H. & Sohma, J. (1968) *J Polym Sci A-2* **6**, 1435–1449.
Neiman, M.B., Fedoseeva, T.S., Chubarova, G.V., Buchachenko A.L. & Lebedev, Ja.S. (1963) *Vys Soedineniya* **5**, 1339–1344.
Neiss, M.A. & Willard, J.E. (1975) *J Phys Chem* **79**, 783–794.
Niekerk, J.N., van, Schoening, F.R.L. & Talbot, J.H.(1953) *Acta Crystallog* **6**, 720–723.
Nilsson, G. & Lund, A. (1984) *J Phys Chem* **88**, 3292–3295.
Pfister, G. & Scher, H. (1978) *Adv Phys* **27**, 747–798.
Phillips, W.A.J. (1972) *J Low Temp Phys* **7**, 351–360.
Platz, M.S., Senthilnathan, V.P., Wright, B.B. & McCardy, C.W. (1982) *J Am Chem Soc* **104**, 6494–6501.
Plonka, A. (1983) *Radiat Phys Chem* **21**, 405–409.
Plonka, A. (1986) *Time-Dependent Reactivity of Species in Condensed Media.* Springer-Verlag, New York.
Plonka, A. (1991) *Progr React Kinet* **16**, 157–333.
Plonka, A. & Kevan, L. (1984) *J Chem Phys* **80**, 5023–5026.
Plonka, A. & Pietrucha, K. (1983) *Radiat Phys Chem* **21**, 439–444.

Plonka A., Kroh, J., Lefik, W. & Bogus, W. (1979) *J Phys Chem* **83**, 1807–1810.
Plotnikov, O.V., Mikhailov, A.I., & Rayavee, E.L. (1977) *Vys Soedineniya* **19A**, 2528–2537.
Pollak, M. (1977) *Phil Mag* **36**, 1157–1169.
Rabinovitch, I.B. (1968) *Isotopic Influence on Physico-Chemical Properties of Liquids*, pp. 3–308. Nauka, Moscow.
Radtsig, V.A. (1976) *Vys Soedineniya* **18A**, 1899–1918.
Radtsig, V.A. & Rajnov, M.M. (1976) *Vys Soedineniya* **18A**, 2022–2030.
Schelimov, B.N., Fok, N.V. & Voevodskii, V.V. (1963) *Kinet Katal* **4**, 539–548.
Scher, H. & Montrol, E.W. (1975) *Phys Rev* **12B**, 2455–2477.
Scwarc, H. (1962) *J Chim Phys* **59**, 1067–1071.
Scwarc, H. (1966) *J Chim Phys* **63**, 137–141.
Senthilnathan, V.P. & Platz, M.S. (1980) *J Am Chem Soc* **102**, 7637–7643.
Serdobov, M.V., Alexandrov, A.I. & Lebevev, Ya.S. (1973) *Khim Vis Energ* **7**, 439–444.
Shimada, S., Hory, Y. & Kashiwabara, H. (1977) *Polymer* **18**, 25–31.
Shimada, S., Hory, Y. & Kashiwabara, H. (1981) *Polymer* **22**, 1377–1384.
Shirom, M. & Willard, J.E. (1968) *J Phys Chem* **72**, 1702–1707.
Sime, J.G., Speakman, J.C. & Parthasarathy, R. (1970) *J Chem Soc A* **11**, 1919–1923.
Smith, C.J., Poole, C.P. & Farach, H.A. (1981) *J Chem Phys* **74**, 993–996.
Smith, W.V. & Jacobs, B.E. (1962) *J Chem Phys* **37**, 141–148.
Smolyakov, B.P. & Khaimovich, E.P. (1982) *Uspekhi Fiz nauk* **136**, 317–343.
Spath, B.W. & Raff, L.M. (1992) *J Phys Chem* **96**, 2179–2185.
Sprague, E.D. (1979) *J Phys Chem* **83**, 849–852.
Stentz, F.B., Taylor, E.D. & Johnsen, R.H. (1972) *Radiat Res* **49**, 124–132.
Stepanov, A.A., Tkatchenko, V.A., Bol'shakov, B.V. & Tolkatchev, V.A. (1978) *Int J Chem Kinet* **10**, 637–648.
Sullivan, P.J. & Koski, W.J. (1963) *J Am Chem Soc* **85**, 384–387.
Sullivan, P.J. & Koski, W.J. (1964) *J Am Chem Soc* **86**, 159–161.
Syutkin, V.M. & Tolkatchev, V.A. (1982) *Radiat Phys Chem* **20**, 281–288.
Syutkin, V.M. & Tolkatchev V.A. (1985a) *Chem Phys* **95**, 115–122.
Syutkin, V.M. & Tolkatchev, V.A. (1985b) *Chem Phys Lett* **122**, 201–204.
Syutkin, V.M. & Tolkatchev, V.A. (1985c) *React Kinet Catal Lett* **29**, 417–421.
Syutkin, V.M. & Tolkatchev, V.A. (1986a) *Khim Fiz* **5**, 1288–1290.
Syutkin, V.M. & Tolkatchev, V.A. (1986b) *Kinet Katal* **27**, 979–983.
Syutkin, V.M. & Tolkatchev V.A. (1989) *Khim Fiz* **8**, 93–97.
Syutkin, V.M., Astashkin, A.V. & Tolkatchev, V.A. (1985) *Chem Phys* **98**, 267–278.
Syutkin, V.M., Dmitryuk, A.V. & Tolkatchev, V.A. (1991) *Fiz Khim Stekla* **17**, 273–281.
Syutkin, V.M., Dmitryuk, A.V. & Tolkatchev, V.A. (1992) *Soviet J Glass Phys Chem* **18**, 224–228 (English translation).
Syutkin, V.M., Garmasheva, N.V. & Tolkatchev, V.A. (1984) *Khim Fiz* **3**, 741–747.
Syutkin, V.M., Tolkatchev, V.A., Dmitrjuk, A.V. & Paramzina, S.E. (1995) *Chem Phys* **196**, 139–147.
Thryion, F.C. & Baijai, M.D. (1968) *J Polym Sci A-1* **6**, 505–507.
Tolkatchev, V.A. (1987) *Chem Phys* **116**, 283–298.
Tolkatchev, V.A., Syutkin, V.M. & Vyazovkin, V.V. (1993) *Chem Phys* **170**, 427–436.
Tolles, W.M., Crawford, L.P. & Valenti, J.L. (1968) *J Chem Phys* **49**, 4745–4749.
Toriyama, K., Nunome, K. & Iwasaki, M. (1980) *J Phys Chem* **84**, 2374–2390.
Tria, J.J. & Johnsen, R.H. (1977a) *J Phys Chem* **81**, 1274–1278.
Tria, J.J. & Johnsen, R.H. (1977b) *J Phys Chem* **81**, 1279–1284.
Tria, J.J. & Johnsen, R.H. (1978) *J Phys Chem* **82**, 1235–1239.
Tria, J.J., Hoel, D. & Johnsen, R.H. (1979) *J Phys Chem* **83**, 3174–3179.
Tsvetkov, Ju.D. & Falaleev, O.V. (1964) *Kinet Katal* **5**, 1119–1120.
Tsvetkov, Ju.D., Lebedev, Ja.S. & Voevodskii, V.V. (1961) *Vys Soedineniya* **3**, 882–890.
Vacek, K. & Schulte-Frohlinde, D. (1968) *J Phys Chem* **72**, 2886–2888.

Vasenkov, S.V. & Tolkatchev, V.A. (1991) *J Chem Biochem Kinet* **1**, 357–368.
Vasenkov, S.V. & Tolkatchev, V.A. (1992) *Kinet Katal* **33**, 1064–1068.
Vasenkov, S.V., Tolkatchev, V.A. & Bazhin, N.M. (1993) *Chem Phys Lett* **207**, 51–54.
Vasenkov, S.V., Bagryansky, V.A., Korolev, V.V. & Tolkatchev, V.A. (1991) *Radiat Phys Chem* **38**, 191–197.
Vorotnikov, A.P., Davydov, E.Ja., Pariisky, G.B. & Toptygin, D.Ja. (1983) *Khim Fiz* **2**, 818–822.
Vorotnikov, A.P., Davydov, E.Ja., Zaitseva, N.I. & Toptygin, D.Ja. (1986) *Kinet Katal* **27**, 62–66.
Vyazovkin, V.L. & Tolkatchev, V.A. (1995) *Chem Phys* **195**, 313–327.
Vyazovkin, V.L., Bol'shakov, B.V. & Tolkatchev, V.A. (1983) *Chem Phys* **75**, 11–16.
Vyazovkin, V.L., Bol'shakov, B.V. & Tolkatchev V.A., (1985a) *Chem Phys* **95**, 93–113.
Vyazovkin, V.L., Tolkatchev, V.A. & Burshtein, A.I. (1985b) *Khim Fiz* **4**, 493–500.
Wang, J.-T. & Williams, F. (1972) *J Am Chem Soc* **94**, 2930–2934.
Wright, B.B. & Platz, M.S. (1984) *J Am Chem Soc* **106**, 4175–4180.
Zaitsev, S.A., Vyazovkin, V.L., Panfilova, E.V., Bol'shakov, B.V. & Tolkatchev, V.A. (1983) *Khim Fiz* **2**, 517–524.
Zanin, A.V., Kiriukhin, D.P., Barkalov, I.M. & Goldanskii, V.I. (1980) *Dokl AN SSSR* **253**, 142–146.
Zamaraev, K.I., Khairutdinov, R.F. & Zhdanov, V.P. (1985) In: *Electron Tunneling in Chemistry*, pp. 3–316. Nauka, Novosibirsk.
Zapadinsky, E.L. & Tolkatchev, V.A. (1988) *Khim Vys Energ* **22**, 305–310.
Zapadinsky, E.L., Zaskul'nikov, V.M. & Tolkatchev, V.A. (1985) *Kinet Katal* **26**, 809–814.
Zapadinsky, E.L., Korolev, V.V., Gritsan, N.P., Bazhin, N.M. & Tolkatchev, V.A. (1986) *Chem Phys* **108**, 373–379.
Zaskulnikov, V.M. & Tolkatchev, V.A. (1979) *Kinet Katal* **20**, 263.
Zaskulnikov, V.M. & Tolkatchev, V.A. (1980) *React Kinet Catal Lett* **14**, 435–438.
Zaskulnikov, V.M. & Tolkatchev, V.A. (1982) *Kinet Katal* **23**, 1063–1070.
Zaskulnikov, V.M., Vyazovkin, V.L., Bol'shakov, B.V. & Tolkatchev, V.A. (1981) *Int J Chem Kinet* **13**, 707–728.
Zeller, R.C. & Pohl, R.O. (1971) *Phys Rev* **B4**, 2029–2041.
Zhuzhgov, E.L., Bubnov, N.N. & Voevodskii, V.V. (1965) *Kinet Katal* **6**, 56–64.
Zumofen, G., Klafter, J. & Blumen, A. (1983) *J Chem Phys* **79**, 5131–5135.

9 Reactivity of Organic Solids: Past, Present and Future

G.R. DESIRAJU and B.S. GOUD
School of Chemistry, University of Hyderabad, Hyderabad 500 046, India

1 Introduction

The most significant recent feature of organic solid state reactions is that they have, in most senses, entered the chemical mainstream. Several reasons have contributed to such a development. The extension of the topochemical principle to organic reactions by Schmidt (1971), that brilliant connection between crystallography and chemistry, paved the way eventually for more sophisticated mechanistic studies of reaction dynamics in a number of topochemical and topotactic transformations. Almost simultaneously, the work of Paul and Curtin (1973) on a wide variety of solid state reactions demonstrated convincingly that 'genuine' organic reactions, i.e. those involving the breaking and making of covalent bonds, take place in the solid state at temperatures at least 50°C or more below the melting point of the reactant. These studies led more or less naturally to work on more complex systems involving several reactant species. Quite independently, reactions of organic solids with other solids and vapours have been investigated and exploited by the chemical industry. Today, as we near the end of the 20th century, industry is more than ever concerned with cleaner and greener chemistry and this factor alone justifies the heightened awareness of organic solid state reactions amongst the chemical community.

It has taken this long for organic solid state reactions to be thus accepted (Desiraju 1987; Pierrot 1990), because unlike in large tracts of inorganic chemistry where 'chemistry' is equivalent to 'solid state chemistry', the overwhelming majority of organic reactions have been carried out in solution, even in the absence of specific reasons. Also, because of this historical reality, solid state organic reactions are judged stringently before serious study by organic chemists. They should be either more selective, more rapid or occur under milder conditions or yield products that are difficult or impossible to obtain in solution. In spite of these handicaps imposed somewhat irrationally by solution chemists, a large number of organic solid state reactions showing considerable synthetic variety have been actively investigated. A number of reviews continue to appear (Desiraju 1984; Scheffer 1987; Ramamurthy and Venkatesan 1987; Seebach 1990; Toda 1993), and here we will concentrate on the latest developments.

2 Photochemical 2 + 2 cycloaddition of alkenes

More work has probably been done on the solid state 2 + 2 cycloaddition reaction of alkenes than on all the other reaction types combined, and our discussion properly begins with a discussion of this category. It is generally accepted that reactions in the solid state tend to occur with a minimum of atomic and molecular motion (topochemical principle). Since this concept was demonstrated by Schmidt (1971) for *trans*-cinnamic acids, it has been refined with several auxiliary ideas, for example the concept of reaction cavity (Cohen 1975), orbital shapes (Kearsley and Desiraju 1985), molecular volume (Gavezzotti 1982) and free space, local stress and steric compression control (Ariel *et al.* 1985). Another spin-off has been the field of crystal engineering which was originally conceived in order that specific crystal packings, which promote specific reaction pathways, might be tailored, but which has now grown to include design strategies for a wide variety of supermolecules and organized structures (Desiraju 1989).

Both intra- and intermolecular 2 + 2 cycloaddition reactions have been studied. The historically important cinnamic acid dimerization has been used by Weber *et al.* (1989), who have prepared new hosts for selective clathration. Using these ideas, we have designed and prepared the unsymmetrical truxinic acid from the 1:1 complex of 3,5-dinitrocinnamic acid and 2,5-dimethoxycinnamic acid and have shown that it clathrates aromatic molecules efficiently (Desiraju and Sharma 1991). The same reaction has been used by Chung *et al.* 1991 to prepare highly strained molecules like [2.2]paracyclophanes. The 2 + 2 cycloaddition reaction has also been utilized in the preparation of ordered polymers and single crystal polymers. Topochemical photoreaction of a molecular complex to afford a perfectly ordered polymer composite has been described recently by Hasegawa *et al.* (1993). In this unique reaction, the reactant contains distinct columns of 2,5-distyrylpyrazine (A) and 4-[2-(2-pyrazinyl) ethyl]cinnamate (B). The solid state reaction occurs within the columns and is of the type $m\text{A} \rightarrow \text{A}_\text{m}$ and $n\text{B} \rightarrow \text{B}_n$, but not of the type $m\text{A} + n\text{b} \rightarrow \text{A}_m\text{B}_n$; in other words, the product polymer is a blend of two polymers A_m and B_n with no mixed polymer A_mB_n. Peachey and Eckhardt (1993) have studied the solid state polymerization of 2,5-distyrylpyrazine (DSP) and 1,4-bis(3-pyridyl-2-vinyl)benzene (P2VB). These workers found that DSP and P2VB are isoelectronic and structurally identical (double bond separations; DSP, 3.94 Å; P2VB, 3.90 Å) and so according to the topochemical principle, identical reactivity should be expected. However, DSP reacts 30 times faster than P2VB. Indeed, Peachey and Eckhardt (1993) argue that the qualitative criteria used for predicting the likelihood of 2 + 2 solid state cycloaddition reaction, derived from geometrical attributes or from a consideration of orbital shapes, may be an oversimplification. A solid state 2 + 2 cycloaddition reaction should not be considered as a collection of individual reactions between pairs of molecules but rather such reactions are necessarily associated with the aggregate behaviour of molecules comprising the crystal. According to Peachey and Eckhardt (1993), a fundamental conceptual problem with solid state reactions, which is usually ignored, is the collective excitation of solids. Solid state reactions may thus be closely related to phase transitions and this more rigorous conception may explain so-called deviations from ideal topochemical behaviour.

A discussion of photochemical topotactic reactions is relevant in this context. Reactions that proceed in a single crystal to single crystal fashion do so because nucleation is homogeneous and isolated reactions take place independently at myriad sites throughout the body of the crystal. In this respect, topotactic reactions may be considered to be those where the interference of the surrounding molecules in the crystal is at a minimum and where the reaction may be considered as essentially taking place between pairs of molecules immobilized in the crystal. Until very recently, it was believed that some 2 + 2 cycloadditions, such as the dimerization of 2-benzyl-5-benzylidenecyclopentanone, are inherently topotactic (Theocharis and Jones 1987), whilst others such as the dimerization of cinnamic acid are not. Therefore, the recent study of Enkelmann *et al.* (1993) is especially noteworthy in that they have shown that the photodimerization of cinnamic acid is topotactic provided the irradiating wavelength is sufficiently far out in the absorption shoulder. Laser irradiation is required because of very low absorbances and topotaxy is believed to follow from this low absorbance, which inhibits surface reactivity thereby preserving crystal transparency and inducing slow uniform reaction in the crystal bulk. These experiments show that there is no real distinction between topotactic and non-topotactic reactions and that these differences may well have more to do with reaction conditions rather than any chemical characteristic. Cooperative effects could gradually build-up with increasing reaction rates, resulting in a failure to maintain single crystal character in the product.

Reactions of >C=C< and >C=N− bonds have also been studied (Fig. 9.1). The azadiene (1), upon irradiation in the solid state gives the intramolecular reaction products (2) and (3) along with the intermolecular 2 + 2 cycloadduct (4) (Teng *et al.* 1991). The formation of (4) is topochemically controlled. The unusual formation of the intramolecular adducts was rationalized on the basis of changes in the molecular environment with changes in the reaction time. As the reaction progresses, the required neighbouring molecular orientation is no longer favourable for the intermolecular reaction because lattice alteration has occurred in the crystal to produce dislocations.

In further confirmation of the fact that ground state crystal geometry need not always lead to correct predictions of solid state reactivity, the effect of temperature on the photopolymerization of *p*-phenylenediacrylic acid diethyl ester has been described (Hasegawa and Shiba 1982). When crystals of this diester were irradiated below 0°C, high yields of crystalline polymer were obtained. Above 250°C, however, crossed polymers were obtained in poor yields. These authors have argued that topochemical reactions occur at optimum temperatures where molecules are neither rigidly immobi-

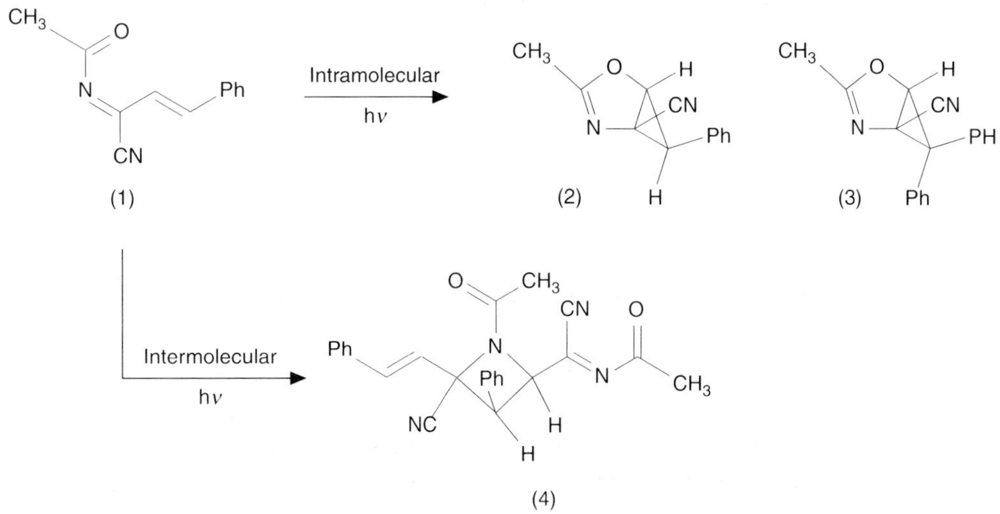

Figure 9.1 Intra- and intermolecular solid state 2 + 2 cycloaddition reaction of azadiene (1).

lized (temperature too low) nor vibrating too fast thereby preventing 'lock in to' a photoreactive position (temperature too high). We have found, for instance, that the difference in the topochemical 2 + 2 photoreactivity of the isostructural 3- and 4-cyanocinnamic acids is caused by differing degrees of ease of molecular motion in the two crystals (Dhurjati *et al.* 1991). The 3-cyano acid is unreactive at room temperature, but is photoreactive when heated to 130°C because topochemical reactivity can only take place after translational motion of molecules along [010]. Such motion is possible in the 4-cyano acid at room temperature itself. Due to a tighter packing in the 3-cyano acid, this motion is possible only at 130°C. Coumarin is unreactive in the solid state but in the presence of antipyrene or phenanthrene, it forms mixed crystals which react upon photoirradiation to give the coumarin dimer (Meng *et al.* 1989). In these examples the antipyrene or phenanthrene acts as a crystal lattice modifier without taking part in the photoreaction. However, cycloaddition of mixed crystals of 1,3-dimethyl-5-formyluracil and indole leads to mixed adducts. This reaction does not appear to be topochemically controlled. Other reactions that cannot be explained by the topochemical principles have been explained on the basis of force field calculations (Angermund *et al.* 1991).

3 Other photochemical reactions

Solid state H-abstraction reactions of enediones and enones, Norrish type II reactions and di-π-methane and tri-π-methane rearrangements have been studied in detail. Scheffer (1987; Scheffer *et al.* 1987) has shown that tetrahydro-1,4-naphthoquinones (enedione) undergo H-atom abstraction by O and C atoms to give cyclobutanones. In the case of 4-hydroxycyclohex-2-en-1-one (enones), however, intramolecular H-atom abstraction by the β-C atom of the enone double bond leads to a diradical which then collapses to a keto-alcohol product.

In the Norrish type II reaction, intramolecular γ-H-atom abstraction by a carbonyl O atom leads to a 1,4-diradical, which collapses to a cyclobutane, or cleaves off an alkene to give an enol, which tautomerizes to the ketone. Such reactions have been studied for α-cycloalkyl ketones in the solid state (Scheffer 1987; Scheffer *et al.* 1987). Most of these substrates exist in the boat conformation with abstracting distances C=O. . .H in the range 2.7–3.1 Å. In the case of α-3-methyl adamantyl-*p*-chloroacetophenone, however, abstraction occurs via the chair form across a distance of 2.7 Å. In general, for intramolecular type II reactions, the abstracting distances are less than 3.1 Å, but Ito *et al.* (1987) have reported that intermolecular H abstraction occurs across a distance of 3.3 Å in 4,4-dimethylbenzophenone. Furthermore, it was observed that 4-methylbenzophenone with a much shorter O. . .H intermolecular distance of 2.7 Å is photostable. These anomalies were ascribed to the inability of the radical pair, produced by abstraction in the latter case, to couple (4.4–4.5 Å separation). In the former case, however, coupling is possible because the corresponding radical pair separation is 3.9 Å. The approximate upper limit for intermolecular solid state H-atom transfer and radical coupling are 3.5 and 4.2 Å, respectively. It is interesting to speculate that these differences may reflect the slight but significant translational freedom that molecules in a crystal possess, so that intermolecular abstraction may actually occur across distances less than the crystallographically determined values. Clearly, further experimental work

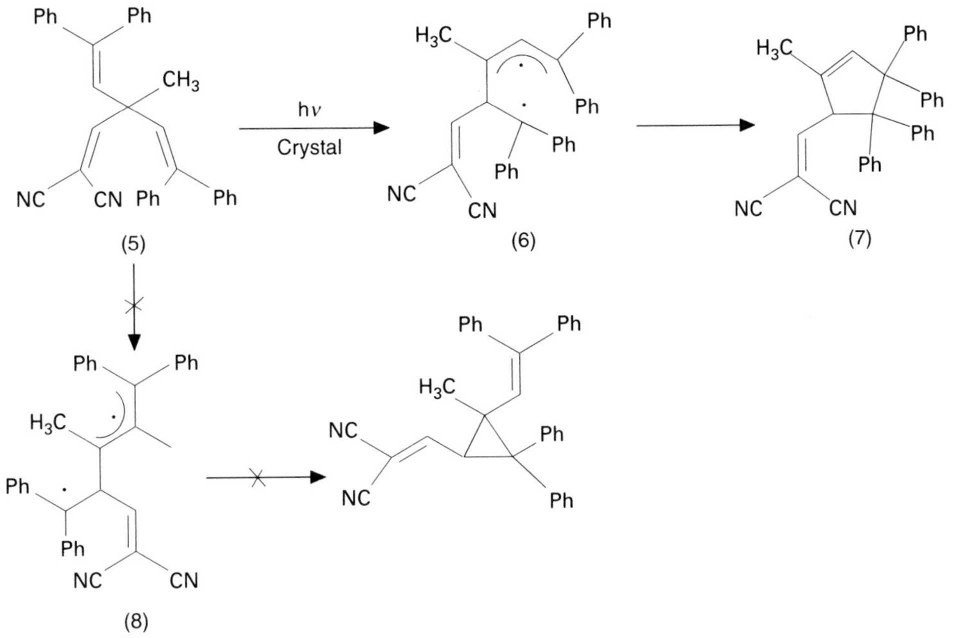

Figure 9.2 Solid state tri-π-methane rearrangement.

is required before firm conclusions can be drawn concerning this question (Ramamurthy 1991).

H abstractions are also common in the solid state reactions of nitro-compounds. The photochemical reactions of nitro-aromatics are amongst the oldest known of unimolecular solid state reactions, and the conversion in the solid state of 2′-nitrochalcone to indigo is a well-known example.

The di-π-methane rearrangement is a very well-known solution photochemical reaction wherein divinylmethanes afford vinylcyclopropanes. This reaction was shown to proceed in the solid state by Zimmerman *et al.* (1973), who pioneered the solution reaction. However, the corresponding tri-π-systems (Fig. 9.2), trivinylmethanes (5), were found to yield only the di-π-methane products, in other words divinylcyclopropanes, in solution. Only when the reaction was carried out in the solid state was the genuine tri-π-product, the vinylcyclopentene (7), obtained (Zimmerman and Zuraw 1989). This product is obtained only in the solid state because the *cis*oid diradical (6) necessary for cyclopentene formation is stabilized by the confines of the crystalline matrix. The formation of the *cis*oid diradical (6) is accompanied by less molecular motion, less molecular volume increase and less overlap with the surrounding molecules than the formation of the *trans*oid diradical, (8), which would have led to the formation of the divinylcyclopropane. This is a significant example of a reaction which proceeds in the solid state due to spatial constraints which do not exist in solution.

4 Asymmetric synthesis

An asymmetric synthesis is one in which achiral molecules are converted into chiral molecules in such a manner that stereoisomeric products are obtained in unequal amounts. Dissymmetry in the solid state environment, which is manifested through the

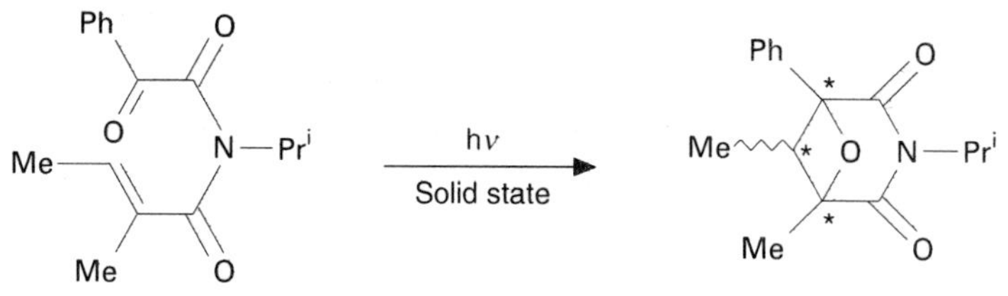

(9) hν Solid state (10)

Figure 9.3 Formation of chiral oxetane (10) from the achiral precursor (9).

adoption of enantiomorphous space groups such as P1, $P2_1$ or $P2_12_12_1$, can be used to induce chirality in such syntheses. Ever since organic solid state reactions were shown to be topochemical, various approaches to solid state asymmetric synthesis have been attempted. For example, the formation of the chiral oxetane (Fig. 9.3 (10)) upon photolysis of crystalline achiral *N*-isopropyl-*N*-tiglylbenzoylformamide (9) via a topochemically controlled Paterno–Buchi reaction has been reported (Sakamoto *et al.* 1993). The diastereoselectivity of the intramolecular 2 + 2 cycloaddition of 4-(3′-butenyl)-2,5-cyclohexadien-1-one has been studied (Schultz *et al.* 1992). Solid state di-π-methane and Norrish type II asymmetric syntheses have been reported (Evans *et al.* 1986). Toda has synthesized a number of chiral aziridines, quinolinones and β-lactams via host–guest chemistry (Toda *et al.* 1989b; Toda 1993). For example, the enantioselective photocyclization of *N,N*-dimethyl phenylglyoxylamide (Fig. 9.4 (11)) to the corresponding β-lactam (12) upon photoirradiation of the 1:1 inclusion crystal of (11) in (13) has been reported.

5 Thermal reactions

Thermal solid state reactions have, in general, not been subjected to as great a degree of crystallographic scrutiny and mechanistic study as have photochemical reactions. However, thermal reactions are more varied, numerous and possibly of greater commercial importance. Many of these reactions are unimolecular and are found to occur when crystals are kept at elevated temperatures (in the dark) for varying time periods. Topochemical control of some sort is probably important in the early stages of these reactions. Several other thermal reactions that involve mixtures of solid state reactants have been studied (Toda 1993). These latter reactions are necessarily more complex because solid–solid diffusion must precede the reaction (Rastogi *et al.* 1963). There is

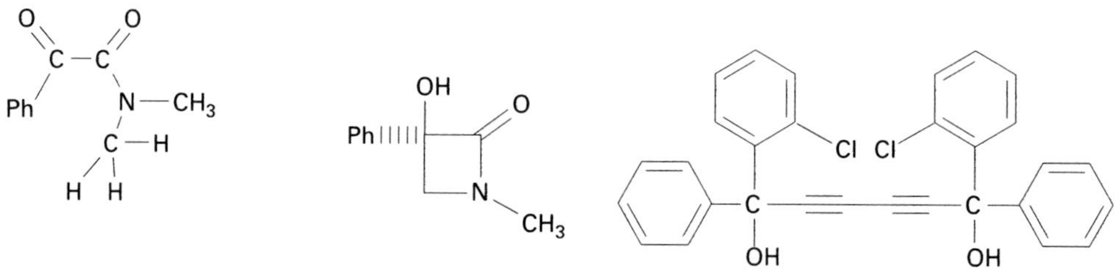

(11) (12) (13)

Figure 9.4 Enantioselective photocyclization of *N,N*-dimethyl phenylglyoxylamide (11).

adequate evidence that reactant vapour pressure is not a significant factor in many of these cases. The efficiency and degree of mixing and grinding are undoubtedly important, and the technique of 'ball-milling' of the reactants has attracted some recent attention (Nixon 1993). This technique improves yields considerably and is believed to be effective because local temperatures at the points of contact between particles could be as high as 1500°C! Rigorous and systematic studies of mechanochemistry are clearly difficult because simultaneous expertise in organic chemistry, physical chemistry and chemical engineering is called for, but there is little doubt that these reactions will become very important in the future. The work of Boldyrev (1993), albeit on inorganic systems, is noteworthy in this respect.

A good example of an intermolecular lattice-controlled thermal reaction has been reported by Nader *et al.* (1985), who found that the light stable butatrienetetracarboxylic ester and its derivatives are thermally converted via 2 + 2 cycloaddition at the central cumulene bonds to [4]-radialene, which is the topochemical product. We have designed a family of crystal structures capable of undergoing an intermolecular solid state Diels–Alder reaction (Kishan and Desiraju 1989). For instance, 3,4-(methylenedioxy) phenylpropiolic acid (Fig. 9.5 (14)) crystallizes with a 4 Å short axis which permits a topochemical 4 + 2 cycloaddition to give the lignan derivative (15). This is a good example of crystal engineering wherein the C–H. . .O bond-forming property of the methylenedioxy group has been utilized to give a crystal packing with a 4 Å short axis. Chapman rearrangement of imino-ethers to *n*-alkylamides has been reported by Dessolin *et al.* (1992), who observed that in 5-methoxy-2-aryl-1,3,4-oxadiazoles the migration of the methyl group from O to N is possible because of the favourable crystal packing. Carbonium ion rearrangements in the solid state have been also studied (Borodkin *et al.* 1983).

Recently, we have investigated a solid state intramolecular Michael-type reaction in which 2′-hydroxy-4′,6′-dimethylchalcone and some of its derivatives are converted to the corresponding flavanones at 50–60°C. This quantitative reaction is not governed by topochemical considerations and presumably takes place after reactant molecules can relax into favourable conformations in voids, caused by initial or surface reaction (Goud *et al.* 1995).

Many thermal reactions are topochemically controlled, but it has been shown that the thermally induced methyl-transfer rearrangement of methyl-2-(methylthio)benzene-

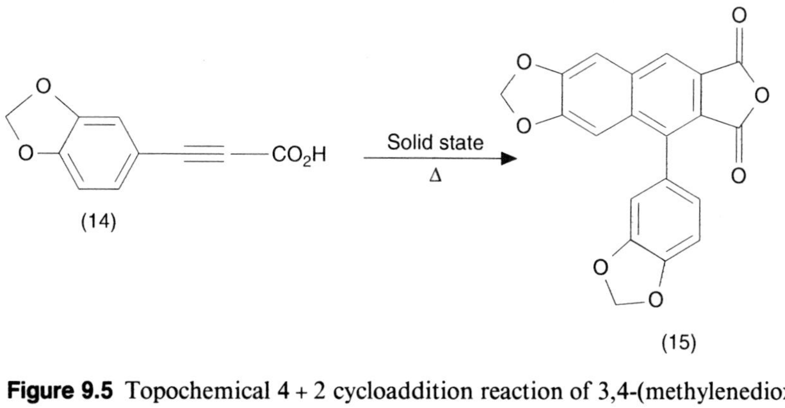

Figure 9.5 Topochemical 4 + 2 cycloaddition reaction of 3,4-(methylenedioxy) phenylpropiolic acid (14) to yield lignam (15).

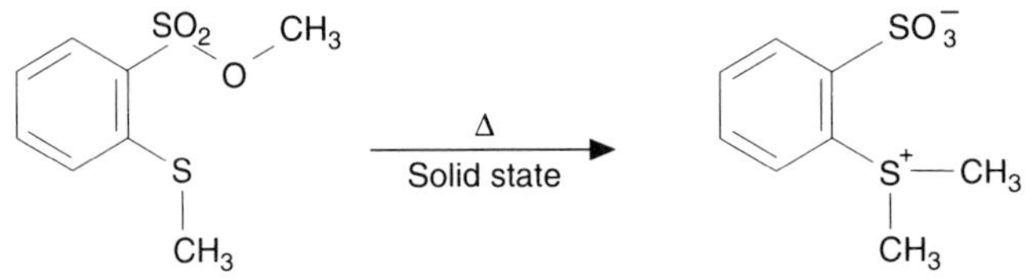

Figure 9.6 Thermally-induced methyl-transfer rearrangement of methyl-2(methylthio)benzenesulphonate (16).

sulphonate (Fig. 9.6 (16)) to the zwitterionic 2-(dimethylsulphonium)-benzenesulphonate (17) occurs at defects such as microcavities, surfaces and other irregularities (Venugopalan *et al.* 1991). This reaction is controlled by defect rather than bulk topochemistry and, accordingly, the transformation in powdered samples is faster than in single crystals but slower than in the melt.

A major contribution to synthetic solid state chemistry has been made by Toda (1993), who has studied a wide variety of synthetic transformations such as the aldol condensation, the Baeyer–Villiger oxidation, coupling reactions, the Grignard, Reformatsky and Luche reactions, the $NaBH_4$ reduction, the pinacol rearrangement and the Wittig reaction. These reactions are usually carried out by keeping a mixture of finely powdered reactants at room temperature. In some cases the solid state reactions are accelerated by heating, shaking, irradiating with ultrasound or by grinding with mortar and pestle. Many of Toda's reactions are characterized by very high yields and selectivities. For instance, the solid state reduction of ketones with $NaBH_4$ is quite selective. Treatment of the powdered 1:1 inclusion crystal of (18) (Fig. 9.7) in host (13) (see Fig. 9.4) with solid $NaBH_4$ gives the keto-alcohol (19) in 100% enantiomeric excess. Crystallographic study showed that the enone carbonyl of (18) is sterically hindered by forming a H bond with the host and that consequently, the unprotected non-conjugated ketone group is selectively reduced (Toda *et al.* 1989a). These and other such reactions surely have potential in solventless scale-up processes and merit further detailed study (Goud and Desiraju, 1995).

6 Gas–solid reactions

A number of gas–solid reactions have been studied from academic and applied viewpoints. The Kolbe–Schmitt reaction is used in the large-scale preparation of salicylic acid and aspirin, and the studies of Paul and Curtin (1975) on reactions of

Figure 9.7 Solid state reduction of dione (18) in diacetylene host (13) (see Fig. 9.4) with solid $NaBH_4$. Only one of the carbonyl groups is selectively reduced to yield (19).

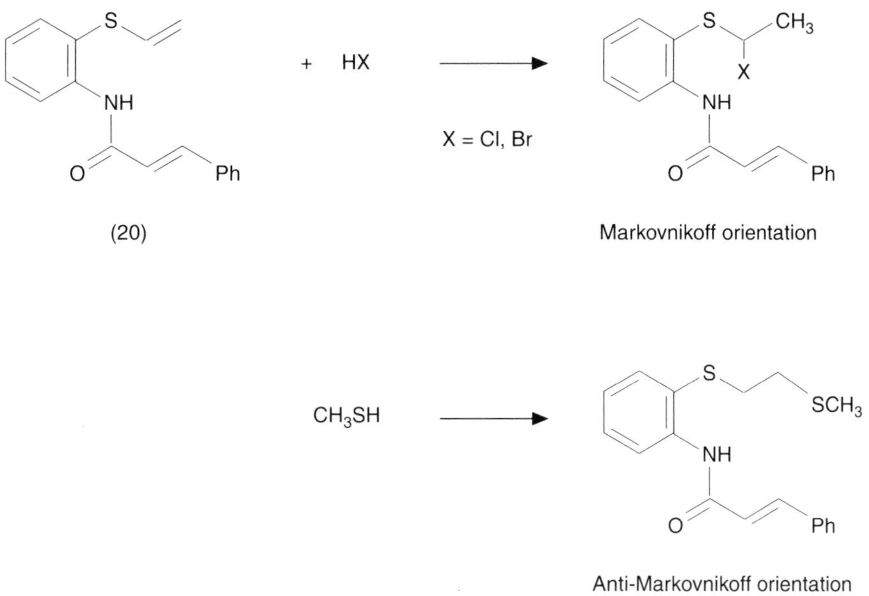

Figure 9.8 A gas–solid reaction of crystalline alkene (20) in the Markovnikoff and anti-Markovnikoff orientation.

ammonia and amine vapours with crystalline carboxylic acids are of fundamental and mechanistic value. The products of these reactions are anhydrous ammonium and alkylammonium carboxylates, which are difficult to prepare by other means. The solid state hydrogenation of thymol at 20°C under H_2 at 1 bar pressure (1×10^5 Pa) and catalysed by 5% Pt/C has been reported (Boudard *et al.* 1969). Very recently, Kaupp and Mathies (1988) have performed the hydrogenation of cinnamic acid doped with Pd at 30°C. We have reported an unusual chlorination–oxidation of 4-phenylthiazole-2(1H)-thione to yield 4-(chloro)phenacyl disulphide (Nalini and Desiraju 1987).

Kaupp and Matthies (1988) state that the particular advantages of organic gas–solid reactions are that no solvent is required, the experimental techniques are very simple and the reactions are often highly regioselective. For example, powdered *N*-vinylphthalimide reacts almost quantitatively with gases like HCl, HBr, HI and Br_2 to yield crystalline 1-chloro-, 1-bromo-, 1-iodo- and 1,2-dibromoethyl phthalimides, respectively. However, in solution the corresponding oligomers and polymers are formed. Kaupp has extended the scope of this gas–solid reaction to other crystalline enamides, enamines and S-vinylthioethers and demonstrated their good synthetic potential. The gases HCl and HBr react at 1 bar (1×10^5 Pa) and at temperatures from 40 to − 80°C, with crystalline (20) (Fig. 9.8) usually in the Markovnikoff orientation. However, the addition of CH_3SH to (20) constitutes the first example of a gas–solid addition in the anti-Markovnikoff orientation. Gas–solid alkylations have also been reported. For example, solid phenol reacts with isobutene gas to give *O*-alkylated tetrabutylphenylether as the main product (Perrin and Lamartine 1990).

7 Radiation-induced reactions

Some solid state reactions are initiated by irradiation with X-rays or γ-rays. The unusual thermal reactivity of sodium *trans*-2-butanoate (Fig. 9.9 (21)), which on

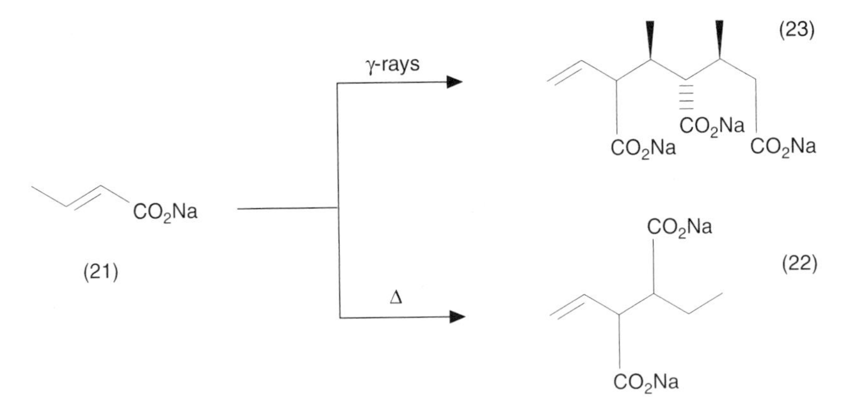

Figure 9.9 Stereospecific conversion of sodium *trans*-2-butanoate (21) to the dicarboxylate dimer (22) on heating.

heating is converted stereospecifically to the dicarboxylate dimer (22), has been discussed (Naruchi and Miura 1987). However, salt (21) was anticipated by Diaz de Delgado *et al.* (1991) to be sensitive to γ-ray irradiation. They found that on γ-irradiation, only one of eight possible diastereoisomers of the sodium salt of 2,4-dimethyl-6-heptene-1,3,5-tricarboxylic acid (trimer) (23) was produced (Diaz de Delgado *et al.* 1991). The formation of this trimer was explained as occurring via loss of an allylic proton to give an anion which adds topochemically to another double bond of a neighbouring molecule at a C=C. . .C=C distance of 3.47 Å.

8 Photochromism and thermochromism

Reversible chemical reactions and/or phase transitions which are light or heat initiated are of great practical interest because such substances can be used in a number of light- and heat-sensitive devices (Durr 1989; Suzuki *et al.* 1989). Generally, these compounds are conjugated or aromatic and the chemical changes are accompanied by colour changes. This topic has been studied extensively. Photochromic changes can be classified into six major groups:

1 triplet–triplet absorption (picene and coronene in polymer films);
2 *cis–trans* isomerization (azomethanes, retinal);
3 tautomerism (salicylidene anilines);
4 dissociation into radicals (octaphenyl-1,1′-bipyrrolyl);
5 pericyclic changes (fulgides);
6 redox photochromism (methylene blue systems).

In spite of all this study, poor reversibility of the photochromic processes is a feature common to most of the systems.

Other solid state reactions that have practical applications are piezochromism and triboluminescence. These effects have been discussed in detail. Piezoelectricity, which is the separation of positive and negative charge on expansion or compression of a crystal, is of technological significance (Maeda and Hayashi 1970; Ohkanda *et al.* 1992). Inorganic materials like quartz and sphalerite show this effect but many piezoelectric organics are also known. An early space group compilation by Rez (1960) was motivated by a search for new piezoelectrics. It was observed that organic materials show a higher proportion (42%) of potential piezoelectrics than do inorganic compounds (17%). Covalency in the bonds, which is a feature characteristic of organic compounds, favours piezoelectric behaviour. A large number of amines and amino acids, and also their salts and complexes, were identified in this connection.

Triboluminescence is the emission of light on the application of mechanical force to a solid (Chandra and Zink 1980). To be triboluminescent the crystal must have a permanent dipole moment. This phenomenon may be seen in substances such as sucrose, tartaric acid, anthranilic acid and acenaphthene.

9 Industrial solid state reactions

Unlike academic laboratories, chemical industries have not shied away from organic reactions in the solid state. The Kolbe–Schmitt synthesis of phenolic acids is an early example of a gas–solid reaction that is used commercially (Perrin *et al.* 1987). In this reaction, solid alkali metal phenoxides, when submitted to the action of CO_2 under elevated conditions of pressure and temperature, yield 2- and 4-hydroxybenzoic acids. Salicylic acid is by far the most important product because it is used in the manufacture of aspirin. Other phenols such as resorcinol also undergo this reaction, the mechanism of which is still not clearly understood. Another example of a solid state reaction practised in industry today is the preparation of sulphanilic acid by the action of heat on anilinium sulphonate. Similarly, homologous C-alkylated derivatives (ortho-, meta-) may be converted to the corresponding sulphonated anilines.

Other solid state (or semi-dry state) processes that are used industrially are the preparation of the dyestuff Colour Index Vat Black 25, (Fig. 9.10 (24)), dry state quaternization of pyridine-azo-β-naphthol and the preparation of melamine. One effective way to prepare melamine is by solid state reaction of urea with CO_2 in a pressurized vessel. It is noteworthy that many of these processes are not new and were being used in Germany during and even prior to World War II. Several newer reactions are undoubtedly in use today but the details of these reactions are shrouded in the patent literature (see, for example, US Patent 2 760 961; US Patent 3 418 321; Ger. Offen. 2 741 395). Industrial solid state reactions certainly have a future in the next century. They are cheaper to run because they do not involve solvent recovery and operators are less exposed to the reaction.

There are industrial and technological uses of organic solid state reactions other than

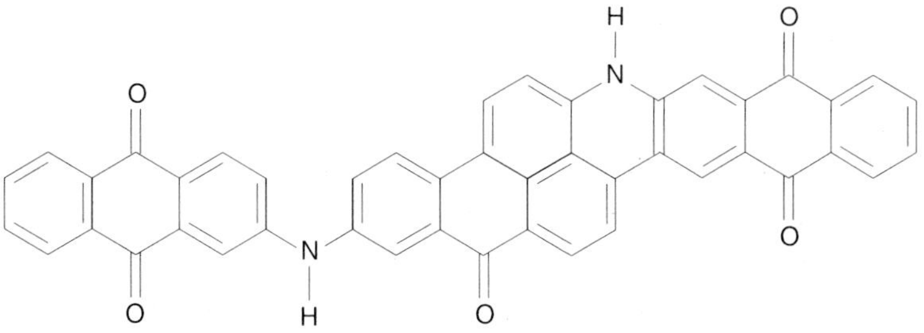

H N O O O N H O

(24)

Figure 9.10 Dyestuff Colour Index Vat Black 25.

the purely synthetic. Solid state reactions can be carried out in mono- and multilayer Langmuir–Blodgett films for various electronic and optical applications. For example, polydiacetylene films have been produced in this manner. The reversible 2 + 2 cycloaddition of acenaphthalene-1,2-dicarboxylic anhydride is used in photochemical information storage systems (Rieke *et al.* 1990). Similar reactions are used in chemical sensors. Photoirradiation of *trans*-syn-3,3′-diphenyl-2,2′-biindenyliden-1,1′-dione in the solid state gives a stable triplet diradical which shows ferromagnetic properties. Solid state photoreactions on organometallic carbonyls and porphyrins are useful in artificial photosynthesis and the photodynamic therapy of cancer.

10 Conclusions

Organic solid state reactions have been known for well over a century. It is generally not known that Wöhler's synthesis of urea from ammonium cyanate, the reaction which symbolizes more than any other the birth of organic chemistry, is a solid state process. The past was a period during which the study of solid state reactions was initiated and formalized. During that time, the topochemical rules were elaborated and a number of reaction types were researched but the full scope of organic solid state chemistry was only a promise. The present is characterized by intense activity, and new solid state reactions with wide synthetic diversity are being developed even as older reactions such as the 2 + 2 cycloaddition are being probed mechanistically and dynamically with the ever increasing array of instrumental techniques at the disposal of the experimental chemist. Whilst it may be risky to make scientific predictions, all indications are that the chemistry of the future will be driven and motivated mainly by the sometimes conflicting criteria of economic viability and environmental concerns. In this regard, organic solid state transformations are ideal candidates for future development because in a solventless regime, the factors mentioned above need not necessarily be in conflict. The future will probably witness the coming together of research activity in academic and industrial laboratories, capitalizing on their respective advantages to design increasingly efficient organic solid state reactions.

11 Acknowledgement

We are indebted to Dr S.M. Gerber for the information provided by him on industrial solid state reactions.

12 References

Angermund, K., Klopp, I., Kruger, C. & Nolte, M. (1991) *Angew Chem Int Ed Engl* **30**, 1354–1356.

Ariel, S., Askari, S., Scheffer, J.R., Trotter, J. & Walsh, L. (1985) *ACS Symposium, Series 278* (ed. M.A. Fox), pp. 243–256. American Chemical Society, Washington, DC.

Boldyrev, V.V. (1993) *J Thermal Anal* **40**, 1041–1062.

Borodkin, G.I., Nagy, S.M., Mamatyuk, V.I., Shakirov, M.M. & Shubin, V.G. (1983) *J Chem Soc Chem Commun* 1533–1535.

Boudard, M., Vannice, M.A. & Benson, J.E. (1969) *Z Phys Chem* **64**, 171–177.

Chandra, B.P. & Zink, J.I. (1980) *Phys Rev* **B21**, 816–826.
Chung, C.-M., Nakamura, F., Hashimoto, Y. & Hasegawa, M. (1991) *Chem Lett* 779–782.
Cohen, M.D. (1975) *Angew Chem Int Ed Engl* **14**, 386–393.
Desiraju, G.R. (1984) *Endeavour* **8**, 201–206.
Desiraju, G.R. (ed.) (1987) *Organic Solid State Chemistry.* Elsevier, Amsterdam.
Desiraju, G.R. (1989) *Crystal Engineering: The Design of Organic Solids.* Elsevier, Amsterdam.
Desiraju, G.R. & Sharma, C.V.K.M. (1991) *J Chem Soc Chem Commun* 1239–1241.
Dessolin, M., Eisenstein, O., Golfier, M., Prange, T. & Sautet, P. (1992) *J Chem Soc Chem Commun* 132–134.
Dhurjati, M.S.K., Sarma, J.A.R.P. & Desiraju, G.R. (1991) *J Chem Soc Chem Commun* 1702–1703.
Diaz de Delgado, G.C., Wheeler, A.K., Snider, B.B. & Foxman, B.M. (1991) *Angew Chem Int Ed Engl* **30**, 420–422.
Durr, H. (1989) *Angew Chem Int Ed Engl* **28**, 413–431.
Enkelmann, V., Wegner, G., Novak, K. & Wagener, K.B. (1993) *J Am Chem Soc* **115**, 10 390–10 391.
Evans, S.V., Garcia-Garibay, M., Omkaram, N., Scheffer, J.R., Trotter, J. & Wireko, F. (1986) *J Am Chem Soc* **108**, 5648–5650.
Gavezzotti, A (1982) *Nouv J Chim* **6**, 443–450.
Goud, B.S. & Desiraju, G.R. (1995) *J Chem Res* **5**, 244.
Goud, B.S, Panneerselvam, K., Zacharias, D.E. & Desiraju, G.R. (1995) *J Chem Soc Perkin Trans* **2**, 325–330.
Hasegawa, M. & Shiba, S. (1982) *J Phys Chem* **86**, 1490–1496.
Hasegawa, M. Kinbara, K., Adegawa, Y. & Saigo, K. (1993) *J Am Chem Soc* **115**, 3820–3821.
Ito, Y., Matsuura, T., Tabata, K. *et al.* (1987) *Tetrahedron* **43**, 1307–1312.
Kaupp, G. & Matthies, D. (1988) *Mol Cryst Liq Cryst* **61**, 119–143.
Kearsley, S.K. & Desiraju, G.R. (1985) *Proc Roy Soc Lond* **A379**, 157–181.
Kishan, K.V.R. & Desiraju, G.R. (1989) *J Am Chem Soc* **111**, 4838–4843.
Maeda, K. & Hayashi, T. (1970) *Bull Chem Soc Jpn* **43**, 429–438.
Meng, J.-B., Fu, D.-C., Yao, X.-K., Wang, R.J. & Matsuura, T. (1989) *Tetrahedron* **45**, 6979–6986.
Nader, F.W., Wacker, C.-D., Irngartinger, H., Huber-Patz, U., Jahn, R. & Rodewald, H. (1985) *Angew Chem Int Ed Engl* **24**, 852–853.
Nalini, V. & Desiraju, G.R. (1987) *J Chem Soc Chem Commun* 1046–1048.
Naruchi, K. & Miura, M. (1987) *J Chem Soc Perkin Trans 2* 113–116.
Nixon, A.C. (1993) *Chem Eng News* **71**, 4–5.
Ohkanda, J., Mori, Y., Maeda, K. & Osawa, E. (1992) *J Chem Soc Perkin Trans 2* 59–63.
Paul, I.C. & Curtin, D.Y. (1973) *Acc Chem Res* **6**, 217–225.
Paul, I.C. & Curtin, D.Y. (1975) *Science* **187**, 19–26.
Peachey, N.M. & Eckhardt, C.J. (1993) *J Phys Chem* **97**, 10 849–10 856.
Perrin, R. & Lamartine, R. (1990) *Structure and Properties of Molecular Crystals* (ed. M. Pierrot), pp. 107–159. Elsevier, Amsterdam.
Perrin, R., Lamartine, R., Perrin, M. & Thozet, A. (1987) In: *Organic Solid State Chemistry* (ed. G.R. Desiraju), pp. 217–326. Elsevier, Amsterdam.
Pierrot, M. (ed.) (1990) *Structure and Properties of Molecular Crystals.* Elsevier, Amsterdam.
Ramamurthy, V. (ed.) (1991) *Photochemistry in Organised and Constrained Media.* VCH, New York.
Ramamurthy, V. & Venkatesan, K. (1987) *Chem Rev* **87**, 433–481.
Rastogi, R.P., Bassi, P.S. & Chadha, S.L. (1963) *J Phys Chem* **67**, 2569–2573.
Rez, I.S. (1960) *Krist* **5**, 63.
Rieke, R.D., Page, G.O., Hudnall, P.M., Arhart, R.W. & Bouldin T.W. (1990) *J Chem Soc Chem Commun* 38–39.
Sakamoto, M., Takahashi, M., Fujita, T. *et al.* (1993) *J Org Chem* **58**, 3476–3477.

Scheffer, J.R. (1987) *Organic Solid State Chemistry* (ed. G.R. Desiraju), pp. 1–41. Elsevier, Amsterdam.

Scheffer, J.R., Garcia-Garibay, M. & Nalamasu, O. (1987) *Organic Photochemistry* Vol. 8 (ed. A. Padwa), pp. 249–346. Marcel Dekker, New York.

Schmidt, G.M.J. (1971) *Pure Appl Chem* **27**, 647–678.

Schultz, A.G., Taveras, A.G., Taylor, R.E., Tham, F.S. & Kulling, R.K. (1992) *J Am Chem Soc* **114**, 8725–8727.

Seebach, D. (1990) *Angew Chem Int Ed Engl* **29**, 1320–1367.

Suzuki, H., Tomoda, A., Ishizuka, M., Kaneko, A., Furui, M. & Matsushima, R. (1989) *Bull Chem Soc Jpn* **62**, 3968–3971.

Teng, M., Lauher, J.W. & Fowler, F.W. (1991) *J Org Chem* **56**, 6840–6845.

Theocharis, C.R. & Jones, W. (1987) *Organic Solid State Chemistry* (ed. G.R. Desiraju), pp. 47–67. Elsevier, Amsterdam.

Toda, F. (1993) *Syn Lett* **8**, 303–312.

Toda, F., Kiyoshige, K. & Yagi, M. (1989a) *Angew Chem Int Ed Engl* **28**, 320–321.

Toda, F., Tanaka, K. & Mak, T.C.W. (1989b) *Chem Lett* 1329–1330.

Venugopalan, P., Venkatesan, K., Klausen, J. *et al.* (1991) *Helv Chim Acta* **74**, 662–669.

Weber, E., Hecker, M., Csöregh, I. & Czugler, M. (1989) *J Am Chem Soc* **111**, 7866–7872.

Zimmerman, H.E. & Zuraw, M.J. (1989) *J Am Chem Soc* **111**, 7974–7989.

Zimmerman, H.E., Boettcher, R.J. & Braig, W. (1973) *J Am Chem Soc* **95**, 2155–2163.

10 Chemical Strategies for the Synthesis of Metal Oxides

C.N.R. RAO

Solid State and Structural Chemistry Unit and CSIR Centre of Excellence in Chemistry, Indian Institute of Science, Bangalore 560012, India

1 Introduction

Tailor-making materials of the desired structure and properties is one of the main goals of materials chemistry, but it is not always possible to do so by conventional methods. A variety of chemical strategies have been employed for the synthesis of solid materials in recent years by making use of subtle relationships between structure and reactivity. Whilst rational synthesis has provided a variety of materials such as SIALON, NASICON and a large number of microporous materials, a rational approach to materials synthesis almost always yields thermodynamically stable materials, but may miss new and novel metastable ones. In this chapter, we shall be mainly concerned with chemical aproaches to the synthesis of oxide materials (Rao 1994) and shall discuss the various methods with examples and specially examine soft chemistry routes to illustrate the control of structure of solids through chemical means. We shall first briefly examine the traditional ceramic method.

2 Ceramic method

The most common method of preparing inorganic solid materials is by the reaction of the component materials in the solid state at high temperatures (Rao and Gopalakrishnan 1986). In some instances, as in the preparation of chalcogenides or of materials where the products or reactants are volatile, the reaction is carried out in sealed tubes. In preparing oxides one generally palletizes the reacting materials and repeats the grinding, palletizing and heating operations several times. Yet, the completion of the reaction or the phasic purity of the product is not assured. The ceramic method suffers from several disadvantages. Various modifications of the ceramic technique have been employed to overcome some of the limitations. An important effort has been to decrease diffusion path lengths. In a polycrystalline mixture of reactants, individual particles are approximately 10 μm in size, representing diffusion distances of roughly 10 000 unit cells. By using freeze–drying, spray–drying, coprecipitation, sol-gel and other techniques, it is possible to reduce the particle size to a few hundred ångströms and thus effect a more intimate mixing of the reactants. In coprecipitation, the required metal cations, taken as soluble salts (for example, nitrates), are coprecipitated from a common medium, usually as hydroxides, carbonates, oxalates or citrates. In actual practice, one takes oxides or carbonates of the relevant metals, digests them with an acid and then the precipitating reagents are added to the solution. The precipitate obtained after drying is heated to the required temperature in a desired atmosphere to produce the final product. The decomposition temperatures of the precipitates are generally lower than the temperatures employed in the ceramic method.

3 Soft chemistry routes

Soft chemistry routes essentially employ simple reactions that can be carried out under mild conditions at relatively low temperatures. Generally, at least one of the steps of synthesis involves a reaction in solution. A few illustrative examples would help to understand the spirit of soft chemistry. Tournoux and co-workers (Marchand *et al.* 1980) prepared a new form (B) of TiO_2 by the dehydration of $H_2Ti_4O_9.xH_2O$, obtained from $K_2Ti_4O_9$ by exchange of K^+ by H^+. We illustrate the kind of transformations involved in Fig. 10.1. Raveau and co-workers (Rebbah *et al.* 1979) prepared $Ti_2Nb_2O_9$ by the dehydration of $HTiNbO_5$, which in turn was prepared by cation exchange with $ATiNbO_5$ (A=K, Rb). $Ni(OH)_2.xH_2O$, with a large intersheet distance of 0.78 nm, has been prepared by the hydrolysis of $NaNiO_2$, followed by the subsequent reduction of NiOOH. Similar transformations of Fe- and Co-doped $Ni(OH)_2.xH_2O$ have been examined (Delmas and Borthomieu 1993). In Fig. 10.2, we show the nature of this kind of soft chemical synthesis. Intercalation and deintercalation reactions are soft chemical routes for the synthesis of many solids. Thus, deintercalation of $LiVS_2$ gives VS_2 which cannot be prepared otherwise; deintercalation of $LiVO_2$ similarly gives metastable VO_2. Recently, we have used deintercalation of amine intercalates of WO_3 to prepare WO_3 in perfect cubic form (ReO_3 type) and also in other metastable forms. In Fig. 10.3 we show the X-ray diffraction patterns of stable WO_3, $WO_3.0.5NEt_3$ and WO_3 (cubic). Heating WO_3 (cubic) at 770 K transforms it to the stable monoclinic form. Acid leaching the NEt_3 intercalate gives another metastable form of WO_3 as shown in Fig 10.4.

Intercalation reactions involve the insertion of a guest species into a solid host lattice without any major rearrangement of the solid structure (Rao and Gopalakrishnan 1986; Jacobson 1992). Many intercalation reactions are known, with a variety of host

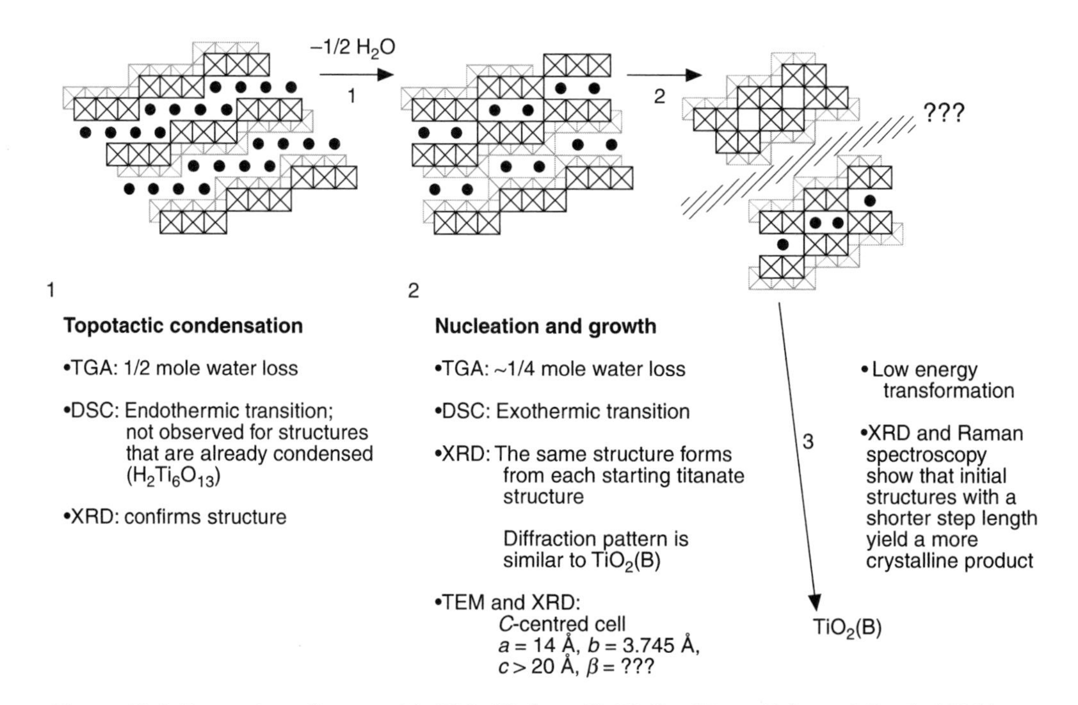

Figure 10.1 Formation of metastable $TiO_2(B)$ from $K_2Ti_4O_9$. (From Feist and Davis 1992.)

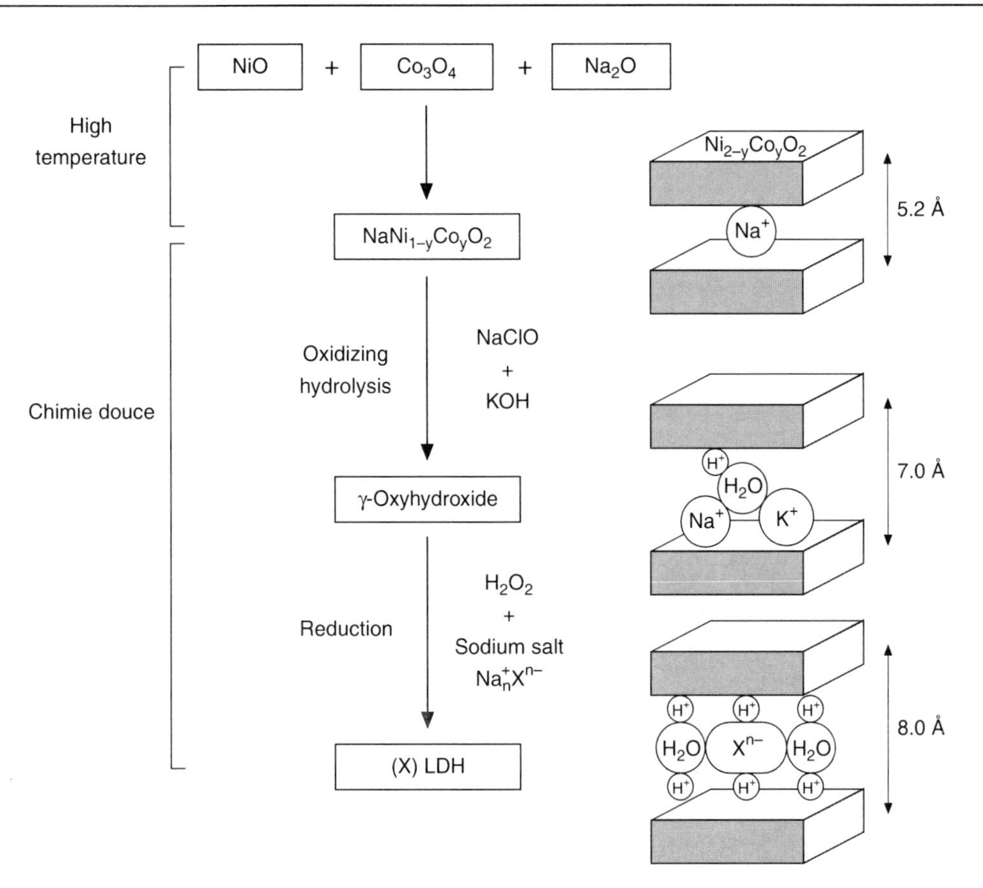

Figure 10.2 Preparation of layered double hydroxides (LDH). (From Delmas and Borthomieu 1993.)

materials which include graphite, oxides and chalcogenides. Thus, Li^+ has been intercalated in a large number of hosts (for example, VS_2, TiS_2, VO_2, Fe_2O_3, Fe_3O_4, MnO_2, etc.). W bronzes may be considered to be intercalation compounds. Chevrel phases are also intercalation compounds since Cu and such cations can be leached out to produce the chalcogenides. Novel materials can be obtained by restacking exfoliated single-layer MoS_2 and other layered materials with molecular units between the layers (Fig. 10.5). A variety of molecules have been incorporated between such single layers of MoS_2, MoO_3 and other layered materials (Divigalpitiya *et al.* 1989). Many of the topochemical reactions are gentle, although they may occur entirely in the solid state.

4 Precursor methods

The best way of reducing diffusion distances in solid state synthesis, in particular by ceramic procedures is to have precursors wherein the relevant metal ions are a short distance apart (Fig. 10.6). Precursor compounds to prepare a variety of perovskite oxides have been known for some time (for example, $Ba[TiO(C_2O_4)_2]$ for $BaTiO_3$, $Li[Cr(C_2O_4)_2.(H_2O)_2]$ for $LiCrO_2$, MCr_2O_4 from $(NH_4)_2M(CrO_4)_2.6H_2O$ and $LaCoO_3$ from $LaCo(CN)_6.5H_2O$). Hydrazinate precursors have been employed to prepare several complex oxides.

Especially noteworthy are the carbonate solid solutions of calcite structure which on

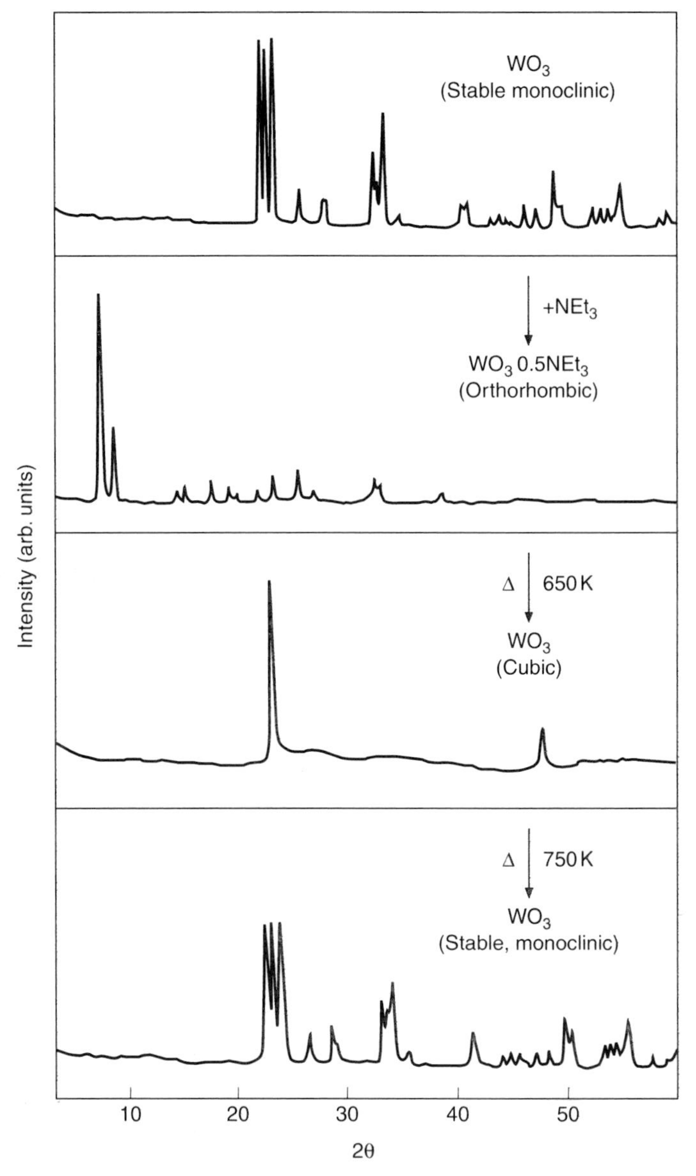

Figure 10.3 X-ray diffraction patterns of stable (monoclinic) WO_3, $WO_3.NEt_3$ and cubic WO_3 (ReO_3 type), obtained by thermal deintercalation of $WO_3.0.5NEt_3$ and WO_3 (monoclinic), obtained by heating the metastable cubic WO_3 phase. (From Ayyappan *et al.* 1995.)

decomposition give the oxide with the desired cation ratios (Rao *et al.* 1986). A variety of complex oxides have been prepared by this method and many of these cannot be prepared otherwise. A large number of carbonate solid solutions of calcite structure containing two or more cations in different proportions and these solid solutions are excellent precursors for the synthesis of oxides since the diffusion distances are considerably lower than in the ceramic procedure. The rhombohedral unit cell parameter, a_R, of the carbonate solid solutions varies systematically with the weighted mean cation radius. (Fig.10.7.) Carbonate solid solutions are ideal precursors for the synthesis of monoxide solid solutions of rocksalt structure. For example, the carbonates are decomposed in vacuum or in flowing dry N_2, to obtain monoxides of the type $Mn_{1-x}M_xO$ (M = Mg, Ca, Co or Cd), of rocksalt structure. Oxide solid solutions for

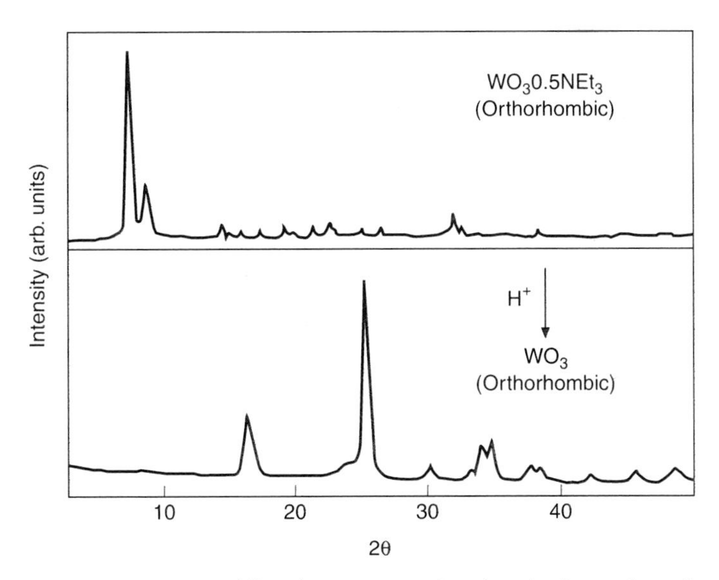

Figure 10.4 X-ray diffraction patterns showing the formation of a metastable orthorhombic phase of WO_3 by the acid leaching of the intercalated amine in $WO_3.0.5NEt_3$ at room temperature. (From Ayyappan *et al.* 1995.)

M = Mg, Ca and Co would require 770–970 K for their formation, whilst those containing Cd are formed at still lower temperatures. The facile formation of oxides of rocksalt structure by the decomposition of carbonates of calcite structure is due to the close (topotactic) relationship between the structures of calcite and rocksalt. The monoxide solid solutions can be used as precursors for preparing spinels and other complex oxides.

A number of ternary and quarternary oxides of novel structures have been prepared by decomposing carbonate precursors containing the different cations in the required proportions. Thus, one can prepare $Ca_2Fe_2O_5$ and $CaFe_2O_4$ by heating the corresponding carbonate solid solutions in air at 1070 (797) and 1270 K respectively, for about 1 hour. $Ca_2Fe_2O_5$ is a defect perovskite with ordered oxide ion vacancies and has the well-known brownmillerite structure with the Fe^{3+} ions in alternate octahedral (O) and tetrahedral (T) sites. Co oxides of similar compositions, $Ca_2Co_2O_5$ and $Ca_2Co_2O_4$, have been prepared by decomposing the appropriate carbonate precursors at around 940 K. Unlike in $Ca_2Fe_2O_5$, anion-vacancy ordering in $Ca_2Mn_2O_5$ gives rise to a square–pyramidal (SP) coordination around the transition metal ion. One can also synthesize quarternary oxides, Ca_2FeCoO_5, $Ca_2Fe_{1.5}Mn_{1.5}O_{8.25}$, $Ca_3Fe_2MnO_8$, etc., belonging to the $A_nB_nO_{3n-1}$ family, by the carbonate precursor route (Fig. 10.8). In the Ca–Fe–O system, there are several other oxides such as $CaFe_4O_7$, $CaFe_{12}O_{19}$ and $CaFe_2O_4(FeO)_n$ (n = 1,2 or 3) which can, in principle, be synthesized starting from the appropriate carbonate solid solutions and decomposing them in a proper atmosphere.

Hydroxide, nitrate and cyanide solid solutions have also been employed as precursors for oxides. Recently, a precursor has been found to prepare Chevrel phases (Nanjundaswamy *et al.* 1987) of the type $A_xMo_6S_8$ by the reaction:

$$2A_x(NH_4)_yMo_3S_9 + 10H_2 \Rightarrow A_{2x}Mo_6S_8 + 10H_2S + 2yNH_3 + yH_2 \quad (10.1)$$

Precursors for the synthesis of $YBa_2Cu_3O_7$ have been described (Rao *et al.* 1993a).

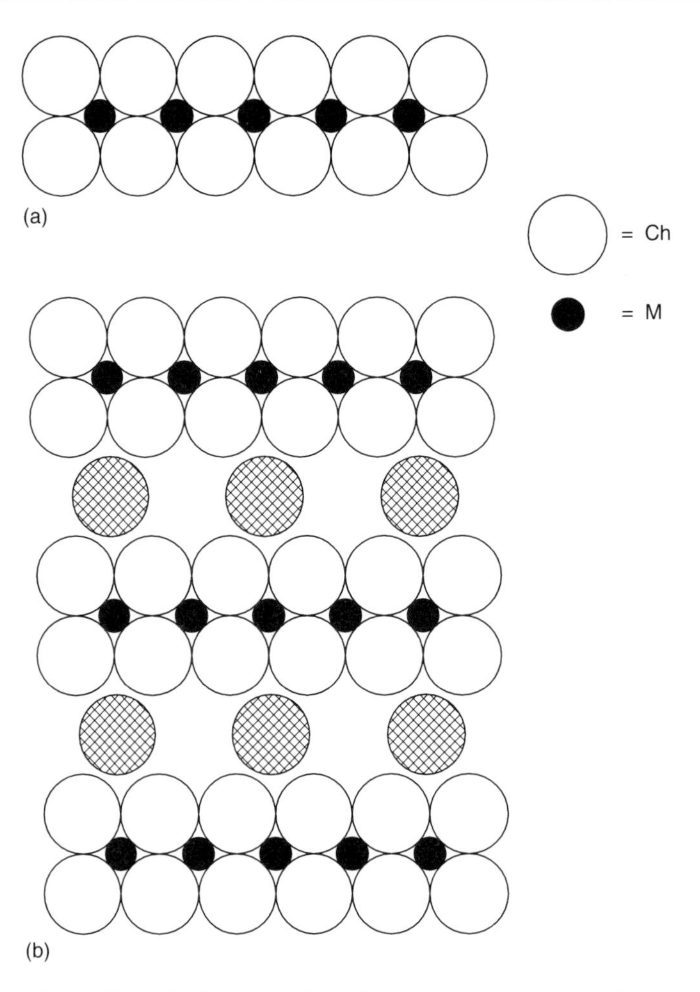

Figure 10.5 Single layers of $MoCl_2$ such as layered materials obtained by exfoliation (a) are employed to obtain new intercalated materials (b). (From Divigalpitiya *et al.* 1989.)

5 Topochemical reactions

Intercalation reactions are generally topochemical in nature. In topochemical reactions the reactivity is controlled by the crystal structure and there is an orientational relationship between the product and the parent just as in Martensitic transformations (Rao and Gopalakrishnan 1986). Dehydration of β-$Ni(OH)_2$ to NiO, as well as the oxidation of $Ni(OH)_2$ to NiOOH, are both topochemical (Figlarz *et al.* 1996). γ-FeOOH topochemically transforms to γ-Fe_2O_3 on treatment with an organic base. Dehydration of many hydrates such as $VOPP_4.2H_2O$ and $MoO_3.H_2O$ is topochemical. The topochemical nature of dehydration has been exploited to prepare MoO_3 in the metal stable ReO_3-type structure (Rao *et al.* 1986). $WO_3.1/3H_2O$ seems to yield WO_3 in different structures depending on the temperature of dehydration (Figlarz *et al.* 1996), as shown in Fig. 10.9. Topochemical reduction and oxidation reactions are known. Thus, the reduction of $YBa_2Cu_3O_7$ to $YBa_2Cu_3O_6$ is topochemical. $La_2Ni_2O_5$ can only be made by the topochemical reduction of $LaNiO_3$.$La_2Co_2O_5$ can similarly be prepared only from $LaCoO_3$ (Vidyasagar *et al.* 1985); in Fig. 10.10 we show the nature of structural transformation involved in the reduction process.

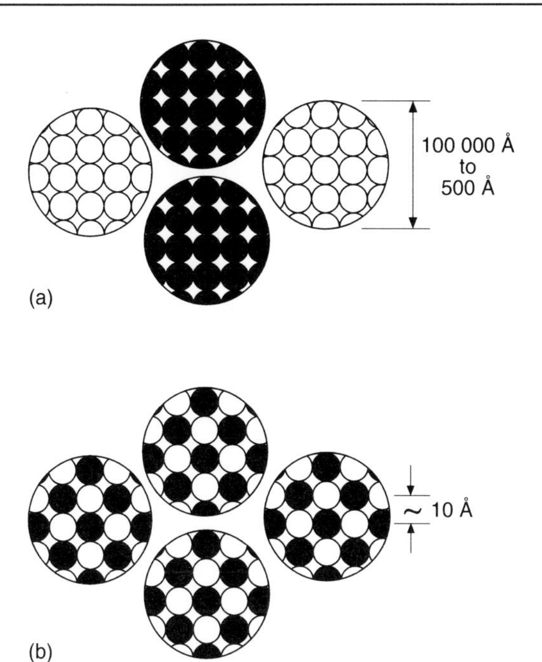

Figure 10.6 Large diffusion distances involved in ceramic preparations involving two types of cations (closed circles and open circles represent two cations) compared to a precursor (the two cations are present in proximity).

6 Ion exchange and alkali flux methods

Ion exchange is an important property of fast-ion conductors such an β-alumina. Ion exchange also provides a means of preparing other materials. Typical examples are the preparation of $LiCrO_2$ from $NaCrO_2$ and $LiNO_3$, $AgAlO_2$ from $KAlO_2$ and $AgNO_3$, and $CuFeO_2$ from $LiFeO_2$ and CuCl. These reactions are carried out in molten state or in aqueous solution. It is through proton exchange with oxide materials containing

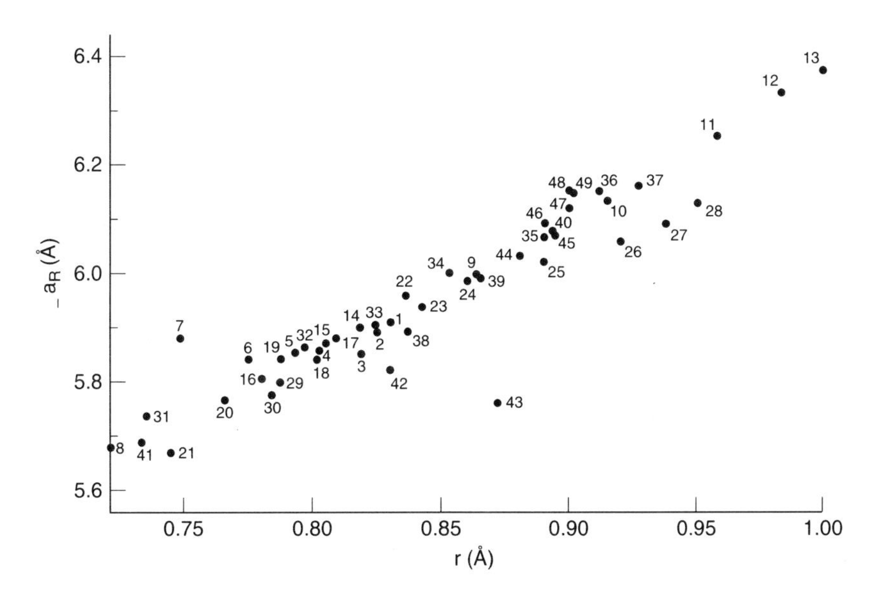

Figure 10.7 Plot of the rhombohedral lattice parameters of a binary and ternary carbonate of calcite structure against the mean cation radius. (From Rao *et al.* 1986.)

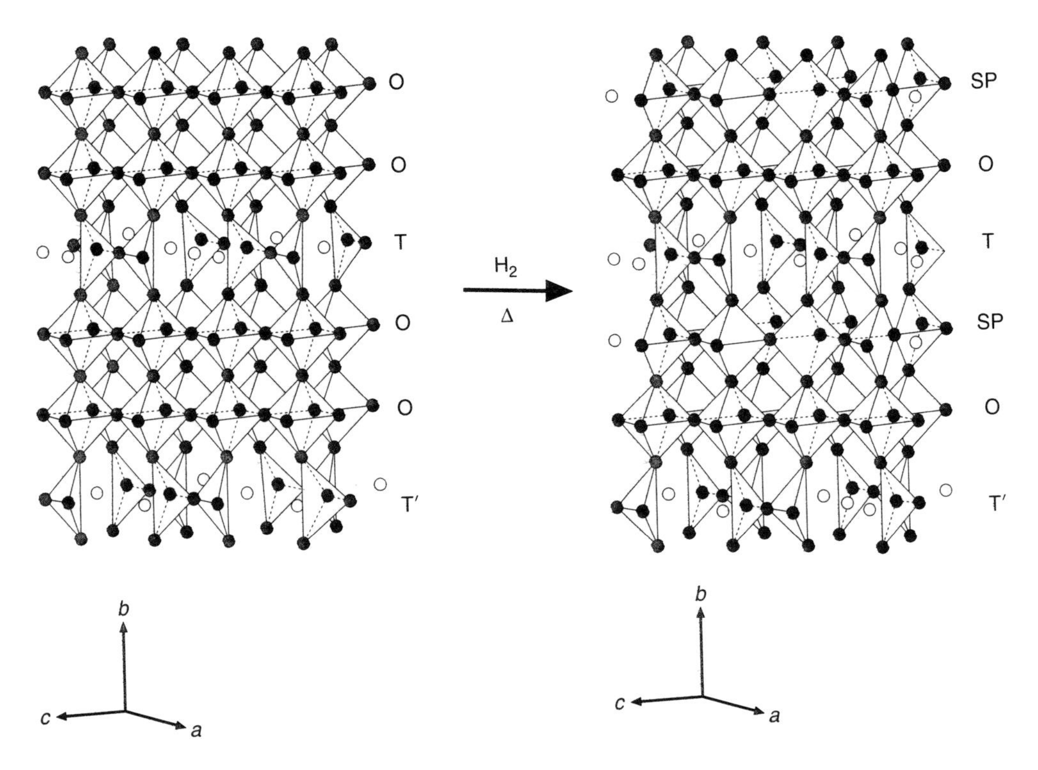

Figure 10.8 $Ca_3Fe_2MnO_{7.5}$ with three types of transition metal coordinations obtained by the topochemical reduction of $Ca_3Fe_2MnO_8$ with only octahedral and tetrahedral coordinations. The latter oxide is obtained by the thermal decomposition of the precursor carbonate, $Ca_2Fe_{4/3}Mn_{2/3}(CO_3)_4$. (From Rao *et al.* 1986.)

alkali metal ions (especially Li^+) that one obtains several materials used as precursors to prepare some of the oxides. We examined some examples whilst discussing soft chemistry routes. Another example is the conversion of $LiNbO_3$ and $LiTaO_3$ to $HNbO_3$ and $HTaO_3$ (Rice and Jackal 1982).

Use of strong alkaline media in the form of solid fluxes or molten solutions is helpful in preparing oxides, especially if one of the metal ions is required in a high-oxidation state. Thus, $Pb_2(Ru_{2-x}Pb_x)O_7{-}y$ with Pb in the 4^+ state is prepared in a highly alkaline media. Molten alkali has been used to prepare superconducting $La_2CuO_{4+\delta}$ (Rao *et al.* 1993a).

7 Electrochemical methods

Electrochemistry in aqueous or molten media has yielded a large number of inorganic materials including carbides, borides, silicides, oxides and sulphides. Many of the oxide bronzes are prepared electrochemically, just as other alkali metal intercalation compounds (for example, Li_xMS_2). The electrochemical method provides the best means of preparing oxides where a transition metal ion is required to be present in a high-oxidation state. In Fig. 10.11 we show the simple electrode systems employed for the purpose. Thus, superconducting $La_2CuO_{4+\delta}$ as well as $SrFeO_3$-type oxides have been prepared electrochemically (Grenier *et al.* 1991). Recently, oxygen-excess $La_8Ni_4O_{17}$ with unusual oxide has been prepared species electrochemically (Demourgues *et al.*

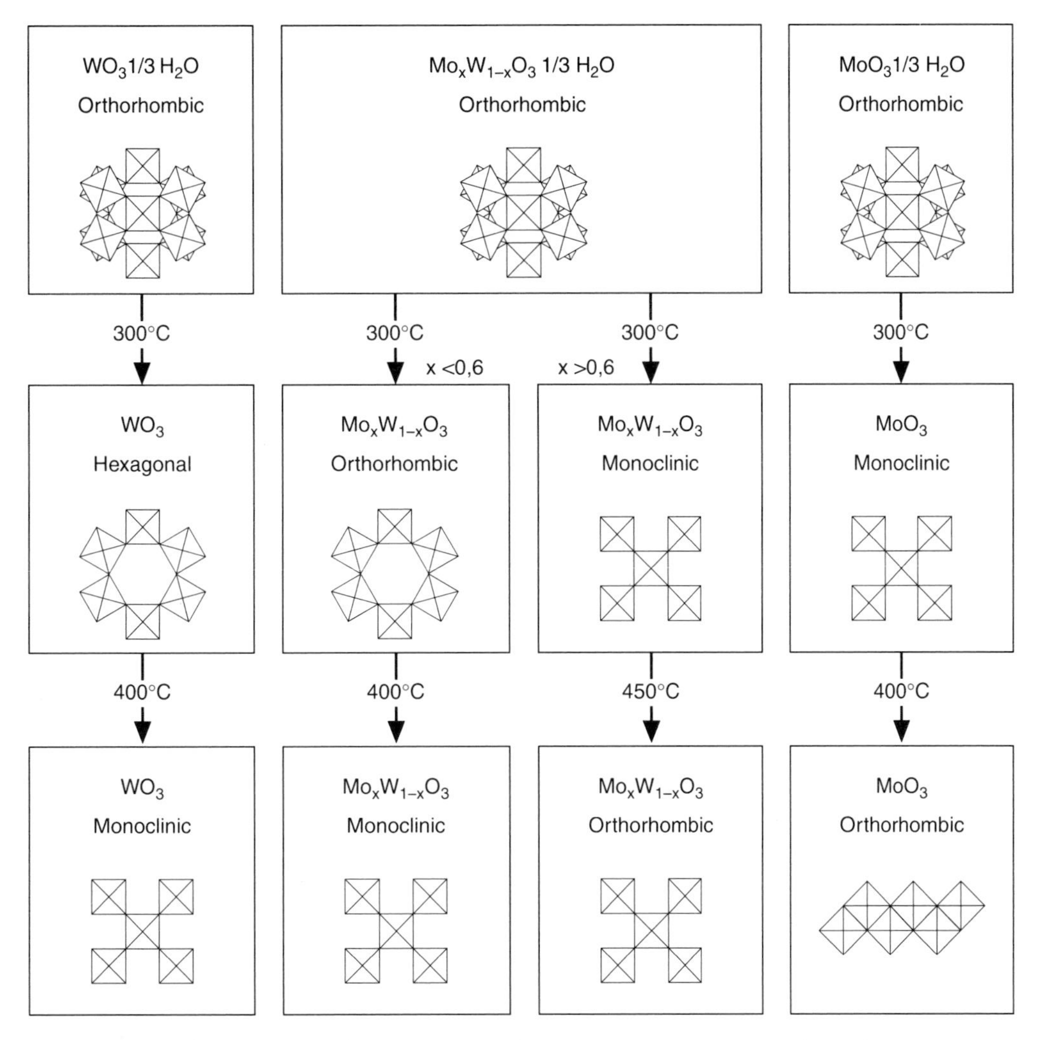

Figure 10.9 Different forms of WO_3 obtained from topochemical dehydration of $WO_3 . 1/3H_2O$ and related hydrates. (From Figlarz *et al.* 1990.)

1993), whilst we have prepared ferromagnetic cubic $LaMnO_{3+\delta}$ with more than 40% Mn^{4+} content. In Fig. 10.12 we show the progressive oxidation of $LaMnO_3$ from the orthorhombic phase to the cubic phase via the rhombohedral phase.

8 Nebulized spray pyrolysis

Pyrolysis of sprays is a well-known method for depositing films. Thus, one can obtain films of oxidic materials such as CoO, ZnO and $YBa_2Cu_3O_7$ by the spray pyrolysis of solutions containing salts (for example,nitrates) of the cations. A novel improvement in this technique is the so-called pyrosol process or nebulized spray pyrolysis, involving the transport and subsequent pyrolysis of a spray generated by an ultrasonic atomizer as demonstrated by Joubert and co-workers (Langlet and Joubert 1992). Xu *et al.* (1990) as well as Raju *et al.* (1995), have employed this method to prepare films of a variety of oxides. When high-frequency (100 kHz–10 MHz range) ultrasonic beam is directed at a gas–liquid interface, a geyser is formed and the height of the geyser is

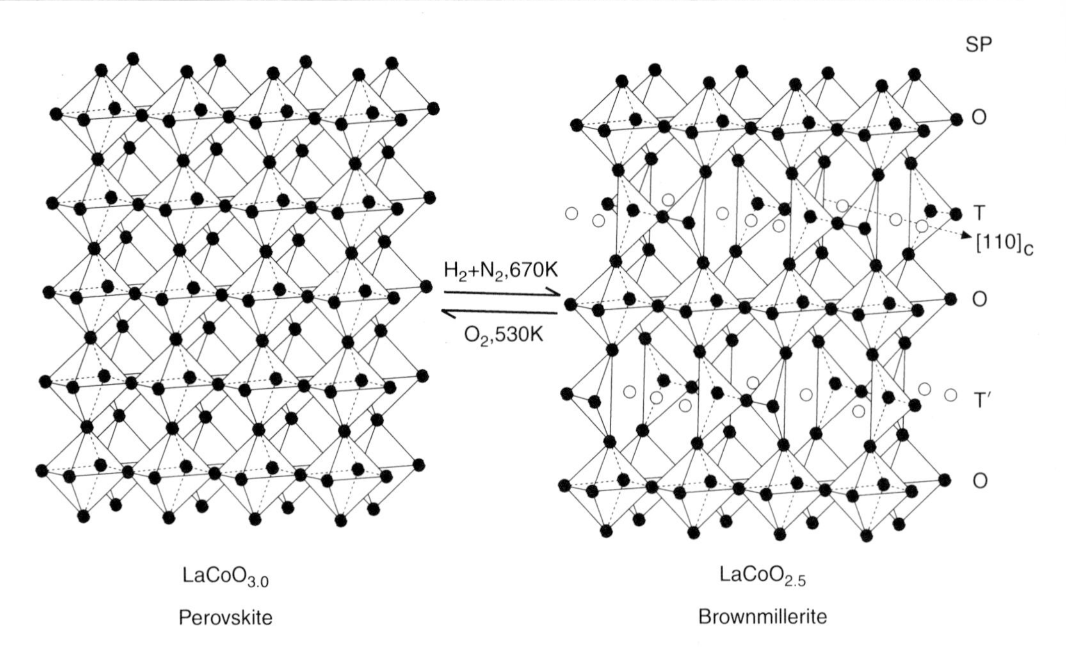

Figure 10.10 Reduction of $LaCoO_3$ to $La_2Co_2O_5$. (From Vidyasagar *et al.* 1985.)

proportional to the acoustic intensity. Its formation is accompanied by the generation of a spray, resulting from the vibration at the liquid surface and cavitation at the liquid–gas interface. The quantity of spray is a function of the intensity. Ultrasonic atomization is accomplished by using an appropriate transducer made of PZT located at the bottom of the liquid container. A 500–1000-kHz transducer is generally adequate. The atomized spray, which goes up in a column fixed to the liquid container, is deposited on a suitable solid substrate and then heat treated to obtain the film of the concerned material. The flow rate of the spray is controlled by the flow rate of air or any other gas. The liquid is heated to some extent, but its vaporization should be avoided. In Fig. 10.13 we show the apparatus employed in the pyrosol method.

The source liquid would contain the relevant cations in the form of salts dissolved in an organic solvent. Organometallic compounds (for example, acetates, alkoxides, β-diketonates, etc.) are generally used for the purpose. Proper gas flow is crucial to obtain satisfactory conditions for obtaining a good liquid spray. Nebulized spray pyrolysis is somewhere between MOCVD and spray pyrolysis, but the choice of source compounds for the pyrosol process is much larger than available for MOCVD. Furthermore, the use of a solvent minimizes or eliminates the difficulties faced in MOCVD. Films of a variety of oxide materials, such as $(Pb,Zr)TiO_3$, $YBa_2Cu_3O_7$ and $LaNiO_3$, have been obtained by the pyrosol method. Nebulized spray pyrolysis is truly inexpensive compared to CVD/MOCVD. The thickness of the films obtained by this method can be anywhere between a few hundred ångströms to a few microns. In Table 10.1, we show typical materials prepared by nebulized spray pyrolysis.

9 Cuprate superconductors

The discovery of a superconducting cuprate with a transition temperature that enabled conduction (T_c) above 77 K created a sensation in early 1987. Wu *et al.* (1987) who

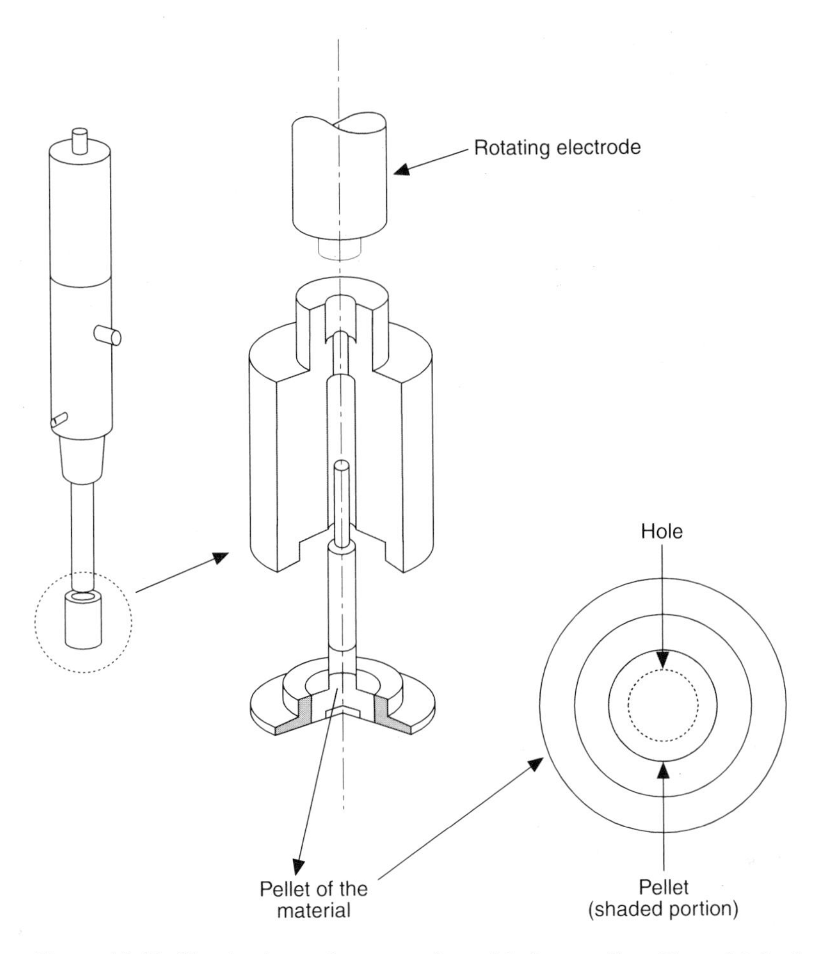

Figure 10.11 Simple electrode system for oxidation studies. (From Mahesh *et al.* 1995.)

announced this discovery first, made measurements on a mixture of oxides containing Y, Ba and Cu, obtained in their efforts to obtain the Y analogue of $La_{2-x}Ba_xCuO_4$. In our laboratory, we worked independently on the Y–Ba–Cu–O system on the basis of solid state chemistry. We knew that Y_2CuO_4 could not be made and that substituting Y by Ba in this cuprate was not the way to proceed. We therefore tried to make $Y_3Ba_3Cu_6O_{14}$ by analogy with the known $La_3Ba_3Cu_6O_{14}$, and varied the Y/Ba ratio as in $Y_{3-x}Ba_{3+x}Cu_6O_{14}$ (Rao *et al.* 1987). By making $x = 1$, we obtained $YBa_2Cu_3O_7$ ($T_c \approx 90$ K). We knew the structure had to be that of a defect perovskite from the beginning because of the route we adopted for the synthesis.

Preparative aspects of the various types of cuprate superconductors have been recently reviewed (Rao *et al.* 1993a). The cuprates are ordinarily made by the traditional ceramic method (mix, grind and heat), which involves thoroughly mixing the various oxides and/or carbonates in the desired proportions and heating the mixture at a high temperature. The mixture is ground again after some time and reheated until the desired product is formed, as indicated by X-ray diffraction. This method may not always yield the product with the desired structure, purity or O_2 stoichiometry. Variants of this method are often employed. For example, decomposing a mixture of nitrates has been found, by some, to yield a better product in the case of the 123 compounds; others prefer to use BaO_2 in place of $BaCO_3$ for the synthesis. One of the

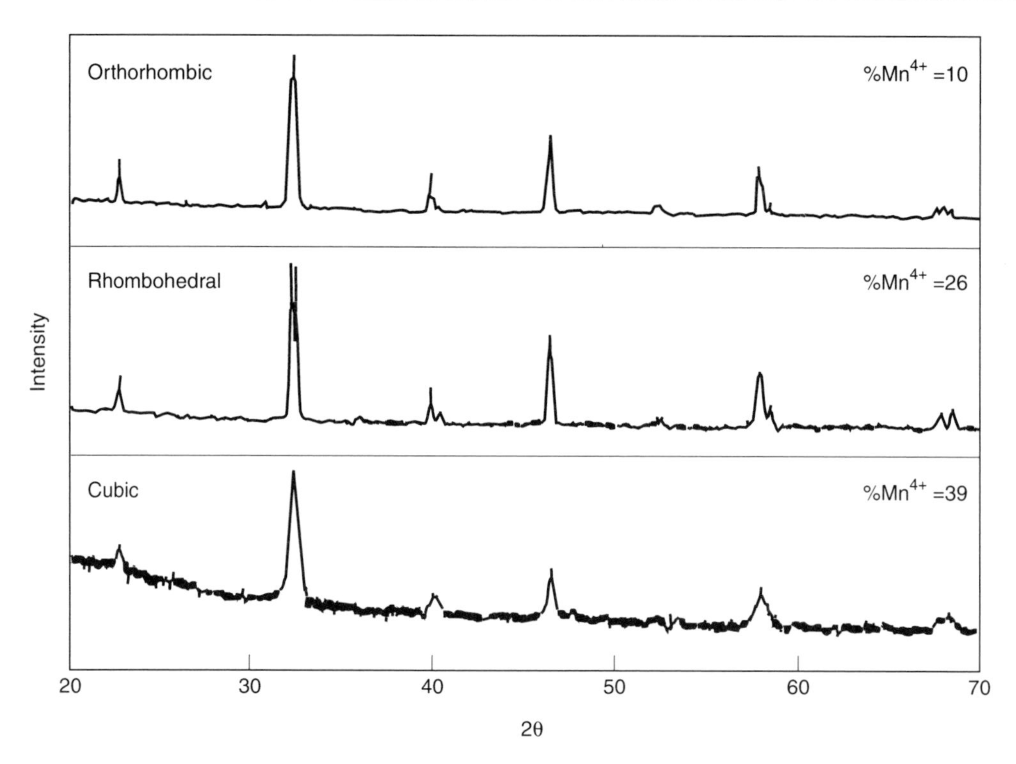

Figure 10.12 Rhombohedral and cubic $LaMn_{1-x}^{3+}Mn_x^{4+}O_{3+\delta}$ from orthorhombic $LaMnO_3$. (From Mahesh *et al.* 1995.)

Table 10.1 Typical films prepared by nebulized spray pyrolysis

Material	Compound used	Solvent	Gas	Substrate*
Pt	Pt acetylacetonate	Acetylacetone	Air	Glass, Al_2O_3,Si (670 K)
ZnO	Zn acetate	Methanol	Air	Glass, Al_2O_3,Si (770 K)
In_2O_3	In acetylacetonate	Acetylacetone	Air	Glass, Al_2O_3,Si (770 K)
SnO_2	$SnCl_4$	Methanol	Air, N_2	Glass, Al_2O_3 (670 K)
$La_4Ni_3O_{10}$,$LaNiO_3$	La + Ni Acetylacetonates	Ethanol	Air/O_2	Si, Al_2O_3 (770 K)
$CdIn_2O_4$	In acetylacetonate, + Cd acetate	Acetylacetone, methanol	Air	Glass, Al_2O_3 (710 K)
TiO_2	Butylorthotitanate	Butanol, acetylacetone	Air, N_2	Glass, steel (670 K)
γ-Fe_2O_3	Fe acetylacetonate	Butanol	Air/argon	Glass, Al_2O_3 (760 K)
(Ni–Zn) Fe_2O_4	Ni, Zn, Fe acetylacetonates	Butanol	Air	Glass (770 K)
Al_2O_3	Al isopropoxide	Butanol	Air	Glass (920 K)
$YBa_2Cu_3O_7$	Dipivaloylmethane derivatives	Butanol	Air/O_2	MgO, $SrTiO_3$ (870 K)

*Substrate temperature is shown.

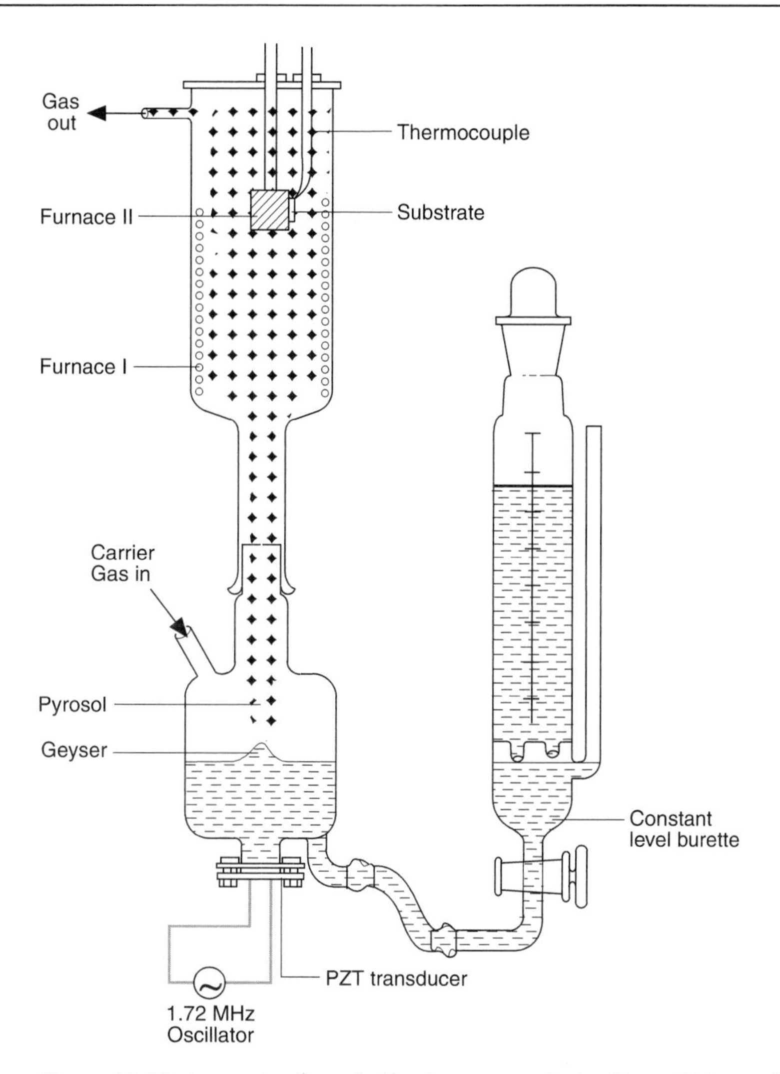

Figure 10.13 Apparatus for nebulized spray pyrolysis. (From Raju *et al.* 1995.)

problems with the Bi cuprates is the difficulty in obtaining phasic purity (minimizing intergrowth of the different layered phases). The glass or the melt route has been employed to obtain better samples. Sealed tube reactions are essential in the case of Tl and Hg cuprates.

Heating oxidic materials under high O_2 pressures or in flowing O_2 often becomes necessary to attain the desired O_2 stoichiometry. Thus, La_2CuO_4 and $La_2Ca_{1-x}Sr_xCu_2O_6$ heated under high O_2 pressures become superconducting, with T_c values of 40 and 60 K, respectively. We have obtained an analogous cuprate with two CuO_2 sheets ($T_c \sim 60$ K) by using $KClO_3$ in the preparation. In the case of the 123 compounds, one of the problems is that they lose O_2 easily. It is necessary to heat these materials in an O_2 atmosphere below the orthorhombic–tetragonal transition temperature. The 124 superconductors were first prepared under high O_2 pressures, but it was later found that heating the oxide or nitrate mixture in the presence of Na_2O_2 in flowing O_2 is sufficient to obtain 124 compounds. Superconducting Pb cuprates, however, can only be prepared in the presence of very little O_2. In the case of the electron superconductor

$Nd_{2-x}Ce_xCuO_4$, it is necessary to heat the material in an O_2-deficient atmosphere. Many of the Tl cuprates have to be heated in vacuum, N_2 or H_2 atmospheres to make them superconducting.

The sol-gel method has been conveniently employed for the synthesis of 123 compounds and Bi cuprates. Materials prepared by such low-temperature methods have to be heated under suitable conditions to obtain the desired O_2 stoichiometry as well as the characteristic high T_c value. 124-cuprates, Pb cuprates and even Tl cuprates have been made by the sol-gel method; the first two are particularly difficult to make by the ceramic method. Coprecipitation of all the cations in the form of a sparingly soluble salt such as carbonate in a proper medium (for example, using tetraethylammonium oxalate), followed by thermal decomposition of the dried precipitate has been employed by many workers to prepare cuprates. Several other novel strategies employed for the synthesis of superconducting cuprates were mentioned whilst discussing the various methods. Strategies where structure and bonding considerations are involved in the synthesis are generally more interesting. One such example is the synthesis of modulation-free superconducting Bi cuprates. Special mention should be made of oxyanion derivatives of cuprates, some of which are superconducting. Anions such as CO_3^{2-}, SO_4^{2-}, PO_4^{3-} and BO_3^{3-} replace the CuO_4 units in the 123 compounds as well as other cuprates (Konishita and Yamada 1993; Ayyappan *et al.* 1993; Huve *et al.* 1993; Maignan *et al.* 1993; Rao *et al.* 1993b; Nagarajan *et al.* 1994).

10 Intergrowth structures

There are several metal oxides exhibiting well-defined recurrent intergrowth structures with large periodicities, rather than forming random solid solutions with variable composition. Such ordered intergrowth structures themselves, however, frequently show the presence of wrong sequences. High-resolution electron microscopy (HREM) enables a direct examination of the extent to which a particular ordered arrangement repeats itself and the presence of different sequences of intergrowth, often of unit cell dimensions. Many systems forming ordered intergrowth structures have come to be known in recent years (Rao and Thomas 1985). These systems generally exhibit homology. The Aurivillius family of oxides of the general formula $Bi_2A_{n-1}B_nO_{3n+3}$ form intergrowth structures of the general formula $Bi_4A_{m+n-2}B_{m+n}O_{3(m+n)+6}$, involving alternate stacking of two Aurivillius oxides with different n values. The method of preparation involves simply heating a mixture of the component metal oxides at about 1000 K. Ordered intergrowth structures with (m,n) values of (1,2), (2,3) and (3,4) have been fully characterized. Intergrowth bronzes and hexagonal Ba ferrites are also examples of ordered intergrowths. What is surprising is that such periodicity occurs even when neither unit involved in the intergrowth is stable by itself.

11 Acknowledegements

The author thanks the EU Science Office and the US National Science Foundation for support.

12 References

Ayyappan, S., Manivannan, V. & Rao, C.N.R. (1993) *Solid State Commun* **87**, 551.

Ayyappan, S., Subbanna, G.N. & Rao, C.N.R. (1995) *Chem Euro J* **1**, 165.

Delmas, C. & Borthomieu, Y. (1993) *J Solid State Chem* **104**, 345–355.

Demorgues, A., Weill, F., Darriet, B. *et al.* (1993) *J Solid State Chem* **106**, 317–338.

Divigalpitiya, W.M.R., Frindt, R.F. & Morrison, S.R. (1989) *Science* **246**, 369–372.

Feist, T.P. & Davis, P.K. (1992) *J Solid State Chem* **101**, 275–287.

Figlarz, M., Gerard, B., Delahaye-Vidal, A. *et al.* (1990) *Solid State Ionics* **43**, 143–155.

Grenier, J.C., Wattiaux, A., Lagueyte, N. *et al.* (1991) *Physica C* **173**, 139–146.

Huve, M., Michel, C., Maignan, A. *et al.* (1993) *Physica C* **205**, 219.

Jacobson, A.J. (1992) In: *Solid State Chemistry* (eds A.K. Cheetham & P. Day). Clarendon Press, Oxford.

Konishita, K. & Yamada, T. (1993) *Nature* **357**, 313–315.

Langlet, M. & Joubert, J.C. (1992) In: *Chemistry of Advanced Materials* (ed C.N.R. Rao), pp. 55–71. Blackwell Scientific Publications, Oxford.

Mahesh, R., Kannan, K.R. & Rao, C.N.R. (1995) *J Solid State Chem* **114**, 294–297.

Maignan, A., Hervieu, M., Michel, C. & Raveau, B. *et al.* (1993) *Physica C* **208**, 116.

Marchand, R., Brohan, L. & Tournoux, M. (1980) *Mat Res Bull* **15**, 1229–1235.

Nagarajan R., Ayyappan, S. & Rao, C.N.R. (1994) *Physica C* **220**, 373–377.

Nanjundaswamy, K.S., Gopalakrishnan, J. & Rao, C.N.R. (1987) *Inorg Chem* **26**, 4286–4288.

Raju, A.R., Aiyer, H.N. & Rao, C.N.R. (1995) *Chem Mater* **7**, 225–231.

Rao, C.N.R. (1994) *Chemical Approaches to the Synthesis of Inorganic Materials*, John Wiley & Sons, Chichester.

Rao, C.N.R. & Gopalakrishnan, J. (1986) *New Directions in Solid State Chemistry*. Cambridge University Press, Cambridge.

Rao, C.N.R. & Thomas, J.M. (1985) *Acc Chem Res* **18**, 113–121.

Rao, C.N.R., Nagarajan, R. & Vijayaraghavan, R. (1993a) *Supercond Sci Tech* **6**, 1–12.

Rao, C.N.R., Nagarajan, R., Ayyappan, S. & Mahesh, R. (1993b) *Solid State Commun* **88**, 757–761.

Rao, C.N.R., Ganguly, P., Raychaudhri, A.K., Mohan Ram, R.A. & Sreedhar, K. (1987) *Nature* **326**, 856.

Rao, C.N.R., Gopalakrishnan, J., Vidyasagar, K., Ganguli, A.K., Ramanan, A. & Ganapathi, L. (1986) *J Mat Res* **1**, 280–292.

Rebbah, H., Desgardin, G. & Raveau, B. (1979) *Mat Res Bull* **14**, 1131–1136.

Rice, C.E. & Jackal, J.L. (1982) *J Solid State Chem* **41**, 308–314.

Vidyasagar, K., Reller, A., Gopalakrishnan, J. & Rao, C.N.R. (1985) *J Chem Soc Chem Commun* 7–8.

Wu, M.K., Ashburn, J.R., Torng, C.J. *et al.* (1987) *Phys Rev Lett* **58**, 908–911.

Xu, W.W., Kershaw, R., Dwight, K. & Wold, A. (1990) *Mat Res Bull* **25**, 1385–1389.

11 Low-Temperature Materials Syntheses

R. ROY

Materials Research Laboratory, The Pennsylvania State University, University Park, Pennsylvania, USA

1 Introduction

Since 1962 the Materials Research Laboratory, Pennsylvania, and I, have categorized materials research under three headings: materials preparation, materials characterization, materials properties. In recent years materials preparation has been referred to as materials 'synthesis and proccessing' and the synthesis of new ceramic materials has become the goal of hundreds of groups across the world. Yet, the record shows that in spite of thousands of person-years of recent effort, not much in the way of really new phases has emerged. Amongst high-temperature super conducting (HTSC) materials, for example, not a single new structural family has emerged—with modest success attained by rather obvious compositional tailoring (for example Bi, Tl and Hg for Ba) of the perovskite plus Cu–O layers. Where some new material has emerged, as in the C_{60} complexes, the discovery is purely serendipitous. The spectacular neodymium iron boride hard magnets constitute perhaps the most significant and also serendipitous example of a discovery of a significant new material phase, The Lanxide or directed metal oxidation process was the equivalent most significant new processing discovery, also made serendipitously. The reasons for this failure to predict new phases or processes 'scientifically' have been treated in a recent detailed review of the status of the content and context of ceramic materials synthesis (Roy 1991).

This chapter reviews the status of ceramic materials synthesis at *low temperature* as practised by the wider community, building on the classics of crystal chemistry and phase equilibrium, because current practitioners are unaware of the rich resources in this increasingly important task. Whilst the vast majority of the current ceramics research effort is devoted to (the properties of) high-temperature materials, such as ZrO_2, SiC and Si_3N_4, there have been few advances in any synthesis aspects of such materials. Contrary to general perception in the materials community, as far as really new materials are concerned (using the measure of the product's value), there have been greater successes of significance in low-temperature materials. In the last three decades completely new families of zeolites with very considerable commercial value have been synthesized, starting with Linde's 4A and 5A and 13X, and moving through the aluminium phosphates to the new large pore phases. New synthetic layer structures (clays) which serve as selective adsorbers have also been synthesized. New cements have radically improved those 'low-tech' materials. Thus, these low-temperature materials are in many ways the leading edge of *new* ceramic materials synthesis. Moreover, the gradual recognition that materials made at low temperatures can be much purer and more perfect structurally are also bringing low-temperature processes into focus. Only very recently, processes such as hydrothermal synthesis, developed extensively by our laboratory in the 1950s, are finally being recognized for their potential.

Still, the vast majority of ceramic materials are made at high temperatures in the 1000–2000°C range. The reason is simple. In solid state reactions, where solid state diffusion controls the kinetics, reaction rates are unworkably slow below such temperatures. To take the most obvious example, SiO_2–quartz and Al_2O_3–corundum are not in equilibrium together in any part of the p–t plane. Yet, an intimate mixture of fine powders would not react at all below nearly 1000°C, although thermodynamically they are less stable together than mullite, andalusite, sillimanite and kyanite. Certainly they would not react with water, even though they should according to the Cu_2O_3–SiO_2–H_2O phase diagram (Roy 1956), to form kaolinite or pyrophyllite. The kinetics are too slow by orders of magnitude.

Hence, the problem confronting the low-temperature materials synthesizer is how to increase the *kinetics* of reaction of ceramics to catalyse reactions, say below 500°C. This is the problem I shall address in this chapter.

2 Traditional and current methods for low-temperature synthesis

In this chapter, I shall attempt to bring together and show the connections amongst many different approaches to catalysing low-temperature reactions amongst solids. Table 11.1 groups the methods into two categories, each subdivided further into several subcategories.

Category A includes five subcategories starting with reactive precursors, fluxes and hydrothermal reactions and ending with radiation catalysis. We will treat each in turn. Then, we will discuss category B methods which include our most current work.

2.1 *Known strategies for catalysis of low-temperature reactions*

2.1.1 REACTIVE PRECURSORS: THE SOL-GEL PROCESS, I

I developed the sol-gel process, starting in 1948 originally, explicitly to respond to the utter lack of reactivity of Al_2O_3 and SiO_2 below 500°C. By creating atomic mixing in solution both from organic or inorganic precursors, it became possible to create highly

Table 11.1 Different methods for catalysing low-temperature reactions

A In widespread use before 1980

1 Utilization of reactive precursors; the sol-gel approach
2 Use of low-melting salts as fluxes
3 (a) The hydrothermal process (high p–t water)
   (b) as for (a) but with electric fields added
   (c) as for (a) but with mechanical grinding added
4 Mechanochemical effects
   (a) fracture alone
   (b) pressure alone
   (c) pressure with shear
5 Use of ionizing and non-ionizing radiation

B Recent research

6 Nanocomposite xerogels via sol-gel II
7 Acoustic wave stimulation (sonochemistry)
8 Hydrothermal with microwave

reactive starting materials of virtually any ceramic composition. This is a very general solution of universal applicability in low-temperature ceramic material synthesis: use solution mixing. By the mid-1950s we had made xerogels of some 5000 such compositions in the most common one-, two-, three-, four- and five-component oxide systems for use as starting materials precisely for this purpose. Indeed, the first review of the sol-gel process (Roy 1956) is titled *Aids in Hydrothermal and 'Wet' (Low Temperature) Phase Equilibrium Studies.* Such materials also made possible the systematic synthesis of low-temperature hydrated or hydroxylated phases. The variety of synthetic hydroxylated phases—dozens of them quite new—ranging from complex clays, zeolites, micas containing 'unnatural' ions such as Ga, Ge, Ni, Mn, etc., to the simple hydroxide and oxyhydroxides (of ions such as Sc and the rare earths), all made via the sol-gel route, is very large and references may be found in a late review (Roy 1987).

2.1.2 THE SOL-GEL PROCESS II: NANOCOMPOSITES

In 1981 I redirected the main thrust of all sol-gel research away from homogeneity of mixing on the atomic scale to heterogeneity on the nanometre scale (see details in Roy 1987). I thereby started work on a class of materials which I first called 'nanocomposites'. The rationale was twofold. *Compositional heterogeneity* provided an additional thermodynamic driving force in the change in the enthalpy of reaction (ΔH_{react}) for lowering temperatures of reaction. This nano-inhomogeneity of starting materials was commercialized early by the Chichibu Cement Company in its mullite powder. In a parallel approach it was also demonstrated that *structural heterogeneity*, specifically for *epitaxial seeding*, provided a lower temperature route for completing a solid state reaction at a lower temperature or directing the reaction to a (metastable) lower temperature phase. Thus, the concept of seeding of gels, first of boehmite gels with Al_2O_3–corundum, which we had first practised widely in 1949–1950 in hydrothermal research, was extended to all ceramic systems under 'dry' firing conditions between 1982 and 1985 (Roy 1982a–c, 1983; Hoffman *et al.* 1984). Based on this work, nanocomposite xerogels and gels have become a major asset and tuneable resource for all low-temperature (and high-temperature) material synthesis in oxides and similar materials.

2.2 *Fluxes*

A flux is typically a low-melting halide or oxysalt mixture. Whilst little sophisticated detailed work has been done on the *mechanisms* of how fluxes work in oxide reactions, the empirical fact that one can use fluxes for inducing subsolidus reactions in oxides goes back at least to Fenner's (1913) work on SiO_2 polymorphism. By using molten Na_2WO_4 (or Li salts) one is able to transform, for example, quartz to tridymite and tridymite to cristobalite reversibly near 870 and 1470°C, respectively. Even today the tungstate flux is the method of choice for synthesizing large amounts of tridymite.

A variety of low-melting halide salts of Al, Zn, Mg and Li have been used as fluxes in a variety of syntheses. In the recent past, fluxes have been used extensively mainly in crystal growth. The use of a flux or low-melting eutectic from which (small) crystals precipitate is often the fastest method to obtain single crystals in the millimetre range for structural characterization.

A search of *Chemical Abstracts* reveals that from 1967 to 1993, out of some 35 638 entries on crystal growth, some 1100 involved growth from fluxes. A scan of these titles reveals that 75% of them describe the attempts to grow even very modestly sized crystals of Y–Ba–Cu superconductors and hard ferrites.

The disadvantage of using fluxes, which has not proved to be a serious deterrent, is the possible inclusion of the flux ions in the growing crystal lattice. To avoid this one uses either a common ion (for example, $BaCl_2$ for $BaTiO_3$) or cations and anions with very different sizes (Cl instead of F) in the flux for growing oxide crystals.

In my opinion, the use of fluxes has been neglected in recent materials synthesis, especially in the low-temperature regime. There are many ternary halide eutectics and even binary eutectics involving Zn and Pb which have melting points from 250 to 350°C which could be explored as fluxes for oxide reactions, not only for crystal growth but also for day to day materials syntheses.

2.3 *Hydrothermal*

By far the most general catalytic tool for materials synthesis of O-lattice phases is the use of modest temperature and high-pressure water—the so-called hydrothermal process. Many regard the water as the limiting case of a flux. In the USA this tool owes its development to Morey and Fenner (1917) and to Morey's group at the Carnegie Institution in Washington. Morey not only designed the pressure vessel which bears his name but did many, many systematic phase diagram studies on the alkali–silicate–water systems. The 'Morey bombs' were used worldwide for low-temperature materials synthesis—including clays and zeolites in the 1930s and 1940s by Hall and Insley, later most extensively by Roy and Osborn and specifically for zeolites by Barrer. The Morey vessels were limited to about 450°C and 1 kbar (100 MPa). Nineteen forty-eight saw the appearance of the Tuttle apparatus and 1949 the test-tube or cold-seal bomb as developed by Roy and Tuttle (1956). The latter has become the worldwide standard for laboratory hydrothermal synthesis for 40 years. It is used in the range of approximately 1000°C and 5 kbar (500 MPa), with special modifications helping it to reach 1200°C and 5 kbar (500 MPa).

Convenient apparatus is essential for exploiting any process for synthesis, and the commercial multiple test-tube vessel facility by Tem-Pres (Roy and Tuttle 1956) provided this for hundreds of laboratories. Using it, an enormous number of syntheses have been carried out. For lower temperatures and pressures, very large autoclaves (3 m × 1 m) are in widespread use, for example for growth of quartz single crystals at about 400°C and 1 kbar (100 MPa). The very important hydrothermal synthesis of commercial zeolites is carried out worldwide in very simple autoclaves, because even lower pressures and temperatures are sufficient.

2.3.1 Low-temperature reactions made possible

The syntheses which could not be done by any process other than the hydrothermal approach include the following.

1 Hydroxylated phases such as clays, zeolites, micas, some oxyhydroxides, etc., since the water must be maintained as a component and modest temperatures are needed for finite rates.
2 Reaction between highly refractory oxide phases through solution in water.
3 Crystallization of glasses and non-crystalline oxides, which is accelerated by several orders of magnitude.
4 Exsolution and order–disorder changes in oxide crystals at low temperatures have so far *only* been catalysed by the use of hydrothermal conditions. The classic example is the work of Roy *et al.* (1953), tracing the exsolution of Al_2O_3 from spinel-Al_2O_3 crystalline solutions, and Datta and Roy (1967) on growing fully ordered spinels.

Since the 1970s, a great deal of the activity in the use of hydrothermal oxide reactions in ceramics moved to the laboratories of Professor S. Somiya, then at the Tokyo Institute of Technology. He specialized in the application of hydrothermal processes to supplement traditional ceramic processes. Excellent reviews of this work are available (Nakamura *et al.* 1977; Somiya *et al.* 1978; Somiya 1991), in which he made not only new phases but ceramic bodies.

2.3.2 HYDROTHERMAL WITH SUPERIMPOSED ELECTRIC FIELD

In 1962, Hawkins and Roy (1962) introduced electrodes into a hydrothermal bomb and studied the influence of an electric field on the synthesis of kaolinite. They showed a very substantial effect ($> 100°C$) of lowering the temperature of crystallization of recognizable kaolinite. More recently, Yoshimura *et al.* (1989) have extended this concept to the synthesis of electroceramic materials such as $BaTiO_3$ and related materials. Several other papers (Somiya *et al.* 1986; Yoo *et al.* 1989) have established that the use of an electric field in a hydrothermal environment can play a useful role in accelerating and changing products in low-temperature synthesis.

2.3.3 HYDROTHERMAL WITH SUPERIMPOSED MECHANICAL FORCES

The concept behind these experiments, first carried out by Dachille and Roy (1960), was that the exposure of fresh highly reactive surfaces, formed by grinding and consequent fracture, to the reactive hydrothermal ambient would give rise to the equilibrium phases.

In fact, we were able to demonstrate quite substantial improvements in reaction kinetics especially in hydration reactions of highly stable oxides to make clays. Again, this was made possible by a simple innovation in apparatus. In effect the Morey bombs were converted into ball mills or rod mills.

2.3.4 MECHANOCHEMICAL EFFECTS INCLUDING HIGH UNIAXIAL PRESSURE WITH AND WITHOUT SHEAR

It is not unreasonable that the process of grinding itself and the attendant repeated fracture with the high 'local' temperatures involved could serve to catalyse reactions. This has proved to be an effective technique for catalysing low-temperature reactions. V.V. Boldyrev in Novosibirsk is the best-known worker who has specialized in the field

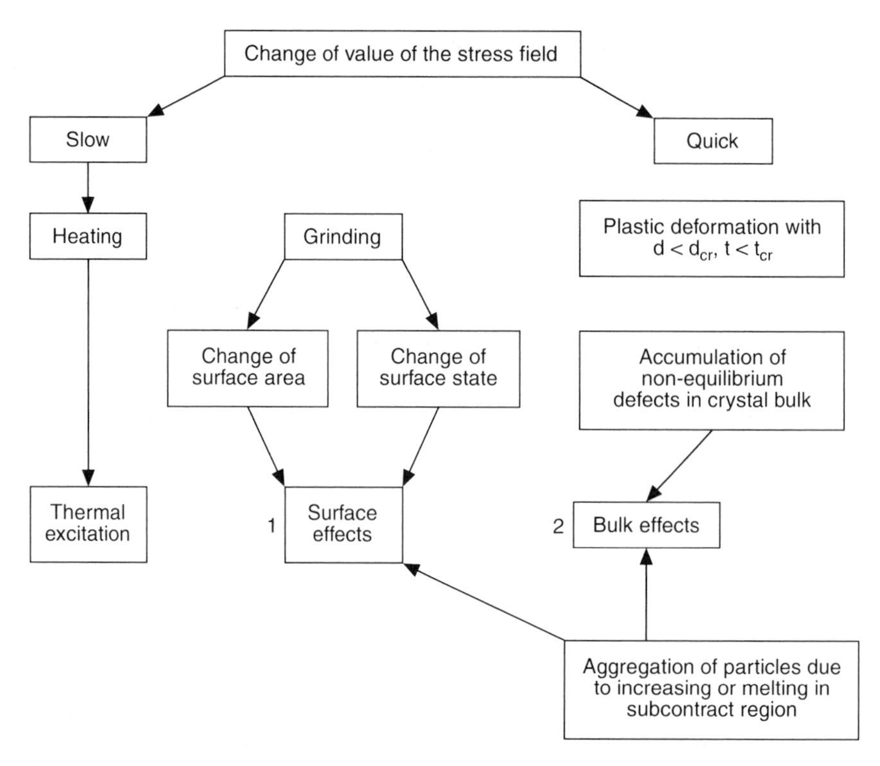

Fig. 11.1 Boldyrev's analysis of how mechanochemical effects work, principally by surface (1) and, to a lesser extent, bulk (2) effects.

and the interested reader is referred to his work (Boldyrev 1986). He summarizes the mechanisms he finds in Fig. 11.1.

At Penn State—at the same time as Boldyrev—Dachille and Roy took a different tack. First, they established that simple uniaxial pressure in the 10–100 kbar (1–10 GPa) regime was sufficient to bring about reactions at low temperatures which could not be accomplished by any other means at all.

The most direct evidence for this is the crystallization of simple glasses such as SiO_2, B_2O_3 or GeO_2 (Dachille and Roy 1959) by the application of pressure alone at much lower temperature than is possible by any other means, specifically in opposed anvil devices. It was argued that the effect may not only be due to pressure, but to the *shearing actions at high pressure*, which are unavoidably present in simple anvil experiments with powders. Nevertheless, it is empirically true that high pressures (> 10 kbar (1 GPa)) alone catalyse low-temperature solid state reactions in ceramic material by orders of magnitude.

Influence of grinding alone. The earliest clean experiment showing the remarkable effects of grinding in ordinary agate mortars was that of Burns and Bredig (1956) who claimed that calcite could be partly transformed to aragonite merely by grinding in a mortar. Dachille and Roy (1961) not only confirmed that observation but extended it to include a whole range of phases such as PbO, PbO_2, $CaCO_3$, MnF_2, Sb_2O_3 and BeF_2. The extraordinary finding was that all the crystalline phases stable at 1 atm (1×10^5 Pa) and room temperature could be transformed by mere grinding for several hours in an automatic agate mortar and pestle, to the stable high-pressure polymorphs. Dachille and Roy estimated from the independently determined p–t equilibria for these phases that the grinding generated pressures of the order of 15–20 kbar (1.5–2.0 GPa) and that the shearing action was responsible for the increase of several orders of magnitude in the *kinetics* of these solid–solid reactions at room temperature, but not measurably in any displacement of the p–t equilibrium.

Influence of the combination of shear and very high pressures. The logical extension of the results reported was to try to separate the effects of pressure from shear. The study of the influence of shearing stresses superimposed upon quasi-hydrostatic pressures of up to 100 000 bar (10 GPa) at temperatures below 550°C was made possible by the development of another simple apparatus by Dachille and Roy (1962). This consists of the Bridgman uniaxial-type apparatus—with a provision for continuous rotation of the bottom piston very slowly back and forth through a 2° arc. The sample is heated externally, and 'displacive-shearing' runs with pressure and temperature automatically controlled can be made for periods exceeding several days if desired. The results clearly separate the influence of 'hydrostatic' pressure itself upon reaction rates from the effect of the added 'displacive-shearing' stresses. Furthermore, from the results it becomes clear that equilibrium relationships between phases are not altered by the shearing stresses. The influence of this type of stress is illustrated for the transformations $SiO_{2\ quartz} \leftrightarrow SiO_{2\ coesite}$; PbO_2 or $MnF_{2\ rutile} \leftrightarrow PbO_2$ or $MnF_{2\ orthorhombic}$; $PbO_{litharge} \leftrightarrow PbO_{massicot}$; $CaCO_{3\ calcite} \leftrightarrow CaCO_{3\ aragonite}$; and the formation of $NaAlSi_2O_6$ (jadeite). Reactions which cannot usually be made to proceed below 300–350°C can be performed at temperatures between 0 and 150°C.

Increases in rates of reaction of two or three orders of magnitude can be attained at a given pressure and temperature. It is not clear whether this should be ascribed to strain energy stored in the lattice or merely to breakage of bonds.

Boldyrev (1986), in his extensive work with mechanochemical effects, comes to virtually the same conclusions for many reactions: it is possible to increase the kinetics by about two orders of magnitude by use of mechanochemical effects.

2.3.5 USE OF IONIZING AND NON-IONIZING RADIATION

The radiation damage caused by uncharged particles (neutrons, α-particles and heavy atoms) has been studied by dozens of workers. However, they have not examined the effects on kinetics of solid state transformations. Roy and Buhsmer (1965) studied a variety of phase transformations in SiO_2, γ-Ca_2SiO_4 and PbO_2, and showed that only modest changes in kinetics were caused by neutron radiation damage.

For ionizing radiation, Boldyrev *et al.* (1989), in work on the CaO–SiO_2 system, was able to increase the kinetics of reaction by two orders of magnitude—albeit at a modest temperature—by using 2 MeV electrons (Figs 11.2 and 11.3).

2.4 *Recent research on low-temperature reaction catalysis*

During the last several years we have reopened the question of the possibility of increasing the kinetics of low-temperature reactions in ceramic systems. Below are the results of three different families of approaches:

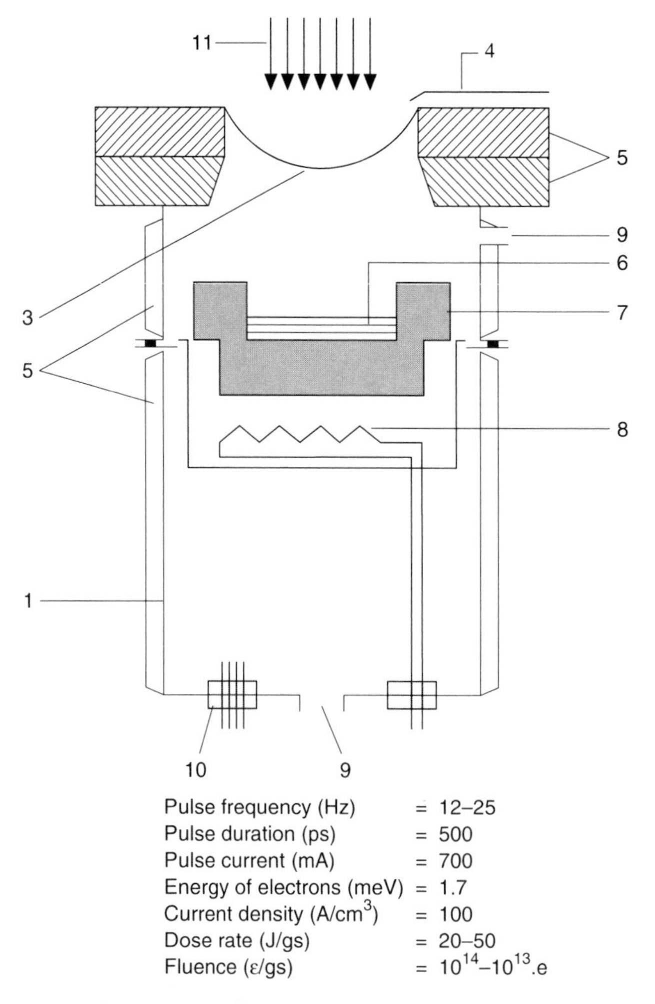

Fig. 11.2 Apparatus for electron beam excitation of solid state reactions.

1 epitaxial nanocomposites as precursors:
2 acoustic wave stimulation (sonochemistry);
3 microwave hydrothermal reactions.

2.4.1 EPITAXIAL NANOCOMPOSITES

In Section 2.1.2 we referred to our extensive work on the remarkable effects on crystallization temperature and kinetics by using nanocomposites in *dry* ceramic systems. We first established these effects in one-component systems such as Al_2O_3, TiO_2, etc. (Suwa *et al.* 1986) and then in binary systems such as Al_2O_3–SiO_2, ZrO_2–SiO_2, etc. (Komarneni *et al.* 1986). Figure 11.4 summarizes what can be achieved in the crystallization of $ZrSiO_4$ (Roy *et al.* 1988). Whereas a homogeneous $ZrSiO_4$ gel will crystallize at 1350°C, if one achieves the same composition by mixing nanoparticles of non-crystalline ZrO_2 and SiO_2; by adding crystalline nanoseeds to a chemically homogeneous gel one can reduce the crystallization temperature by nearly 300°C. The case of $ThSiO_4$ (Vilmin *et al.* 1987), with its two polymorphs was, however, by far the most instructive. Not only could one lower the temperatures of reaction by

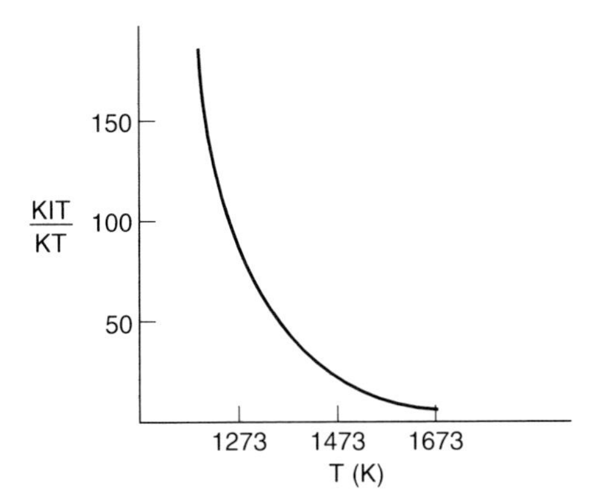

Fig. 11.3 Change of reaction rate by a factor of 20 at lower temperatures for the reaction $2CaO + SiO_2 = Ca_2SiO_4$.

200–300°C, but one could *totally* direct the $ThSiO_4$ gel to form (and retain at 1400°C for days) the huttonite or thorite structures merely by nanocompositing with 1–5% seeds of the respective phases.

The most spectacular example of epitaxial seed catalysis is one very recent work showing that albite glass, which could not be crystallized at 1050°C for 2 years, could be crystallized in 24 hours if seeded with crystalline albite (Takahashi *et al.* 1990).

In a study designed to test whether these effects could be realized at lower temperatures, we studied crystallization in the Al_2O_3–P_2O_5–water system (Liu *et al.* 1992). Our results clearly showed that below 400°C one can *direct* the phase formation by seeding, as summarized in the following.

Crystal growth, under hydrothermal conditions in the binary join SiO_2–$AlPO_4$, was investigated up to 400°C using several starting materials made by a partial solution route. Precursors used were boehmite (AlOOH), H_3PO_4, non-crystalline silica (Ludox, Cab-O-Sil) and quartz. Studies up to 400°C showed that SiO_2 $AlPO_4$ and its hydrate, were the only crystalline and non-crystalline phases present along the binary join, and no substantial crystalline solution or any ternary phase was observed. Three polymorphic forms of $AlPO_4$, i.e. berlinite, tridymite and cristobalite form, existed at as low as 75°C. The nature of the SiO_2 precursors greatly influenced the development of the polymorphic phases of $AlPO_4$. The low-quartz precursor suppressed the formation of

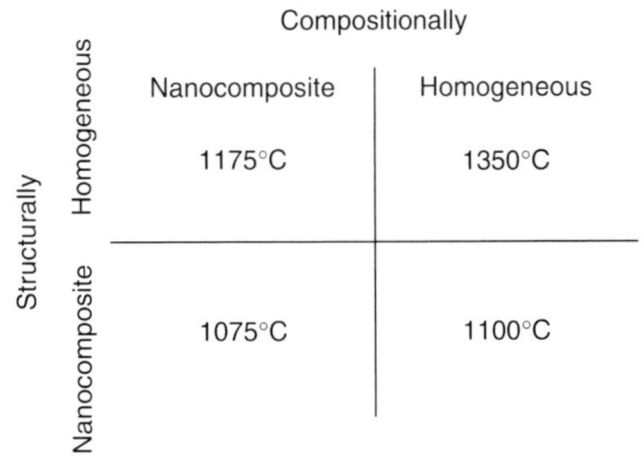

Structurally	Compositionally: Nanocomposite	Compositionally: Homogeneous
Homogeneous	1175°C	1350°C
Nanocomposite	1075°C	1100°C

Fig. 11.4 Effect of using compositional and structural nanocomposite gels of the $ZrSiO_4$ composition compared to a single-phase homogeneous gel.

the cristobalite form of $AlPO_4$ and favoured berlinite production. On the other hand, non-cystalline SiO_2 with a cristobalite-like broad X-ray diffraction (XRD) peak suppressed the formation of berlinite and enhanced that of the cristobalite form of $AlPO_4$. The SiO_2 precursors acted as structural seeds for the growth of $AlPO_4$. These precursor effects indicate that heteroepitaxy is very significant during the nucleation and growth of $AlPO_4$ phases on the surface of SiO_2 particles, even in these low-temperature reactions. Clearly, the potential for exploitation of final phase direction and the kinetics of its formation exists in the low-temperature hydrothermal regime.

2.4.2 ACOUSTIC WAVE (SONOCHEMICAL) STIMULATION (AWS)

In the last decade considerable work has appeared in the field of 'sonochemical' breakdown of liquid phases. This work was limited, almost exclusively, to organic materials and to decomposition of homogeneous liquids. Excellent reviews by Suslick (1986, 1990) give an overview of the field and Suslick did in fact show that metallic phases may be melted and/or corroded in aqueous suspension at 'room temperature'. The general explanation is that the collapsing bubble during cavitation could generate temperatures in the order of 5000 K and modest local pressures. We decided to see if *novel* materials synthesis or combination reactions could be accelerated by the AWS route. In the event, it became clear that the kinetics of many reactions can be increased by about two orders of magnitude. In attempting to understand the fundamentals of the process we changed the frequency from 20 to 500 kHz and even 1 MHz. This had virtually no effect. The more interesting effect was the size effect (Srikanth *et al.* 1992). The relevant study was the leaching of K^+ into water from a series of micas. The leaching at 60 °C was enhanced by more than a factor of 10. But, to our surprise, for the 20 kHz field, it was clear that very small particles ≈ 1 μm), which on the face of it should react faster, did not. Instead, particles of the various micas near 20–40 μm in size coupled better with the cavitation process and reacted faster.

In another example (Fang *et al.* 1992), ultrasonic energy was used to accelerate the formation of hydroxyapatite (HAp). The experiments were carried out in aqueous systems on two different sets of reactants: (i) a mixture of $Ca_4(PO_4)_2O$(TCP) and $CaHPO_4.2H_2O$ (brushite); and (ii) α-$Ca_3(PO_4)_2$(α-C_3P). The reaction systems were exposed to ultrasound of 20–kHz for various times ranging from 5 to 80 min. The products were characterized by XRD and scanning electronmicroscopy. Parallel experiments without ultrasound were carried out for calibration. The results show that the ultrasound substantially accelerates both reactions. With ultrasound, the time required for the TtCP–brushite system to complete the reaction forming HAp was reduced from 9 hours to 25 min at 25 °C, and from 3 hours to 15 min at 38 °C. At 87 °C, α-C_3P does not hydrolyze within 1 hour in deionized water unless the pH is adjusted. Hydrolysis of α-C_3P was induced by sonication in less than 20 min, and longer treatment results in the formation of a homogeneous sol of HAp.

Also, in a preliminary study of the use of AWS of cementitious materials (Srikanth *et al.* 1991), we have demonstrated that occasionally a new reaction can be made to occur. In the systems $Al_2O_3 + H_2O$ and $Al_2O_3 + P_2O_5 + H_2O$, it can be concluded definitely that the high-power ultrasound influences inorganic reactions near room temperature. The rates were highly dependent on the particle size of the precursor powder used. In the case of Rhone-Poulenc $Al_2O_3 + P_2O_5 + H_2O$ it was observed that just 20 min exposure of ultrasound had produced a highly crystalline hydrated phase of $AlPO_4$.

2.4.3 HYDROTHERMAL PROCESSES WITH SUPERIMPOSED MICROWAVE FIELDS

The empirical observation had been made in industry that the oxides in common minerals, which are very resistant to dissolution in acids, bases and even molten carbonates, could be dissolved in relatively dilute acids at very modest temperatures in short times if, and only if, the slightly superheated systems were placed in a microwave field. Again, this principle has been incorporated into a simple convenient commercial apparatus and today microwave hydrothermal is in use the world over for rapid sample *dissolution* for chemical analysis.

Our approach was to reverse the direction of the reaction and try to show that we can *synthesize* materials by increasing the kinetics of reactions in the same temperature regime.

Our preliminary study (Komarneni *et al.* 1992, 1993, 1994, 1995) clearly demonstrates that microwave–hydrothermal synthesis enhances the kinetics of crystallization of various ceramic oxides such as TiO_2, ZrO_2, Fe_2O_3, $KNbO_3$ and $BaTiO_3$ by one or two orders of magnitude. In addition to catalysing the reactions, this novel processing may lead to novel phases as we have demonstrated here with a new layered alumina phase. The crystal size, morphology and level of agglomeration of the different ceramic oxides can be controlled by parameters such as concentration of the chemical species, pH, time and temperature. Submicron powders of TiO_2, ZrO_2, Fe_2O_3, $KNbO_3$ and $BaTiO_3$ have been prepared by optimizing the above parameters. It is suggested that microwave–hydrothermal processing may prove to be a valuable process in the low-temperature production of fine ceramic powders including clays and zeolites.

3 Production of useful low-temperature materials: ordered phases, hydroxylated materials, nanopowders, nanocomposites

The value of low-temperature methods is just starting to be recognized by solid state physicists who often performed ultraprecise measurements on one-of-a-kind specimens of a material. In oxide materials where perfect ordered crystals are needed, low-temperature materials offer the only route available. Obviously, there is no other way to synthesize clays, zeolites, oxyhydroxides, etc. In addition to their huge and permanent dominant (in volume) role in materials technology, their scientific value is just being recognized.

Nanosized powders of virtually any two-, three- or four-component combination of major ceramic oxides—about 5000 compositions—have been routinely produced in our laboratory by the sol-gel process for 40 years. These are very largely non-crystalline as produced. *Only* the low-temperature material processes ('hydrothermal plus') can yield well-crystallized nanopowders (Komarneni *et at.* 1992).

I introduced the term 'nanocomposities', and I believe the major potential for nanosized particles lies here, not in single-phase nanopowders, because one takes advantage of an intrinsic chemical energy in such intimate mixtures as explained earlier (Komarneni *et al.* 1986; Suwa *et al.* 1986; Vilmin *et al.* 1987; Roy *et al.* 1988).

4 Acknowledgements

The author acknowledges financial support from the Office of Naval Research, Advanced Research Projects Agency and the National Science Foundation for various studies involved in this review.

5 References

Aramaki, S. & Roy, R. (1962) *J. Am Ceram Soc* **45**, 229–242.

Boldyrev, V.V. (1986) *J Chim Phys* **83**, 821–829.

Boldyrev, V.V., Voronin, A.P., Gribkov O.S. & Tkachenko, E.V. (1989) *Rev Solid State Sci* **3**, 193–202.

Burns, J.H. & Bredig, M.A. (1956) *J Chem Phys* **25**, 1281–1285.

Dachille, F. & Roy, R. (1959) *J Am Ceram Soc* **41**, 78–80.

Dachille, F. & Roy, R. (1960) *Nature* **186**, 451–455.

Dachille, F. & Roy, R. (1961) *Reactivity of Solids* (ed. J.H. De Boer), pp. 502–512. Elsevier Science Publishers, Amsterdam.

Dachille, F. & Roy, R. (1962) *The Physics and Chemistry of High Pressures*, pp. 77–84. Society for Chemical Industry, London.

Datta, R.D. & Roy, R. (1967) *J Am Ceram Soc* **50**, 578–586.

Fang, Y., Agrawal, D.K., Roy, D.M., Roy, R. & Brown, P.W. (1992) *J Mat Res* **7 (8)**, 2294–3000.

Fenner, C.N. (1913) *Am J Sci* **36**, 331–350.

Hawkins, D.B. & Roy, R. (1962) *J Am Ceram Soc* **45**, 507–508.

Hoffman, D.W., Roy, R. & Komarneni, S. (1984) *J Am Ceram Soc* **67**, 468–476.

Komarneni, S., Li, Q.H., Roy, R., (1994) *J Mat Chem* **4 (12)**, 1903–1909.

Komarneni, S., Roy, R. & Li, Q.H. (1992) *Mat Res Bull* **27**, 1393–1395.

Komarneni, S., Suwa, Y. & Roy, R. (1986) *J Am Ceram Soc* **69**, C155–161.

Komarneni, S., Li, Q.H. Stefansson, K.M. & Roy, R. (1993) *J Mat Res* **8 (12)**, 3176–3185.

Komarneni, S., Pidugu, R., Li, Q.H. & Roy, R. (1995) *J Mat Res* **10 (7)**, 1–8.

Liu, C., Komarneni, S., & Roy, R. (1992) *J Am Ceram Soc* **75 (10)**, 2665–2670.

Morey, G.W. & Fenner, C.N. (1917) *J Am Chem Soc* **39**, 1173–1179.

Nakamura, K., Hirano, S.I. & Somiya, S. (1977) *J Am Ceram Soc* **60**, 513–520.

Roy, R. (1956) *J Am Ceram Soc* **49**, 145–148.

Roy, R. (1982a) *Bull Am Ceram Soc* **61**, 374–378.

Roy, R. (1982b) *Mat Res Soc Ann Meet Abstr*, p. 370. MRS, Pittsburgh Pa.

Roy, R. (1982c) *Mat Res Soc Ann Meet Abstr*, p. 377. MRS, Pittsburgh Pa.

Roy, R. (1983) *Bull Am Ceram Soc* **62**, 375–381.

Roy, R. (1987) *Science* **238**, 1664–1668.

Roy, R. (1991), *New Materials: Fountainhead for New Technologies and New Science, International Science Lecture Series.*, pp. 1–37. U.S. National Academy Press, Washington, DC.

Roy, R. & Buhsmer, C.P. (1965) *J Appl Phys* **46**, 331–334.

Roy, R. & Tuttle, O.F. (1956) *Phys Chem Earth* **1**, 138–147.

Roy, R., Komarneni, S. & Yarbrough, W. (1988) *Ultrastructure Processing of Advanced Ceramics*, Ch. 42 (eds J. MacKenzie & D. Ulrich). Wiley Interscience.

Roy, D.M., Roy, R. & Osborn, E.F. (1953) *Am J Sci* **251**, 337–348.

Somiya, S. (1991) In: *Ceramics Today; Tomorrow's Ceramics* (ed. P. Vincenzi), p. 997. Elsevier, Amsterdam.

Somiya, S., Hirano, S.I. & Yoshimura, M. *et al.* (1978) In: *Proceedings of the International Symposium on Densification of Oxide and Nonoxide Ceramics* (eds S. Somiya & S. Seito), pp. 276–277. Gakujutsu Bunken Fukyu Kai, Tokyo.

Somiya, S. (1986) *Hydrothermal Reactions* **I**, p. 117. Uchida Rokakuho Pub. Co., Tokyo.

Srikanth, V., Agrawal, D.K. & Roy, R. (1991) *J Mat Res* **6 (11)**, 2412–2419.

Srikanth, V., Roy, R. & Komarneni, S. (1992) *Mat Lett* **15**, 127–130.
Suslick, K.S. (1986) *Adv Organomet Chem* **35**, 73–79.
Suslick, K.S. (1990) *Science* **247**, 1439–1443.
Suwa, I.S., Komarneni, S. & Roy, R. (1986) *J Mat Sci Lett* **5**, 21–25.
Takahashi, T., Agraval, D. & Roy R. (1990) *J Ceram Soc Jpn Int* **98**, 46–52.
Vilmin, G., Komarneni, S. & Roy, R. (1987) *J Mat Res* **2 (4)**, 489–498.
Yoo, S.E., Yoshimura, M. & Somiya, S. (1989) *J Mat Sci Lett* **8**, 530–534.
Yoshimura, M., Yoo, S.E. & Hayashi, M. (1989) *Jpn J Appl Phys* **28**, 62 007–62 014.

12 Reactivity of Solids and New Technologies

V.V. BOLDYREV

Institute of Solid State Chemistry, Siberian Branch of the Russian Academy of Sciences, Kutateladze, 18, Novosibirsk-128, 630128, Russia

1 Introduction

Many technological processes used presently for mineral raw processing or for the synthesis of modern materials include the following sequence of steps:

$$\text{Starting raw material} \Rightarrow \begin{array}{c}\text{melting}\\ \text{evaporation}\\ \text{dissolution}\end{array} \Rightarrow \begin{array}{l}\text{chemical reaction}\\ \text{in the gaseous,}\\ \text{liquid or solid state}\end{array} \Rightarrow \text{final desired product} \quad (12.1)$$

Starting solid raw materials are first transferred into the liquid or gaseous state by evaporation, fusion or dissolution. Then, the necessary chemical reactions are carried out in the gas mixture in the melt or in solution. The final product is then separated from by-products, from the rest of the starting reactants and or from the solvent. Quite often, this final product is solid, as were the starting reactants.

Obviously, this transfer of the reactants into the liquid state and the subsequent transfer of the products back into the solid state is not the most elegant solution of the problem. First, it results in additional consumption of energy required for the gasifying or melting or reactants, or for the regenerating the solvent. Second, the process becomes more time-consuming, since more technological steps are required. Finally, a transfer of all (or part of) the reactants in the gaseous or liquid state increases the danger of environmental pollution.

Modifications and modernizations of such a technology are usually aimed at the optimizing of individual steps, and not of the sequence of steps as a whole. For example, various heat exchangers can be applied in order to utilize the heat evolved during the exothermal strages to intensify the endothermal processes. As other examples one can mention the introduction of recycling wherever and whenever possible, or various methods of making the system more isolated. In general, this approach can be termed as a 'tactical' one, not affecting the 'strategy' of production. In many cases this tactical approach has proved to be fruitful. However, its possibilities are limited. We realize more and more that the time has come when further technological progress requires changes in the very strategy of the technology. The sequence of steps in the technological chain itself must be changed. One of the possible variants of such a change in the strategy could be to exclude the chemical processes in the gaseous and liquid states, when both reactants and products are solids:

$$\text{Starting raw materials} \Rightarrow \text{solid state reaction} \Rightarrow \text{final desired product} \quad (12.2)$$

Those new 'solid state technologies' are often economically and ecologically advantageous. Besides, in some cases they make it possible to get an unusual product, which cannot be obtained otherwise, or to carry out an unusual technological operation.

This chapter deals with some examples of new solid state technologies. In order to be able to control a technological operation in such technologies one should know how to control the reactivity of solids.

2 Thermal decomposition of solids and new technologies

Thermal decomposition is one of the most common and best studied types of solid state reactions (Garner 1955; Brown *et al.* 1980; Galwey 1996). In the early 1960s we suggested some methods of controlling the course of thermal decomposition, i.e. of controlling the rate of the reaction and/or the characteristics of the products. The main idea of these methods was to control the types and concentrations of the defects in the solid, since thermal decomposition of solids was shown to be very sensitive to the presence of these defects. Moreover, a particular reaction was shown to be selectively sensitive to some particular type of imperfection (Boldyrev 1960, 1965, 1973, 1975, 1976, 1993b; Boldyrev *et al.* 1979). The type and concentration of the imperfections in a crystal can be controlled by varying the techniques of crystal growth and the storage conditions, as well as by preliminary treatment of the crystals, such as, for example, the preliminary irradiation or the preliminary mechanical treatment. For example, by varying the conditions of crystal growth and applying the surface active compounds one can modify the crystal habit. Mechanical treatment affects the concentration and the distribution of dislocations (Boldyrev V.V., Avvakumov E.G. 1971). Doping of crystals (introduction of impurities) may result in a change in the concentration of intrinsic point defects (vacancies and interstitial ions). Preliminary irradiation can affect the concentration of point defects, and results in the formation of heterophase inclusions.

It was shown that all the reactions of thermal decomposition of solids can be roughly subdivided into two large classes (Boldyrev 1965).

1 To the first class belong the reactions that can be approximated by a local cleavage of the chemical bond in the crystal. These reactions were shown to be especially sensitive to the imperfections, which change the ratio between the volume and the surface of the crystal. Internal surfaces included: dislocations, vicinales, grain boundaries. The change of crystal habit was also shown to affect the reactions included in the first class.

2 The reactions belonging to the second class require a charge or a mass transfer at distances considerably larger than the distance between the nearest neighbours in the crystal or a cell parameter. These reactions were shown to be especially sensitive to point defects: impurities, vacancies, interstitial ions.

2.1 *Application of solid state thermal decomposition reaction in the manufacturing of printed circuit boards*

As an example, I would like to consider the application of one of the solid state thermal decomposition reactions in the manufacturing of printed circuit boards, i.e. in the 'metallization of the dielectric plates'. This process is used in electronics, radio and electrical engineering. The annual sales of printed circuits in the industrially developed countries is comparable with the sale of catalysts in the chemical industry. The process is one of the main sources of pollution of natural water with harmful copper ions.

The conventional industrially used method of the metallization of dielectrics was suggested in the 1950s. It involves a sequential treatment of a half-finished polymer board in a series of tanks containing water solutions. The process includes the following stages (Table 12.1):

1 sensibilization of the surface with a tin chloride solution and the activation of the surface by the deposition of palladium particles;

2 deposition of copper on the surface with formaldehyde in the presence of palladium (so-called 'chemical copper plating');

3 electrochemical growth of the copper layer.

These stages alternate with the stages of washing of the half-finished board. Washing waters and waste solutions are the main sources of environmental pollution by heavy metals: copper, tin, palladium (Lomovsky *et al.* 1991).

The main technological solutions in this process are 'chemical copper plating' and those of 'electrochemical metallization'. The waste solutions remaining after the electrochemical metallization can be regenerated somehow. The situation with the waste solutions remaining after the chemical copper plating is much worse: no efficient methods of their regeneration are available. One can hardly expect such a method to be developed in the nearest future. The prediction is pessimistic, since these waste solutions are so diluted and contain such a low concentration of copper ions that one can hardly suggest a regeneration technology which would be economical. Besides, copper in these solutions forms stable complex compounds, and this makes the regeneration of the solutions especially difficult.

Washing waters remaining after the etching of the boards, as well as after the activation, chemical copper plating and electrochemical deposition, also contribute to environmental pollution. More than 0.1 g of palladium are lost with the washing waters, calculated for each 1 m^2 of the board. Such a waste of palladium is not only harmful because of the environmental problems, but it is also quite a loss from the economical point of view. Attempts to accumulate palladium in a special collector allow the problem to be solved to some extent only; more than 30% of palladium is still

Table 12.1 Technological stages of metallization of dielectrics

Conventional technology	Non-palladium metallization
Preparation of surface Removal of oxide layer Washing	Preparation of surface
Treatment in tin (IV) chloride solution	Wetting in a solution of thermosensitive copper salt solution
Hydrolysis of tin (II) chloride	Crystallization and decomposition of the salt layer upon thermal treatment
Treatment with a palladium salt solution	
Washing in collectors	
Chemical deposition of copper	
Washing	Washing
Galvanization	Galvanization

lost with the washing waters. The main danger related to the manufacturing of printed circuit boards from the ecological point of view is, however, the pollution not with palladium but with copper ions. Copper is known to be a cumulative poison inducing harmful changes in a human body. Up to 18 g of copper polluters get into the waste and washing waters per 1 m^2 of manufactured boards. The contributions of various stages of traditional technology to the overall environmental pollution is summarized in the left-hand column of Table 12.2 (Lomovsky *et al.* 1991). Taking into account the scale of the industrial production of printed circuit boards, one can easily imagine the real scale of environmental pollution related to this process.

Recently, a qualitatively new technology of the metallization of dielectrics was proposed at the Novosibirsk Institute of Solid State Chemistry belonging to the Siberian Branch of Russian Academy of Sciences (Lomovsky and Boldyrev 1989, 1994). This new technology allows one to decrease the number of 'wet' stages, does not require expensive palladium at all and decreases the contamination of waste waters. The basic stage of the new technology is a solid state reaction, i.e. the thermal decomposition of copper hypophosphite. Active copper formed during the thermal decomposition precipitates at the surface of the dielectric, and then a thicker copper layer can be deposited electrochemically. Such stages of the traditional technology as

Table 12.2 Technological wastes of heavy metal ions in the metallization by conventional and non-palladium methods

Conventional technology stages	Ion content (g/m^2)	Non-palladium technology, stages	Ion content (g/m^2)
Deprived of fat	Copper, 0.01–0.02	Deprived of fat	Copper, 0.01–0.02
Etching	Copper, 5–10		
Activation insolutions of tin and palladium salts	Tin, 10–15	Wetting in a thermosensitive copper salt solution	–
Electrolytic polishing	Copper, 3–5	Thermal treatment at 110–140°C	–
Chemical copper plating	Copper, 6–8	Cleaning of the surface from sludge	
Pickling	Copper, 0.01–0.02	Pickling	Copper, 0.01–0.02
Electrolytic copper plating	–	Electrolytic copper plating	–
Washing in a tank collector	Copper, 3–4	Washing in a tank collector with an electrolyzer	–
Washing in running water	Copper, 1–2	Washing in the collector of a neutralizer	–
	Total copper no less than 18		Total copper ~ 0.04

the preliminary treatment of the surface by the solution of tin chloride, the activation of the surface by the solution of a palladium salt with subsequent washings, as well as the stage of chemical copper plating, are no longer required (see Table 12.1). In general, more than 50% of the wet stages are no longer required by the new technology. Not only has it decreased the environmental pollution (the right-hand column of Table 12.2), but the process became much more economical (manufacturing of each plate became almost two times cheaper). A study of the mechanism of thermal decomposition of copper hypophosphite made it possible to find the methods of controlling both the rate of the technological process and the properties of the product of decomposition (copper), and its catalytic and electrophysical characteristics. As a result, the process could be totally automated.

2.2 *Application of thermal decomposition to get materials with desired properties (precursor technique)*

In many reactions of thermal decomposition the structure of products was shown to be related to the structure of the starting reactant (Dankov 1946; Bernal 1960; Feitknecht 1964). The phenomenon was termed *topotaxy*. To preserve at least some structural patterns of the reactant in the structure of the product is often beneficial from the energetical point of view. Topotaxy and topotactical reactions are often used to get materials with desirable properties, which are predetermined by the structure of the starting reactant. The idea is that if there is more than one possible precursor for the synthesis of the final compound, one should choose the precursor with the crystal structure that is most similar to the desirable product structure. By varying precursors, one can get polymorphs of the same compound, including the unusual and metastable ones.

The study of Nitschmann (1938) seems to be one of the first examples of a successful application ofthe precursor technique to the solution of a technological problem. Nitschmann, from the laboratory of Kollschütter, was working in 1935 for the company Farbenindustrie trying to get a yellow pigment from iron shlam. He used what would be called today the precursor technique in order to get an unusual yellow polymorph of iron hydroxide, which cannot be synthesized in any other way (Nitschmann 1938).

Later on, the ability of the product of the solid state reaction to 'memorize' the structure of the parent reactant was repeatedly used in the organic and inorganic solid state synthesis. It is not possible to consider all of them in the present chapter, and I will limit the discussion to two examples only.

The studies of Delmon and coworkers (Delmon 1986; Haruta and Delmon 1986) can be considered as a classical example. They have clearly shown the importance of the coupling effects during solid state reactions for the synthesis of catalysts, in particular, for the control of the morphology of the catalyst.

Oswald and Reller (1989) have used the coordination compounds as precursors for the synthesis of oxide materials with unusual structures and properties. As an example, one can consider the synthesis of the amorphous V_2O_5, which is a widely used catalyst of the oxidation of SO_2 to SO_3. The traditional method of producing V_2O_5 is by the thermal decomposition of NH_4VO_3. In the structure of NH_4VO_3 the VO_3 tetrahedra form continuous infinite chains. The thermal decomposition of NH_4VO_3 proceeds topotactically and the chains are preserved also in the final product, V_2O_5, which remains crystalline. Additional efforts are required in order to transfer the crystalline V_2O_5 into the amorphous form, in which the VO_3 are separated from each other. Oswald and Reller (1989) decided to solve the problem of the synthesis of amorphous V_2O_5 by choosing other precursors in which the VO_3 tetrahedra do not form chains, but are separated. They have compared the thermal decomposition of two coordination compounds, i.e. of the ethylenediammonium vanadate $(NH_3–CH_2–CH_2–NH_3)^{2+}(VO_3)_2^{2+}$ and of the propylenediammonium vanadate $(NH_3–CH_2–CH_2–CH_2–NH_3)_2^{2+}(V_2O_7).^{4-}3H_2O$. In the structure of the ethylenediammonium vanadate the VO_3 tetrahedra form chains, and the thermal decomposition of the salt gives crystalline V_2O_5. In the structure of the propylenediammonium vanadate the $V_2O_7^{4-}$ anions form individual groups, separated from each other by large propylene diamine cations. Thermal decomposition of the propylenediammonium vanadate results in the formation of high-active pseudoamorphous V_2O_5 with average block size about 100 nm.

2.3 *Mechanochemistry and 'Dry' technologies*

Plastic deformation of a solid usually results not only in changes in the shape of the solid, but also in the accumulation of various types of the defects. These defects affect the physical and chemical properties of the solid. Therefore, plastic deformation can be used as a method of changing the reactivity of solids.

One of the preculiar features of the process of mechanical activation is that the activation does not takes palce immediately after the mechanical treatment of the solid has started, but only after the sample has been dispersed to such an extent that the particles have reached some critical size and further dispersion is no longer possible (Schönert 1990; Goldberg and Pavlov 1993). The main result of the mechanical activation is not an increase in the specific surface, but the accumulation of the defects drastically changing the reactivity in the whole bulk of the solid (Thiessen *et al.* 1967; Boldyrev and Avvakumov 1971; Avvakumov 1979; Boldyrev and Heinicke 1979; Heinicke 1984; Tkacova 1984; Boldyrev 1986; Boldyrev *et al.* 1990; Bokhonov *et al.* 1993).

Increasing reactivity of solids by mechanical activation can be considered as one of the methods of obtaining solids in a metastable 'active' form (Hüttig 1943). Different solid state reactions are sensitive to different types of defects (Boldyrev 1960). Therefore, the aim of mechanical activation is not just to increase the overall concentration of the defects in general, but to accumulate selectively the defects of that type to which the particular reaction is most sensitive. This can be achieved by modifying the type, the duration, the intensity of the mechanical treatment, the temperature of the sample or the partial pressure of gases overthe sample. One should also take into consideration the crystal structure, the types of chemical bonds and the mechanical properties of the sample (Boldyrev 1993b).

As one of the examples I would like to consider the development of a new mechanochemical method of the manufacturing of phosphatic fertilizers from natural phosphate rocks (Boldyrev *et al.* 1977; Kolosov *et al.* 1979). The most natural of phosphate rocks (various forms of the tricalcium phosphate) are poorly soluble in the humic acids present in soil. Therefore, before the rock can be used as a phosphatic fertilizer it should be subjected to a special treatment. In the laboratory, when the results of the various types of treatment are studied, one usually uses standard solutions to imitate the action of humic acids present in the soil: solutions of citric acid (imitation of acid soils) or of ammonium citrate (imitation of neutral or alkaline soils). If the soil is acid, the so-called 'secondary phosphate' can be applied as phosphatic fertilizers. They are more reactive than the metamorphic crystalline phosphate rocks. Therefore, a simple comminution of the phosphatic rock, giving grains of such a size that the time of the penetration of the acid inside the grain at the distance equal to its radius becomes comparable with the time during which the plants need phosphorous for their development, is sufficient to produce a phosphatic fertilizer, known as a 'phosphorite meal'. The situation is more complicated if the soil is neutral or basic, and if metamorphic phosphatic rocks with low intrinsic solubility in weak organic acids are used. In the traditional technology, the metamorphic phosphatic rocks are treated by sulphuric acid. At the first stage of this treatment a 'superphosphate pulp' is formed as a result of the interaction between the acid and the apatite. Seventy per cent of the apatite react with sulphuric acid at this stage giving *o*-phosphoric acid and calcium sulphate. A by-product is hydrogen fluoride, since apatite rocks usually contain fluorides. The second stage of the process lasts from 5 to 20 days. During this stage the *o*-phosphoric acid reacts with the remaining 30% of the apatite to give monocalcium phosphate $Ca(H_2PO_4)_2.H_2O$. The mixture of the monocalcium phosphate and gypsum (the hydrate of calcium sulphate) is formed, and is known as the commercial product 'superphosphate'.

The technology is far from being perfect, both from economical and ecological points of view. Amongst the main disadvantages of this technology one can mention, for example, the necessity to use large amounts of sulphric acid (68–70%), the atmospheric pollution with hydrogen fluoride and the formation of gypsum as an unwanted diluent. The process is also very time-consuming. It is possible to improve the technology by substituting the sulphuric acid with phosphoric acid. In this case the process becomes less time-consuming and no gypsum is formed. The resulting valuable product is known as a 'double or triple superphosphate'. However, the application of phosphoric acid instead of sulphuric acid makes the process much more expensive. Besides, the necessity of treatment of the rock by any acid does not seem to be very attractive from the ecological point of view. The application of acids is especially problematic for the districts in which these acids are not manufactured, and their transportation from other places is expensive and dangerous for the environment.

The Institute of Solid State Chemistry in Novosibirsk, in collaboration with German researchers, has suggested a new technology of manufacturing phosphatic fertilizers starting from natural raw minerals (Paudert *et al.* 1981). The new method does not require the treatment of the ore by acids at all, and is based on the mechanical activation of the natural phosphates. The control (citrate acid and citrate ammonium) tests of the solubility of the fertilizer produced by this new technology have shown its quality to be comparable with that of commercial phosphatic fertilizers such as superphosphate.

The new technology is especially attractive for agriculture in such regions as, for example, Siberia. First, one does not need acid any longer (in traditional technology, 0.7 tons (700 kg) of acid per 1 ton (1000 kg) of the superphosphate are required).

Second, the new technology makes it commercially beneficial to exploit small (< 1×10^6 ton) (1×10^9 kg) natural deposits of the phosphates, which are widely spread in Siberia. These deposits are too small to make their exploitation profitable if the traditional 'acid' technology is used.

The dry technology of the manufacturing of phosphatic fertilizers is also ecologically attractive (Frolov *et al.* 1986).

1 The application of the mechanically activated phosphates does not increase the acidity of soil. On the contrary, they were shown to meliorate the soil, to improve its structure and, in some cases, to decrease the acidity.

2 In contrast to the superphosphate, mechanically activated phosphates do not react with the oxides of iron and aluminium present in soil. Therefore, no phosphorous is transferred in forms that cannot be utilized by the plants.

3 The mechanically activated phosphates improve the content of active phosphorous in the soil for a period as long as 5 or even 7 years. (for a comparison, traditionally applied water-soluble phosphate fertilizers should be introduced into soil every 2 years.) This is also good since the soil is more rarely subjected to the compressive action of the wheels of the machines when the fertilizers are being introduced.

4 The pollution of the environment with hydrogen fluoride is excluded. The fluorine present in the starting rock is accumulated in the soil in the form of the non-solvable and low-active fluorite.

In Germany, the mechanically activated phosphates were termed 'tribofos'. The field tests of tribofos have shown it to be better than commercial phosphatic fertilizers such as 'phosphorite meal' or 'hyperfos', especially for basic soils. The German chemists have also suggested a method of the treatment of the phosphate rocks, combining the mechanical activation and the partial chemical dissolution of the apatite. The product 'tribofos T' was shown to be no worse than superphosphate (Harenz *et al.* 1979).

The mechanical activation of the phosphate ores is not the only one example of a successful application ofthe dry technology instead of a traditional one. Similarly, mechanical activation can be used to intensify the leaching of the wolframite and sheelite (Zelikman *et al.* 1979). According to the traditional hyrometallurgical technology, tungsten is obtained in the course of a two-stage treatment of the wolframite by a large (two to three times as compared with the stoichiometrically required amount) excess of soda in the autoclave at temperatures of about 473 K. Preliminary mechanical activation of wolframite makes it possible to obtain sodium tungstate from the wolframite in one stage. The productivity of the leaching apparatus becomes 2.5 times higher, and the consumption of soda becomes 15–20% smaller (Zelikman *et al.* 1979).

Another example is the hydrometallurgical processing of chalcopyrite, giving copper from sulphide ores (Biangarai and Pietsch 1979). Mechanical activation is used for the acid leaching of copper ores. Copper is deposited from the solution obtained by leaching directly at the place of processing of the ore. The technology is known as the 'process of Lurgi–Mitterberg'. It is one of the examples of the use of the mechanochemical activation in an environmentally cleaner process. The leaching of sulphide copper ores is intensified if the ore is mechanically activated preliminarily (Habashi 1982; Balaz *et al.* 1983).

Mechanical activation of graphite was shown to be helpful in decreasing its consumption in metallurgy as a cover and in improving the quality of the cast in the mould (Mamina *et al.* 1984).

Kurakbaeva *et al.* (1992) have described a mechanochemical method of the neutralization of the impurities in the granulated electrothermophosphorous slag by the additives of oxides (CaO), oxidants ($Ca(ClO)_2$, $KMnO_4$, CrO_3) and also of the catalysts of the oxidation (MnO, CuO, Fe_2O_3) present in the sulphide and phosphide slags.

Mechanical activation of dolomites and of magnesites—the natural minerals containing carbonate ions, which can be used to absorb the radionuclides ^{90}Sr from natural mineral waters—increases the capacity of the sorbent to a noticeable extent (Kornilovich *et al.* 1992).

Mechanical activation was shown to increase the reactivity of solids in the heterogeneous catalytical reactions. Thus, as a result of the mechanical activation of zinc ferrite during 12–24 min, its ability to take up sulphur from gases containing H_2S increases more than two times. Moreover, the mechanical activation also facilitates the regeneration of the catalyst. Thus, to regenerate a non-activated catalyst one must heat it at 600 °C for 24 hours. The same catalyst, preliminarily subjected to mechanical treatment, regenerates after 6 hours heating at 500 °C (Sepelak *et al.* 1994).

2.4 *Application of the mechanical activation in inorganic and organic synthesis*

Synthesis of new compounds is one of the main aims of chemistry. At the very beginning of this chapter we considered a typical sequence of stages used nowadays to produce a solid product from the solid raw minerals, using the liquid or gaseous state as an intermediate. It always seemed to be attractive to exclude these gas or liquid state stages and to make the reaction proceed in the solid state. However, a solid state synthesis is usually characterized by an extremely low technological rate. This low rate is due to the small initial contact area between the two solid reactants and to the poor mixing. Besides, the layer of solid product is formed at the contact between the two solid reactants. The layer separates the reactants from each other, and the process becomes limited not by the kinetics of the reaction itself, but by the diffusion through this product layer. The standard techniques of intensifying solid state synthesis are the increase in the reaction temperature and also the repeated grinding of the reaction mixture in the course of the process. However, application of these techniques is also related to a number of problems. The increase in the temperature can stimulate not only the reaction, but also undesirable thermal transformations of the solid product (thermal decomposition, deactivation, a phase transformation into another, high-temperature polymorphism, etc.). Therefore, alternative techniques of intensifying solid state synthesis are being developed. One of these techniques is the mechanical activation of the mixture of solid reagents. The preliminary mechanical activation can facilitate the subsequent thermal reaction between the components of the mixture. Alternatively, the reaction can take place at the very moment of the mechanical treament of the mixture. The mechanical activation of the mixture increases the number of contacts between the particles of the reactants and, more importantly, the contact surface area. The shear stresses arising at the contacts between the particles facilitate the removal of the product from the contact zone and the regeneration of the contacts between the reactants. The process becomes limited by the kinetics of the reaction itself, and not by the diffusion through the product layer. The heat evolved as a result of the relaxation of the mechanical stresses in the contact zone can induce the 'contact melting'. This can also be one of the main reasons of the intensifying of solid state reactions, especially for the metallic systems, as well as for the reactions between the molecular solids.

The examples of a successful application of the mechanochemistry to intensify the synthesis in the mixtures of inorganic and organic solids are already quite numerous. I would like to limit the discussion to a few of them.

2.4.1 MECHANICAL ACTIVATION AND INORGANIC SYNTHESIS

Mechanochemical synthesis in metallic systems. In the last few years the solid solutions of metals with unusual composition and the new intermetallic compounds are being widely produced by the so-called technique of 'mechanical alloying' (Koch 1991). By applying mechanical activation and exploiting the topotactical solid state reactions one can produce new solid materials with unusual structures, composition and properties. For example, mechanical alloying of nickel, cobalt and aluminium gives, respectively, $Ni_{35}Al_{65}$ and $Co_{40}Al_{60}$ crystallized in the body-centred cubic lattice. The alkali leaching of these intermetallic compounds gives metallic nickel and cobalt with very high catalytic activity, the so-called Nickel–Reney catalyst (Ivanov *et al.* 1988). Mechanical alloying of the powders of Nd, boron and iron allows one to synthesize $Nd_2Fe_{14}B$, which is used to manufacture permanent magnets (Schultz 1992). Mechanical alloying can be used to synthesize hydrogen-storage materials (Song *et al.* 1985; Ivanov *et al.* 1987; Konstanchuk *et al.* 1987). Mechanical alloying of copper, mercury and tin makes the processing of stopping amalgam (used in childrenes dental practice) environmentally clean (Ivanov 1990, 1993). Relatively recently, mechanical alloying was used to synthesize the icosahedral phase of Mg_{32} $(Zn,Al)_{49}$ (Ivanov *et al.* 1991). Previously, the icosahedral phases were obtained only by other techniques, for example by a super-rapid cooling of the melt.

Mechanochemical synthesis in the non-metallic systems. The mechanochemical synthesis of the yellow cadmium pigment can serve as an example (Pajakoff 1985). Usually, yellow cadmium pigment is synthesized following one of two techniques. According to the first technique, the homogenized mixture of cadmium and sulphur is heated at 500–600°C in a sealed vessel. In this case one gets not only the desirable cadmium sulphide, but also a number of undesirable by-products, including the volatile compounds of sulphur, which are dangerous for the environment. According to the second technique, the cadmium sulphide is precipitated from the neutral solutions of cadmium salts by sulphides of alkaline metals. The precipitate is decanted, washed and dried at 80°C. After that, in order to transfer the precipitate from the amorphous non-stable into the crystalline stable form, it is annealed at 400°C. The main disadvantage of this method is a large consumption of water and a large amount of waste waters contaminating the environment. One needs special apparatus for filtering and drying the precipitate. The prolonged high-temperature (400°C) annealing of the sample is also not very attractive.

A simple, safe and environmentally cleaner mechanochemical method of synthesis of the yellow cadmium pigment was suggested as an alternative to the two traditional techniques mentioned above. A solid mixture of a cadmium salt (usually cadmium carbonate) and sodium sulphide is subjected to the mechanical treatment during 2–4 hours in a standard ball mill. If 1% of ammonium sulphate is added to the mixture, the time of the mechanical treatment can be even shorter. X-ray and chemical analyses of the product, as well as technological tests, have shown its composition, its degree of crystallinity and its stability to light, as well as its stability with respect to the action of alkali, to be not worse than those of the commercial product obtained following traditional technology.

Another pigment, i.e. the red cadmium pigment, based on the solid solution of sulphide and selenide of cadmium, can also be synthesized mechanochemically, by the mechanical treatment of the solid mixture of cadmium carbonate, sodium sulphide and metallic selenium. By varying the concentration of selenium in the mixture one can get pigments with the colour varying from orange to deep red. Mechanochemically synthesized red cadmium pigment was shown to be of high quality and could be used to colour ceramic surfaces. The main advantage of the mechanochemical synthesis of this pigment, as compared with the traditional themal technique, is that the mixture is not heated and, therefore, the harmful effect of poisonous SO_2 and SeO_2 on the environment is excluded.

In many publications the mechanochemical technology was reported to be advantageous for the synthesis of various types of functional ceramics, making the manufacturing process easier, cheaper and cleaner. For example, the preliminary mechanical activation of zircon considerably decreases the temperature of the synthesis of the lead zirconate, and also improves its piezoelectric properties (Biggers *et al.* 1978). Arai *et al.* (1974) have shown the preliminary mechanical activation of the mixture of barium carbonate and zircon allows one to decrease the temperature of the synthesis of barium zirconate by several hundreds of degrees centigrade. The reason is in the homogenization of the mixture and in the generation of the defects in the solids.

The barium titanate ceramics synthesized by mechanical activation of powders obtained by the decomposition of titanyl oxide was shown to have higher electrical stability and anomalously high dielectric permeability (Boldyrev *et al.* 1989).

Pauli *et al.* (1992) and Isupova *et al.* (1994) have shown that lanthanum cobaltites and manganites with high catalytic activity can be synthesized mechanochemically from oxides or carbonates. Preliminary mechanical activation of the mixture of the starting reactants (during 5 min only), made it possible to decrease the temperature of the subsequent synthesis down to 700°C. The low temperature of the synthesis made it possible to get the desirable product with a 100% yeild. The specific surface of the catalyst was from 7 to 15 $m^2 g^{-1}$. The catalytic activity of the catalyst obtained with preliminary mechanical activation of the reactant mixture was shown to be three times higher than that of the same catalyst synthesized by a traditional thermal technology.

It was shown that high-temperature ceramic superconductors can also be synthesized mechanochemically (Awano *et al.* 1990; Naito *et al.* 1990; Hainovsky *et al.* 1991). Further numerous examples of the successful applications of the mechanochemical techniques in the inorganic synthesis can be found, for example, in the *Proceedings of the Conferences on the Mechanochemistry* and in a relatively recent monograph edited by Avvakumov (1991).

Sometimes it is preferable to carry out the mechanical treatment of the mixture and its heating simultaneously. For example, the lithium–zinc–manganese ferrite can be obtained in one stage by mechanical activation at the temperature 450–500°C (Tkachenko *et al.* 1983). For a comparison, the traditional method of synthesis of this compound requires several repeated stages of heating at the temperatures 900–1000°C and grinding of the mixture.

In some cases additional external heating of the sample is not required, since enough heat is evolved at the contacts between the particles as a result of the mechanical treatment itself. The reaction takes place at the movement of the mechanical activation. As examples one can mention the mechanochemcial synthesis of the apatite from the mixture of calcium oxide, dicalcium phosphate and calcium fluoride (Chaikina *et al.* 1978), or the synthesis of the barium–lanthanum–tungstate starting from the mixture of the carbonates of barium and lanthanum, and the tungsten trioxide (Urakaev *et al.* 1985).

It is worth noting that these latter examples are interesting not only because the synthesis takes place directly in the mechanocemical activator, but also because the reactant mixtures are not only two, but also three and more component. These reactions were shown to give first the intermediate product in an amorphous state. The stability of this state and, consequently, the reaction rate, are strongly dependent upon the environment, especially the presence of water vapours over the sample (Shapkin *et al.* 1989).

The so-called 'soft mechanochemical synthesis' was also described (Avvakumov 1994). The acid–base pairs are taken as the starting reactants. For example, calcium titanate is synthesized from: the mixture of $Ca(OH)_2$ and $TiO_2.1.2H_2O$; metasilicate, from the mixture of CaO and H_2SiO_3; and wolframate, from CaO and H_2WO_4. The water formed in the course of the mechanochemical reaction between the solid components of the mixture accelerates the synthesis: it improves the rheological properties of the mixture and also acts as a catalyst of the chemical reaction (Boldyrev 1972; Juhaz 1989).

To optimize the inorganic solid state synthesis it is often necessary that the same sample is subjected to not one, but, subsequently, to several different types of the mechanical treatment. These different types of treatment require not only different regimes of the exploitation of the same mechanical activator, but often the application of different activators. This is one of the main problems of the mechanochemical inorganic synthesis, which is far from being solved now, but should be solved if we want to achieve some noticeable progress in the future. It is not uncommon nowadays that one tries to use the same activator for all the operations. Obviously, this is not efficient, if at all possible, since dispersion, mixing and activation of the solid components require quite different types of mechanical treatment.

Dispersion and mixing of the components can be carried out not only mechanically, but also using some alternative techniques, for example the sol-gel technique, coprecipitation from the solution or condensation from the vapour. The subsequent mechanical treatment of the mixture can be carried out to make the components of the mixture interact (Isobe and Senna 1992). The interaction may result, first of all, in the formation

Table 12.3 Preparation of sodium salicilate

Traditional liquid technology used in industry	Mechanochemical technology
1 Dissolution of the acid in water (suspension) **2** Neutralization of $NaHCO_3$	**1** Mixing of the acid and neutralizing agent
3 Filtration	**2** Mechanical treatment
4 Drying	**3** Packing
5 Milling	
6 Packing	
Note: the duration of the process is 50–70 hours. The yield is 450–500 kg of the product per cycle. Two chemical reactors 2–3 m^3 in volume, equipment for filtration, a vacuum roller drier and milling equipment are requried.	Note: the process is carried out in continuous conditions at a rate of up to100 kg/hour^{-1}. The necessary equipment is easily mounted in one technological line occupying a floor space of 4 × 6 m, and 4 m in height

of the contact between the reactants, and then either in the chemical reaction or in the formation of the mechanocomposite.

Another possible solution is the separating of the space of the mechanical activator into three parts. The activator is constructed in such a way so that in the first part of the reactor the components are comminuted, in the second one mixed, and in the third one the chemical reaction itself takes place (Boldyrev 1993a).

2.4.3 MECHANOCHEMICAL ORGANIC SYNTHESIS

Development of dry technologies, in which no solvents are used, is also important for organic synthesis. I would like to consider some examples from the practice of the Institute of Solid State Chemistry (Novosibirsk).

One of the important products of pharmaceutical industry is sodium benzoate. Traditionally it is produced by the neutralization of the benzoic acid by soda in aqueous solution. A standard technological cycle consists of six stages. Production of 500 kg of the benzoate requires 3000 l of water. A standard duration of the cycle is 60 hours. The same amount of sodium benzoate can be produced by a mechanical treatment of the mixture of solid powders of benzoic acid and soda in an activator for 5–8 hours only. The consumption (and the contamination) of enormous amounts of water is excluded. The acceleration of the technological rate of the process is also obvious.

A similar example is provided by the synthesis of sodium salicylate. The traditional scheme of the production of this compound includes six technological stages and requires 70 hours. One needs 500 l of water and 100 l of ethanol to produce 500 kg of sodium salicylate. The same amount can be produced in a mechanochemical activator in one stage during 7 hours from the solid starting reactants without any solvent at all. The traditional and the mechanochemical methods of synthesis are compared in Table 12.3.

An example of cyclization reaction is the formation of the sodium salt of oxazepam after mechanical activation of sin-oxime-2-benzoil-2.4-dichloroacetanilide in the presence of solid NaOH (Fig. 12.1).

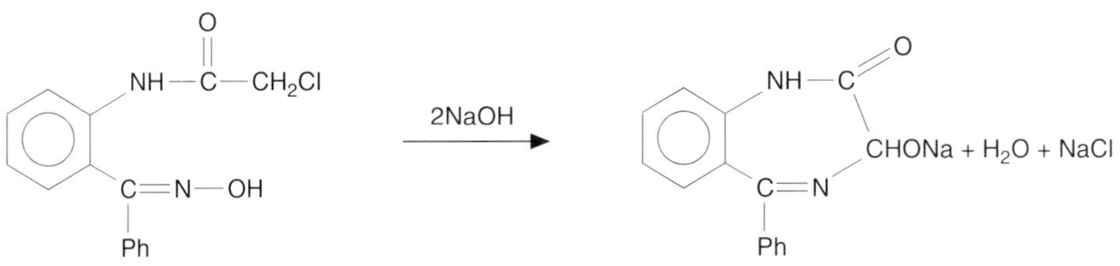

Figure 12.1 Synthesis of sodium acedypiril.

The reaction mixture was dissolved in i-PrOH : H_2O = 70% : 30% neutralized by HCl and analyzed using HPLC. In the activators used, the reaction time is no longer than 5 min and the product yield is as high as 80% of its theoretical value. In the industrial synthesis of oxazepam, the similar conversion is performed in the C_2H_5OH medium with heating for several hours.

The hydrogenolysis of 'Gibberilline' A3 is used when synthesizing stimulants of growth of plants. The reaction is carried out in methanol in the presence of sodium borohydride as a reducing agent, and a catalytic additive of nickel(II) chloride (Yang and Pan 1987). Alternatively, the hydrogenolysis can be carried out by a mechanical treatment of the solid mixture of Gibberilline A3 with Mg_2NiH_4 and Mg_2CoH_4 (acting as a source of hydrogen).

In some cases the dry mechanochemical technologies are advantageous as compared with the traditional solution ones, not only because they decrease the time required for the synthesis, exclude the consumption of solvents and reduce the environmental pollution, but also because they give a product of higher quality.

For example, the standard method of producing phthalozole in the pharmaceutical industry is to heat sulfathiazole with phthalic acid or phthalic anhydride (acting as acidylate agents) in ethanol or in a water–ethanol mixture (Chiang *et al.* 1957) (Fig. 12.2). The product (III) is unavoidably contaminated by phthalozolimid (IV) and the esters of phthalic acid (V) formed in the course of by reactions. Mechanochemical synthesis gives pure phthalozole with no admixtures of impurities (Chuev *et al.* 1989).

The contacts between reactants are important for a solid state synthesis. the ease of formation of these contacts depends on the physical properties of the reactants; first of all, on their plasticity. It is not uncommon that the rate of a solid state synthesis is determined not by the rate of the chemical reaction itself, but by the rate of the formation of the contacts between the reactants (Dushkin *et al.* 1991, 1994). For example, the oxidation of the methylimidazol in solution proceeds faster than that of the phenylimidazol. However, the solid state mechanochemical reaction of a solid derivative of methylimidazol with potassium persulphate is slower than the coresponding reaction of the phenylimidazol derivative since the latter is more plastic.

2.4.4 DISPERSE SYSTEMS

The synthesis of the so-called 'disperse systems' is of special importance for the pharmaceutical industry. It is used to increase the dissolution rate (and, sometimes, also the solubility) of slowly and poorly soluble drugs. The method of producing

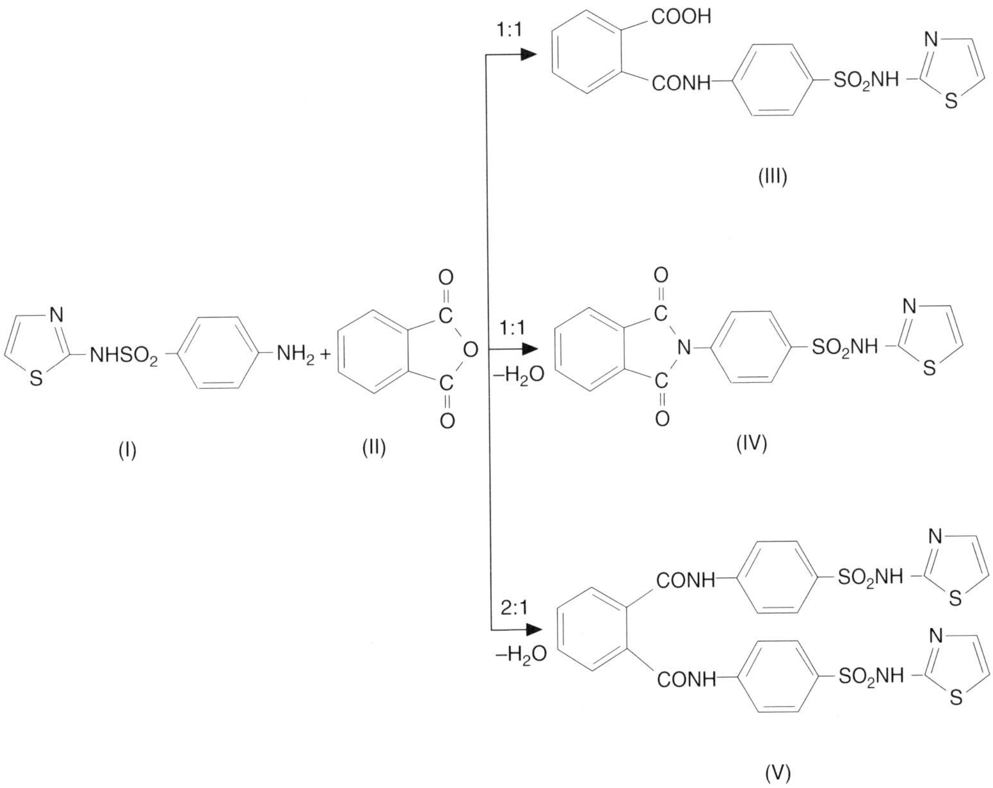

Figure 12.2 Synthesis of phthalozole.

disperse systems was first suggested by Sekiguchi and Obi (1961). The medically active substance and the cosolvent (a physiologically inert, but easily soluble support) are either melted together and then crystallized from the melt or, alternatively, are codissolved in some solvent with subsequent recrystallization (Chiou and Riegelman 1971). The aim of these procedures is to prevent the aggregation of the 'micronized' particles (the phenomenon very common for molecular crystals). If the aggregation is suppressed, the specific surface area of the drug increases, which is favourable for the interaction between the drug and the solvent. Sometimes, the interaction between the drug and the inert support is strong enough, so that an unstable complex or a micelle are formed (Sheffer 1981). This also facilitates the dissolution of the drug and, hence, the biological activity of the drug increases. Both methods of producing the disperse systems are not perfect. The main problems are the partial decomposition of the drug during fusion (melting) and the difficulty of separating the crystals from the solvent.

Dispersal of easily soluble systems can be produced by an alternative mechanochemical technique, similar to the procedure of mechanical alloying described above. During the mechanical treatment the drug particles are simultaneously comminuted, subjected to plastic deformation, undergo transitions into metastable polymorphs and are steadily covered by the cosolvent (Ikekawa and Hayakawa 1981; Huttenrauch *et al.* 1985; Yamamoto and Nakai 1989; Boldyrev *et al.* 1990b; Yokoyama *et al.* 1990). The mechanochemical procedure was successfully applied to increase the solubility of aspirin (decstrine or cellulose acting as cosolvents) (Nakai 1987), of sulfathiazole (polyvinylpyrrolidone as a cosolvent) (Boldyrev *et al.* 1994a, b) or of ibuprofen (polyethylene glycol as a cosolvent) (Boldyrev *et al.* 1994a, b). The main advantages of the mechanochemical procedure of the synthesis of disperse systems are the following: the drugs are not heated and no solvents are required.

3 Concluding remarks

The first attempts to use solid state reactions in industry date back to the beginning of this century. However, the early expectations that these reactions would soon be applied widely in industry turned out to be too optimistic. The main reason of this was the lack of theoretical background. We are just starting to understand the reactivity of solids and working out the methods of the control of the solid state reactions. Without a theory one can hardly develop a good technology. The 'trial and error' method is too expensive and time-consuming, and developing a technology becomes an art rather than a science.

Until now, considerable progress was achieved in the development and the optimization of liquid state technologies, for example by optimizing the reactors, or by making the process automatic and controlled by computers. However, the possibilities of these tactics are also limited and may soon be exhausted. At the same time, the requriements of the technological processes become stricter and stricter. The attitude towards environmental problems has changed since the beginning or even the middle of this century. The rich natural deposits of raw minerals became exhausted. A new strategy, and not just a new tactic is required to develop a new technology meeting modern economical and ecological demands. One of the possible solutions of the problem may be in the development of dry technologies based on the solid state reactions and the knowledge of the theory of the reactivity of solids.

4 References

Arai, Y., Yame, T., Takiguchi, H. & Kubo, T. (1974) *Nipp Kadaku, Kaishi* **N9**, 1611–1616.
Avvakumov, E.G. (1979) *Mechanical Methods of Activation of Chemical Processes*; Nauka, Novosibirsk.
Avvakumov, E.G. (ed.) (1991) *Mechanochemical Synthesis in Inorganic Chemistry*, p. 1–263. Nauka, Novosibirsk.
Avvakumov, E.G. (1994) *Chem Sustain Develop* **1, N2**, 101–130.
Awano, M., Takagi, H. & Kuwahara, Y. (1990) *Proceedings of Second World Congress on Particle Technology, September 19–22,* pp. 369–347. Kyoto, Japan.
Balaz, P., Tkacova, K., Jusko, F. & Bochkarev, G.R. (1983) *CSSR Patent N 223537.*
Bernal, J.D. (1960) *Schweiz Arch* **26**, 69–75.
Biangarai, S. & Pietsch, H. (1979) *Erzmetall,* **29, N2**, 73–76.
Biggerk, Hankey, D. & Tarhay, L. (1978) *Processing of Crystalline Ceramic Materials Science Research Serie* Vol. 11. (ed. H. Palmour, R. Davis, T. Hare). pp. 335–342. Plenum Press, NY.
Bokhonov, B., Rykov, A., Paramzin, S., Pavlukhin, Y.T. & Boldyrev, V.V. (1993) *J Mat Synth Proces* **1, N5**, 341–347.
Boldyrev, V. (1960) *Kinet Catal* **1**, 203.
Boldyrev, V. (1965) *Kinet Catal* **6**, 934.
Boldyrev, V.V. (1972) *Kinet Katal* **13**, 1414.
Boldyrev, V. (1973) *Russ Chem Rev* **42**, 1161.

Boldyrev, V. (1975) *J Thermal Anal* **7**, 685.
Boldyrev, V. (1976) *J Thermal Anal* **8**, 175.
Boldyrev, V. (1986) *J Chim Phys* **83, N11/12**, 821–829.
Boldyrev, V. (1993a) *Proceedings of the First International Conference on Mechanochemistry, Kosice*, Vol. 1 (ed. K Tkacova), pp. 18–26. Cambridge Interscience Publishing, Cambridge.
Boldyrev, V. (1993b) *J Therm Anal* **40**, 1041–1062.
Boldyrev, V.V. & Avvakumov, E.G. (1971) *Uspekhi Khim* **60, N10**, 835.
Boldyrev, V. & Heinicke, G. (1979) *Zeitsch Chem* **19**, 353.
Boldyrev, V., Boulens, M. & Delmon, B. (1979) *The Control of the Reactivity of Solids*, pp. 1–228. Elsevier Science Publishers Amsterdam.
Boldyrev, V., Kolosov, A. & Chaikina, M. (1977) *Dokl Akad Nauk SSSR* **233**, 892.
Boldyrev, V.V., Lapshin, V.I., Fokina, E.L. & Yarmarkin, V.K. (1989) *Dokl Akad Nauk SSSR* **305, N4**, 852–854.
Boldyrev, V.V., Markel, A.L., Yagodin, A.Y. & Dushkin, A. (1990) *Riform Med* **N3**, 49–53.
Boldyrev, V.V., Shakhtshneider, T.P., Burleva, L.P. & Severzev, V.A. (1994a) *Drug Develop Pharm* **20 (6)**, 1103–1114.
Boldyrev, V., Shakhtshneider, T., Burleva, L., Vasiltchenko, M. & Severzev, V.A. (1994b) *Chem Sustain Develop* **2 (1)**, 455.
Brown, M.E., Dollimore, D. & Galwey, A.K. (1980) Reactions in the solid state. In: *Comprehensive Chemical Kinetics*, Vol. 22 (eds C.H. Bamford & C.F. Tipper) pp. 1–340. Elsevier Publishing Co., Amsterdam.
Chaikina, M.V., Shapkin, V.L., Kolosov, A.S. & Boldyrev, V.V. (1978) *Proc Siber Branch Acad Sci USSR Ser Chem* **3**, 96.
Chiou, W.L. & Riegelman, S. (1971) *J Pharm Sci* **60**, 1281.
Chiang S.-C., Juan C.J., Chang, C.-F. & Sen, M.J. (1957) *Acta Chimica Pharm* **5, N4**, 351–355.
Chuev, V.P., Lyagina, L.A., Ivanov, E.Y. & Boldyrev, V.V. (1989) *Dokl Akad Nauk SSSR* **307, N6**, 1429.
Dankov, P.D. (1946) *Russ J Phys Chem* **20**, 853.
Delmon, B. (1986) *J Chim Phys* **83**, N11/12, 875–883.
Dushkin, A.V., Negovitsina, E.V., Boldyrev, V.V. & Druganov, A.G. (1991) *Sibirsky Khim Z* **5**, 75.
Dushkin, A.V., Rykova, Z.V., Shakhtshneider, T.P. & Boldyrev, V.V. (1994) *Intern J Mechanochem Mech Alloy* **1, N1**, 48–55.
Feitknecht, W. (1964) *Pure Appl Chem* **9 (3)**, 423–440.
Frolov, A., Ignatov, V. & Shinkarenko, S. (1986) *Phosphorite Powder Methods of Increasing Quality*, pp. 1–32. INGEKHIM, Moscow.
Galwey, A. (1996) In: *Reactivity of Solids: Past, Present and Future* Ch. 2 (ed. V.V. Boldyrev), pp. 15–72. Blackwell Science Limited, Oxford.
Garner, W.E. (ed.) (1955) *Chemistry of Solid State*. Buttersworth, London.
Goldberg, E.L. & Pavlov, S. (1993) *Proceedings of the First International Conference on Mechanochemistry, Kosice*, Vol. 1 (ed. Tkacova K.), pp. 68–70. Cambridge Interscience Publishing, Cambridge.
Habashi, F. (1982) *Chem Eng News* **60**, 46.
Hainovsky, N., Pavlukhin, Y.T. & Boldyrev, V.V. (1991) *Mat Sci Eng* **B8**, 283–287.
Harenz, H., Paudert, R., Muller, S. & Linke, E. (1979) *Arch Acker Pflaunzenbau Bodenkunde* **23**, 707.
Haruta, M. & Demon, B. (1986) *J Chim Phys* **83, N11/12**, 857–867.
Heinicke, G. (1984) *Tribochemistry Akad Vlg Berlin*, 1–955.
Hutenrauch, R., Fricke, S. & Zielke, P. (1985) *Pharm Res* **6**, 352.
Hüttig, G. (1943) *Handbuch der Katalyse*, Bd. 6 (ed. G.M. Shwab), p. 490. Springer Verlag., Vien.
Ikekawa, A. & Hayakawa, S. (1981) *Bull Chem Soc Jpn* **54**, 2587.
Isobe, T. & Senna, M. (1992) *Mat Sci Forum* **88–90**, 752–754.
Isupova, L.A., Sadykov, V.A., Avvakumov, E.G., Pauli, I.A., Andryushkova, O.A. & Polu-

boyarov, V.A. (1994) *Proceedings of the 6th International Symposium on Scientific Bases for the Preparation of Heterogeneous Catalysts, September 5–8*, pp. 333–341. Louvain-la-Neuve, Belgium.
Ivanov E. (1993) In: *Proceedings of the First International Conference on Mechanochemistry*, Vol. 1 (ed. K. Tkacova), pp. 49–56. Cambridge Interscience Publishers, Cambridge.
Ivanov, E., Bokhonov, B. & Konstanchuk, I. (1991) *J Mat Sci* **26**, 1409–1411.
Ivanov, E., Konstanchuk, I., Stepanov, A. & Boldyrev, V.V. (1987) *J Less-common Met* **131**, 25–29.
Ivanov, E., Grigorjeva, T., Golubkova, G. *et al.* (1988) *Mat Lett* **7, N1,2**, 55–57.
Ivanov, E.Y. (1990) Synthesis of metastable high reactive intermetallides and solid solutions during mechanical alloying. Thesis, Institute of Solid State Chemistry, Novosibirsk.
Juhaz, A.Z. (1989) *Colloid Polym Sci* **267**, 1036.
Koch, C. (1991) Mechanical milling and alloying. In *Materials Science and Technology. A Comphrensive Treatment. Processing of Metals and Alloys* Vol. 15, (ed. R.W. Cahn). V.C.H. Weinhaim, pp. 194–241.
Kolosov, A.S., Chaikina, M., Boldyrev, V. *et al.* (1979) Patent USSR 712407, 26.07.1979.
Konstanchuk, I.G., Ivanov, E.Y., Pezat, M., Dariet, B., Boldyrev, V.V. & Hagenmuller, P. (1987) *J Less-common Met* **131**, 181–189.
Kornilovich, G.Y., Spasenova, L.N., Korsukov, A.A., Pshinko, G.I. & Masko, A.N. (1992) *Chem Technol Water* **14, N1**, 48–52.
Kurakbaeva, R.H., Dorfman, E.A., Polimbetova, G.S., Shokorova, L.A., Kutabaev, K.K. & Kozlovsky, V.A. (1992) *Zhurn Prikl Khimi* **65, N3**, 487.
Lomovsky, O.I. & Boldyrev, V.V. (1989) *Russ J Appl Chem* **11**, 2444–2455.
Lomovsky, O.I. & Boldyrev, V.V. (1994) *J Mat Synth Process* **2, N4**, 199–207.
Lomovsky, O.I., Revzin, G.E. & Boldyrev, V.V. (1991) *Zhurn Vsesoyznogo Khimicheskogo Obshestva Mendeleeva* **N3**, 340.
Mamina, L.I., Kozlov, B.I. & Lukjanova, T.A. (1984) *Izv Sib Otd Akad Nauk SSSR Ser Khim Nauk* **64**, 34–7.
Naito, M., Yoshikawa, K., Yotsuya, T., Sekine, H., Asano, T. & Maeda, H. (1990) *Proceedings of Second World Congress on Particle Technology, September 19–22*, pp. 424–430. Kyoto, Japan.
Nakai, Y. (1987) *Izv Sib Otd Akad Nauk SSSR Ser Khim Nauk* **5**, 31–36.
Nitschmann, H. (1938) *Helvet Chim Acta* **21 (6)**, 1609–1618.
Oswald, H.R. & Reller, A. (1989) *Pure Appl Chem* **61, N8**, 1323–1330.
Pajakoff, S. (1985) *Österreich Chem Zeitsch* March, 48–51.
Paudert, R., Harenz, H., Heinicke, G. (1981) Patent GDR 147772, 22.04.81 COB/205660.
Pauli, I.A., Avvakumov, E.G., Isupova, L.A. *et al.* (1992) *Siber Chem J* **3**, 133–137.
Schönert, R. (1990) In: *Proceedings of Second World Congress on Particle Technology* Part II (ed. G. Jimbo), p. 257. Japanese Society of Powder Technology, Kyoto.
Schultz, L. (1992) *Mat Sci Forum*, **88–90**, 687–694.
Sekiguchi, K. & Obi, N. (1961) *Chem Pharm Bull*, **9**, 866.
Sepelak, V., Jancke, K., Rochfer-Mendau, J., Schteinke, U., Uecker, D. & Rogachev, A. (1994) *Kona* **12**, 87–94.
Shapkin, V.L., Urakaev, F.H., Vakhrameev, A.M. & Boldyrev, V.V. (1989) *Proc Siber Branch acad Sci USSR Ser Chem* **5**, 125–128.
Sheffer, E. (1981) In: *Techniques of Solubilization of Drugs* (ed S.H. Yalkowsky), pp. 170–178. Marsel Dekker Inc., New York.
Song, M.Y., Ivanov, E.Y., Darriet, B., Pezat, M. & Hagenmuller, P. (1985) *Intern J Hydrogen Energy* **10, N3**, 169–178.
Thiessen, P.A., Meyer, K. & Heinicke, G. (1967) *Grundlagen der Tribochemie*, pp. 1–183. Akademie Verlag, Berlin.
Tkachenko V.A., Letuk, P.M. & Bashkirov, L.A. (1983) *Proc Siber Branch Acad Sci USSR Ser Chem* **5**, 30.

Tkacova, K. (1984) *Mechanical Activation of Minerals*, pp. 1–50. Elsevier Science Publishers, Veda.

Urakaev, F.H., Avvakumov, E.G., Chumachenko, Y.V. & Boldyrev, V.V. (1985) *Izv Sib Otd AN SSSR Ser Khim* **5**, 59–63.

Yamamoto, K. & Nakai, Y. (1989) In: *Proceedings of the Second Japan–Soviet Symposium on Mechanochemistry* (ed. G. Jimbo), p. 189. Japanese Society of Powder Technology, Tokyo.

Yang, B. & Pan, X.-Ju. (1987) *Synth Commun*, **17**, 997.

Yokoyama, T., Yoshida, Y., Kondo, A., Hamiya, H. & Jimbo, G. (1990) *Proceedings of the Second World Congress on Particle Technology, September 19–22*, pp. 548–555. Kyoto.

Zelikman, A.N., Aranisova, F.A., Ermilov, A.G. & Rakova, N.D. (1979) *Izv Sib Otd Akad Nauk SSSR* **29**, 14–17.

Index

Page references to figures are in *italic* and those for tables are in **bold**